DNA Synthesis

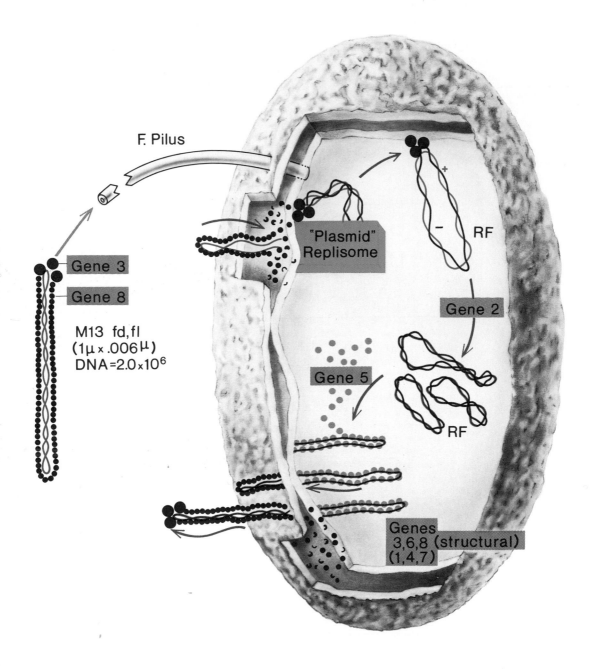

Life cycle of a filamentous virus, a bacterial parasite. Penetration of the viral DNA is coupled to its replication. Understanding replication of this small, well-defined viral chromosome illuminates the DNA synthetic processes of the cell itself (see Chapter 8, Section 3).

DNA Synthesis

Arthur Kornberg

STANFORD UNIVERSITY

W. H. FREEMAN AND COMPANY
San Francisco

751214

COVER: DNA polymerase bound to interruptions in a DNA molecule (from a photograph courtesy of Dr. Jack Griffith). See Fig. 4-6 for more detail.

Library of Congress Cataloging in Publication Data

Kornberg, Arthur, 1918–
    DNA synthesis.

    Includes bibliographical references.
    1. Deoxyribonucleic acid synthesis. I. Title.
[DNLM: 1. DNA—Biosynthesis.   QU58 K84d 1974]
QP624.K67   574.8'732   74–3005
ISBN 0–7167–0586–9

Printed in the United States of America

SECOND PRINTING

2   3   4   5   6   7   8   9

Contents

Preface

The rapid flow of facts and ideas in biochemistry makes it difficult to write an article, let alone a book. But such turbulence submerges useful facts and ideas, which become too specialized for general textbooks and are lost sight of even in detailed annual reviews. That this was true of the biochemistry of DNA synthesis became clear during preparation of the Robbins Lectures, given at Pomona College in April 1972, and it prompted me to undertake this effort.

This book emphasizes biochemical rather than physiological aspects of DNA synthesis. The scope has been broadened beyond that of an earlier book (*Enzymatic Synthesis of DNA, 1962*) to include topics clearly pertinent to DNA synthesis: precursors, repair, recombination, restriction, and transcription. It is hoped that with an enlarged scope and simplified language the book will serve students with a beginning interest in DNA synthesis as well as those working directly on the subject. Citations of the literature favor reviews and recent papers, which will in turn give the interested reader more complete bibliographies.

At the conclusion of writing this book, I am surprised and embarrassed at the large number of people whom I have enlisted in the preparation of this relatively modest effort. I am most indebted to my wife, Sylvy, who helped me write this book and do the early work on DNA synthesis. I am also grateful to Fred Robbins for the initial stimulus, to Charlene Levering for the illustrations, to Inge Loper for the typing, to Stephanie Lee Rowen for a careful early reading of the text, to I. Robert Lehman, David S. Hogness, A. Dale Kaiser, and R. David Cole for a critical reading of the entire text, and to colleagues too numerous to mention who gave me information and advice in preparing many sections of the book. ·The best requital for all these contributions is a useful book, and this has been my primary goal.

February 1974   *Arthur Kornberg*

DNA Synthesis

1

Structure and Functions of DNA

1. DNA: Past and Present[1,2]

The recorded history of DNA can be divided into three periods

1869–1943. This age opened with the discovery of a new organic phosphate compound in cells rich in nuclear substance. At first called nuclein and later chromatin, this compound was subsequently shown to consist of deoxyribonucleic acid (DNA) and protein. Other discoveries followed. Analysis of DNA showed it to contain four kinds of building blocks, called nucleotides. DNA was distinguished from ribonucleic acid (RNA) in having a different sugar, deoxyribose, in place of ribose; and a distinctive base, thymine, in place of uracil.

1. Fruton, J. S. (1972) *Molecules and Life*, John Wiley & Sons, N. Y.
2. McElroy, W. D. and Glass, B. (eds.) (1957) *The Chemical Basis of Heredity*, Johns Hopkins Press, Baltimore.

Although there was some reason to believe DNA might be the genetic material, there was even more reason to assign this role to proteins. Each molecule of DNA was thought to be a repeating polymer of one kind of tetranucleotide unit. Since there are only 256 possible arrangements of the four bases in a tetranucleotide, there could be only 256 kinds of DNA molecules. How could the complexity of genes be expressed using such limited informational capacity? Proteins with greater size and composed of twenty different amino acids were thought more suitable for a genetic role. One must also recall that in the 1930s DNA was still called thymus nucleic acid, and widely believed to occur only in animal cells: RNA had been isolated only from plant cells and was called yeast nucleic acid. In fact, plant and animal cells were sometimes distinguished on the basis of this chemical feature.

1944–1960. This age begins with the first important evidence that DNA is the genetic substance: the discovery reported in 1944[3] that DNA prepared from one strain of pneumococcus could "transform" another strain. The purified DNA carried a genetic message which could be assimilated and expressed by cells of another strain. DNA was further recognized to be a molecule far more complex than a repeating tetranucleotide, varying in composition from organism to organism[4]. Yet, since traces of protein could not be ruled out as contaminants in the transforming DNA, some doubt remained that DNA was the genetic material, and the impact of these discoveries on genetics remained relatively small.

Two persuasive discoveries were eventually made. The first was the demonstration in 1952[5] that infection of *E. coli* by T2 bacteriophages involved the injection of the DNA of the virus into the host cell. The viral protein structures appeared to serve merely to inject the DNA into the bacteria and then to be largely discarded outside the cell. The DNA from the virus thus directed the bacterial cell to produce many identical copies of the infecting virus. This experiment dramatized the role of DNA as the carrier of information for producing the unique proteins of the virus and for duplicating its DNA many times over.

A second, remarkable event was the discovery, in 1953[6], of the complementary, double-stranded (duplex) structure of DNA, and with it the recognition of how the molecule can be replicated. Complementary pairing of the nucleotide constituents of one strand to those of the second strand was postulated to explain in a simple way how one DNA duplex can direct the assembly of two molecules identical to itself. In this model, each strand of the duplex serves as a template upon which the complementary strand is made.

3. Avery, O. T., MacLeod, C. M. and McCarty, M. (1944) *J. Exp. Med.* **79**, 137; Hotchkiss, R. D. **in** McElroy, W. D. and Glass, B. (eds.) (1957) *The Chemical Basis of Heredity*, Johns Hopkins Press, Baltimore, p. 321.
4. Chargaff, E. (1950) *Experientia* **6**, 201.
5. Hershey, A. D. (1953) *CSHS* **18**, 135.
6. Watson, J. D. and Crick, F. H. C. (1953) *CSHS* **18**, 123.

These discoveries and other important ones that followed led to the realization that DNA has two major and discrete functions. One is to carry the genetic information that brings about the specific phenotype of the cell; DNA is transcribed into RNA, and the RNA is then translated into the amino acid language of the proteins. The other major function of DNA is its own replication. For duplicating the genotype of the cell, DNA serves as a template for converting one chromosome into two identical chromosomes.

1960–present. The beginning of this age is not marked by a specific event. It is an age in which the generally held conceptions of both the structure and the dual functions of DNA have not been challenged but rather have been expanded. Without epochal discoveries, the justification for distinguishing this third age lies in a recent radical change in viewpoint toward DNA. Genetics and DNA have become a branch of chemistry. Despite its chemical complexity, DNA is being modified, dissected, analyzed, and synthesized in the test tube. There are insights into a metabolic dynamism of DNA that had not been anticipated. DNA suffers lesions and is repaired. DNA molecules exchange parts with one another. DNA molecules are specifically modified and degraded, twisted and relaxed, transcribed in reverse from RNA as well as directly into RNA. DNA functions not only in the nucleus but also in mitochondria and chloroplasts. There is now a stimulus to determine the total base sequence of DNA and to resynthesize it. There is also a confidence that the metabolic gyrations of DNA in the cell can be understood in as explicit detail as those of, say, glucose or glutamate.

2. The Primary Structure[7]

The two kinds of nucleic acid, the ribonucleic acids, RNA, and the deoxyribonucleic acids, DNA, are polymers of nucleotides.

A nucleotide (Fig. 1-1) has three components: (i) a purine or pyrimidine base, linked through one of its nitrogens by an N-glycosidic bond to (ii) a 5-carbon cyclic sugar (the combination of base and sugar is called a nucleoside) and (iii) a phosphate, esterified to carbon 5 of the sugar. Nucleotides occur also in activated di- and triphosphate forms, in which one or two phosphates are linked to the nucleotide by phosphoanhydride (pyrophosphate) bonds.

In each of the two main kinds of nucleic acids there are only four types of nucleotides. These are distinguished by their bases: adenine (A), guanine (G), uracil (U), and cytosine (C) in RNA, and adenine, guanine, thymine (T), and cytosine in DNA. The bases and their nucleoside and nucleotide forms are listed in Table 1-1. In base

---

7. Sober, H. A. (ed.) (1970) _Handbook of Biochemistry, Selected Data for Molecular Biology_, 2nd edition, The Chemical Rubber Co., Cleveland (Section G).

4

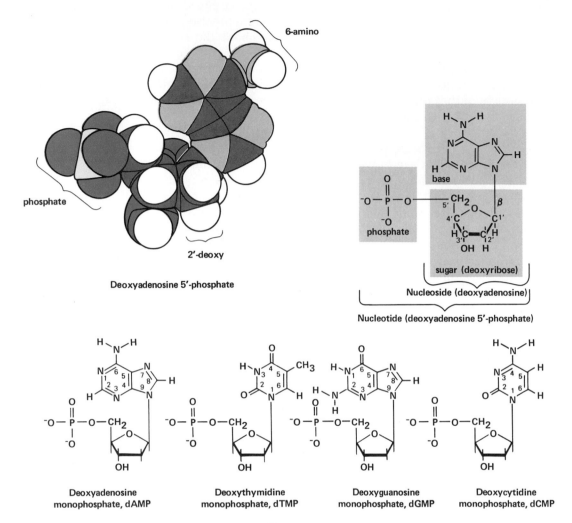

FIGURE 1-1

composition, there is only one distinction between RNA and DNA: RNA contains uracil whereas DNA contains thymine (5-methyl-uracil). Exceptions to this rule are found in transfer RNA and the DNAs of certain phages. The significance of the structures of the various bases becomes apparent when considering the secondary structure of nucleic acids (see next section). The only other distinction between RNA and DNA is in their sugar-phosphate backbones: RNA contains only ribose, DNA contains only 2-deoxyribose.

Designating deoxyribo-containing nucleosides and nucleotides with the prefix deoxy- or deoxyribo- has been accepted in all cases except for those containing thymine. Because thymine was originally thought to occur only in DNA, it seemed redundant to use the prefix, and the terms thymidine, thymidine monophosphate, and thymidylate were accepted. However, now that the natural occur-

TABLE 1-1
Nucleic acid nomenclature

Base	Nucleoside[a]	Nucleotide[a]	Nucleic acid
PURINES			
Adenine	Adenosine	Adenylate	RNA
	Deoxyadenosine	Deoxyadenylate	DNA
Guanine	Guanosine	Guanylate	RNA
	Deoxyguanosine	Deoxyguanylate	DNA
⎰Hypoxanthine	Inosine	Inosinate	Precursor of ⎱ adenylate and⎰ guanylate
PYRIMIDINES			
Cytosine	Cytidine	Cytidylate	RNA
	Deoxycytidine	Deoxycytidylate	DNA
Thymine	Thymidine	Thymidylate	DNA
Uracil	Uridine	Uridylate	RNA

[a]*Nucleoside* and *nucleotide* are generic terms which include both ribo and deoxyribo forms. Deoxyribonucleosides and deoxyribonucleotides are designated as deoxynucleosides and deoxynucleotides, respectively, to make the names less cumbersome. Hypoxanthine nucleotides are bracketed because they are precursors of the purine nucleotides but not common components of nucleic acids.

rence of ribothymidylate has been recognized as one of the unusual nucleotides in transfer RNA and the thymine ribonucleotide (and nucleoside) have been made available by chemical and enzymatic syntheses, nomenclatural confusion does arise. In current practice, the terms thymidine and deoxyribothymidine (deoxythymidine) are used interchangeably; the ribonucleoside and ribonucleotides of thymine bear the prefix *ribo*.

Polynucleotide chains are long, unbranched polymers formed by bridges between the 5'-phosphate of one nucleotide and the 3'-hydroxyl of the sugar of the next (Fig. 1-2). The schematic diagram of nucleotide chains (Fig. 1-2, lower right) is frequently useful, and is viewable either in the horizontal or vertical direction. Thus the backbone of the polymer is a chain of:

$$\text{phosphate} \diagup \underset{5'\ \text{sugar}\ 3'}{} \diagdown \text{phosphate} \diagup \underset{5'\ \text{sugar}\ 3'}{} \diagdown \text{phosphate}$$

The important linkage is the 3',5'-phosphodiester bridge. This linkage is especially vulnerable to hydrolytic cleavage, chemically and enzymatically. It should be clear that such cleavage, depending on which of the phosphate ester bonds is broken, may yield either the naturally occurring 5'-phosphonucleoside or its 3' isomer (Fig. 1-3).

Because of free rotation about the phosphate bonds, polynucleotide chains were originally considered to be highly flexible and to assume essentially random conformations. However, detailed

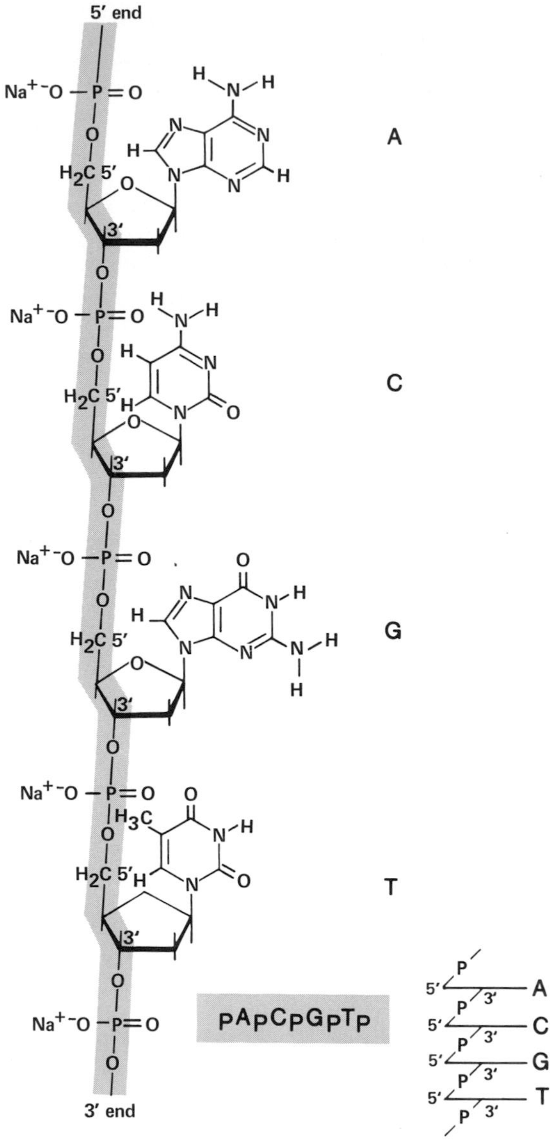

FIGURE 1-2
Segment of a polydeoxynucleotide as a sodium salt.

studies[8] now indicate that the main degrees of rotational freedom are limited to the two O–P bonds in the phosphodiester linkage and the glycosyl bond of base to sugar. Thus, there are numerous restraints imposed on the nucleic acid backbone that permit it relatively few conformations and which impart a considerable stiffness to it. To regard the single-strand polynucleotide chain as a completely random coil is therefore unwarranted.

8. Levitt, M. (1972) in *Polymerization in Biological Systems, Ciba Foundation Symposium* **7**, Elsevier, N.Y. p. 147.

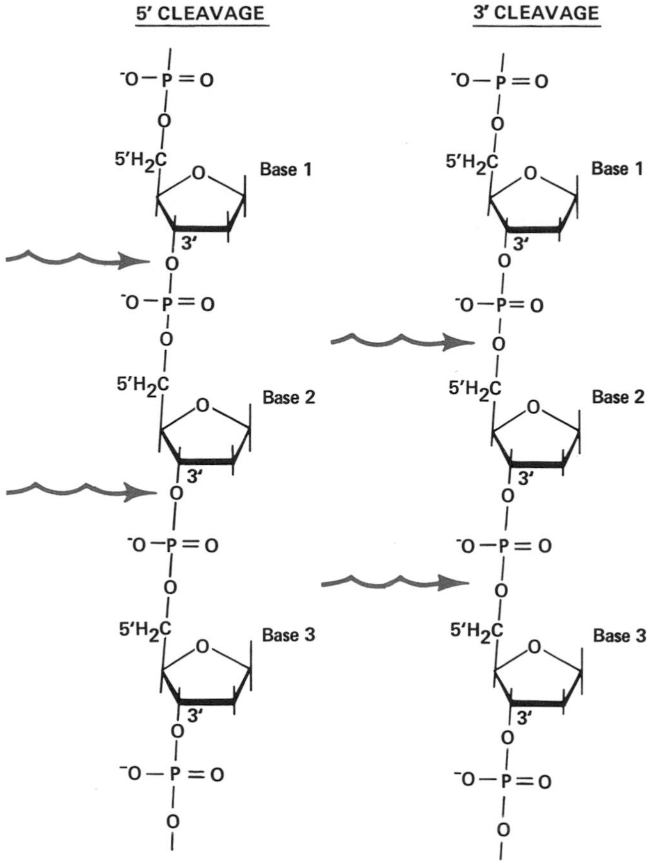

5′ CLEAVAGE **3′ CLEAVAGE**

FIGURE 1-3
Cleavages of polynucleotide chains. 5′-Cleavage
yields 5′ mononucleotides (5′-phosphoryl termini);
3′-cleavage yields 3′ mononucleotides (3′-phosphoryl
termini).

3. The Double Helix[9,10]

Two distinctions among the nitrogenous bases of the nucleotides
are crucial to the secondary structure of nucleic acids. One rests on
the presence of keto and amino groups that provide opportunities
for hydrogen bonding. On this basis T or U, both keto compounds,
can pair with A, an amino compound, by a hydrogen bond. G and
C, each having both keto and amino groups, can form two hydrogen
bonds. In fact, an additional hydrogen bond can be formed between
the ring nitrogens in the A–T pair and also in the G–C pair (Fig. 1-4).

The second important distinction among the bases is that they
come in two sizes: the pyrimidines T or U, and C are smaller than
the purines A and G. However, the base pairs A–T and G–C prove

9. Watson, J. D. and Crick, F. H. C. (1953) *CSHS* **18**, 123.
10. Arnott, S., Wilkins, M. H. F., Hamilton, L. D. and Langridge, R. (1965) *JMB* **11**, 391.

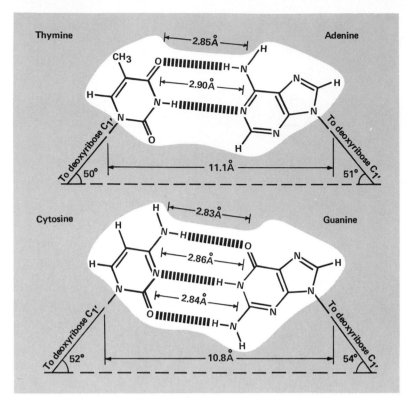

FIGURE 1-4
Hydrogen-bonded base pairs. Interatomic distances and angles are nearly the same for A–T and G–C pairs.

to be identical in size: a pair of pyrimidines would be much smaller, a pair of purines much larger. It turns out that the A–T and G–C base pairs have not only the same size but the same shape!

These base-pairing characteristics of the nucleotides are responsible for the pairing of two chains to form a rigid, strongly stabilized duplex (Figs. 1-5, 1-6). When two chains align themselves in this way, they assume a double-helical structure, as worked out in detail by X-ray diffraction studies[11] of DNA fibers and by model building. The DNA double helix has several remarkable features.

(i) The sugar-phosphate backbones of the two chains form right-handed helices with the same helix dimensions and a common helix axis. The diameter of the helix is 20 Å. The helix has two grooves, one deep and one shallow.

(ii) The sugar-phosphate backbones of the two chains are connected by hydrogen bonds between the bases, whose surface planes are perpendicular to the helix axis (Fig. 1-7). The bases are on the inside; the backbone is on the outside. The planes of neighboring base pairs are 3.4 Å apart. There are ten base pairs for each turn of

11. Arnott, S., Wilkins, M. H. F., Hamilton, L. D. and Langridge, R. (1965) *JMB* **11**, 391.

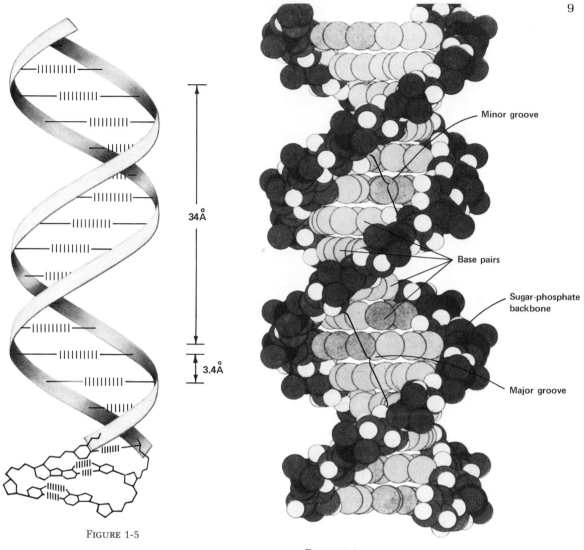

FIGURE 1-5

FIGURE 1-6
Space-filling model of the DNA double helix.

Minor groove

Base pairs

Sugar-phosphate
backbone

Major groove

34Å

3.4Å

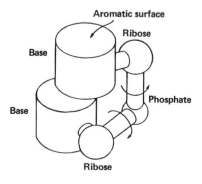

Aromatic surface

Ribose

Base

Base

Phosphate

Ribose

FIGURE 1-7
Schematic model of a dinucleotide. This
model shows flat aromatic surfaces of the
bases, rigidity of the backbone, and the
main degrees of freedom in the bonds.
(Courtesy of Dr. M. Levitt.)

the helix; thus each base pair is rotated 36° relative to its neighbor and each full turn has a length of 34 Å.

(iii) Stacking of the bases through hydrophobic interactions between their flat aromatic surfaces (Fig. 1-7) stabilizes the helical structure against the electrostatic repulsive force between the negatively charged phosphate groups. This stabilizing energy may be equal to or greater than that of hydrogen bonding of bases between chains.

(iv) The two chains are antiparallel. Chemically, they are arranged in opposite directions, that is, the structure . . . P–5′-sugar-3′–P . . . opposes the structure . . . P–3′-sugar-5′–P . . . . When the chains are traced from 5′ end to 3′ end (5′ → 3′) the direction is up the helix on one chain and down it on the other (Fig. 1-8).

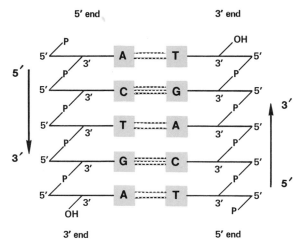

FIGURE 1-8
Segment of a DNA duplex showing antiparallel orientation of the complementary chains.

(v) The only base pairs allowed by the structure are A–T and G–C. All base pairs have the same shape and size. The glycosidic bonds have the same positions and orientations in these pairs. Modified bases (see Chapter 2, Section 10) can also enter into base pair formation provided they form hydrogen bonds with a partner.

(vi) An important element of symmetry in the helix is a two-fold (dyad) axis of rotation passing through the plane of each base pair and relating the N–C glycosidic bonds. Rotation of a base pair by 180° allows the sugar and phosphate groups of the two antiparallel chains to have the same conformations. As a result, the connection between the glycosidic carbon atoms can be made by any one of the four base pairs: A–T, T–A, G–C and C–G.

(vii) Because all four base pairs fit equally well, any sequence within a strand is possible and the double helix has a uniform diameter throughout its length.

(viii) Rotation about the N-glycosidic bond permits various geo-metrical relationships of the base to the sugar. The range of con-formations designated _anti_ occurs more frequently than those conformations that are 180° opposite, designated _syn_ (Fig. 1-9)[12].

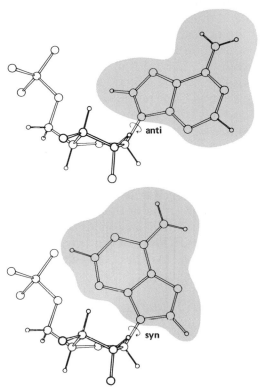

FIGURE 1-9
Model for rotation about the N–C glycosidic bond. The _anti_ conformation is favored over the _syn_. (Courtesy of Professor M. Sundaralingam.)

(ix) Antiparallel double helices can also be formed between a DNA chain and an RNA chain. These are called DNA-RNA hybrids. Examples of hybrid base pairs are deoxy A·ribo-U, deoxy T·ribo A, deoxy C·ribo G, and deoxy G·ribo C. These base pairs occur in the transcriptional hybrids generated by RNA polymerase (see Chapter 10).

A significant consequence of this structure is that any sequence of bases is tolerated in a given chain. DNAs of identical base composi-tion can be entirely different in sequences. Given the sequence and orientation of one chain, however, base-pairing rules and anti-parallelism determine the sequence and orientation of the comple-mentary chain.

---

12. Sundaralingam, M. (1969) _Biopolymers_ **7**, 821.

The most important consequence of the duplex model for DNA structure is the introduction of the concept of complementarity[13]. It provided the explanation for accurate replication of a very long chain (Fig. 1-10). This inherent feature of DNA structure is the basis not only of its replication but also of its capacity to transmit information. Complementarity has come to explain transcription and translation and thus the entire sequence of events in the expression of genetic functions.

There is an interesting exception to the rule that naturally occurring DNA is a duplex. The DNA of certain small bacterial viruses such as ϕX174 and M13 is a closed single-stranded loop about 2 μ long, containing about 6000 residues. However upon entering the host, E. coli, the viral DNA is converted to a duplex circle. Subsequently, the complementary strand synthesized on the viral DNA template is itself used repeatedly as a template in the production of viral strands for new phages. Although single-stranded DNA may be found in solution as a random coil, parts of the coil can form helical duplexes (Fig. 1-11). The resistance of about 1 percent of ϕX174 and M13 DNAs to nucleases specific for single strands indicates the presence of one or more hairpin-like duplexes[14], each containing about twenty base pairs. These may prove to be important in the replication and transcription of these viral chromosomes (Chapter 8).

4. Denaturation and Renaturation

A most remarkable physical feature of the DNA helix, and one that is crucial to its functions in replication and transcription, is the ease with which its component chains can come apart and rejoin. Many techniques have been used to measure this melting and reannealing behavior. Nevertheless, important questions remain about the kinetics and thermodynamics of denaturation and renaturation and how these processes are influenced by other molecules in the test tube and cell.

Denaturation

The chains of the DNA duplex come apart when the hydrogen bonds between the bases are broken. This melting of the secondary structure can be brought about in solution by increasing the temperature or by titration with acid or alkali. Acid protonates the ring nitrogens of A, G, and C; alkali deprotonates ring nitrogens of G and T. The unusual lability of the purine glycosidic bonds at low pH makes the use of acid undesirable in denaturation experiments.

Stability of the duplex is a direct function of the number of triple-hydrogen-bonded G–C pairs it contains; the larger the mole fraction of G–C pairs, the higher the temperature or pH of melting. In addition to hydrogen bonds, other forces stabilizing the helix are base

13. Watson, J. D. and Crick, F. H. C. (1953) CSHS **18**, 123.
14. Schaller, H. Voss, H. and Gucker, S. (1969) JMB **44**, 445.

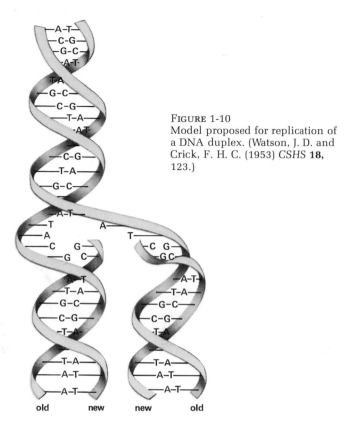

FIGURE 1-10
Model proposed for replication of a DNA duplex. (Watson, J. D. and Crick, F. H. C. (1953) *CSHS* **18,** 123.)

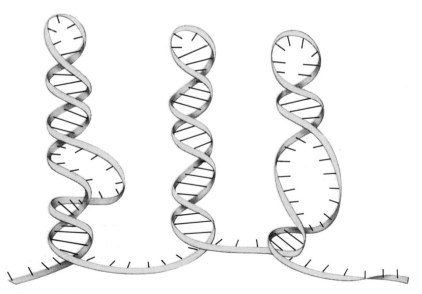

FIGURE 1-11
Secondary structure in single-stranded DNA. Duplexes are formed by annealing of complementary regions.

stacking and related hydrophobic interactions, and neutralization of the electrostatic repulsive forces of the phosphate groups by salt, divalent metal ions, and polyamines.

In thermal melting (Fig. 1-12)[15], unwinding of the chains begins in regions high in A–T base pairs[16] and proceeds to regions of progressively higher G–C content. Melting is conveniently moni-

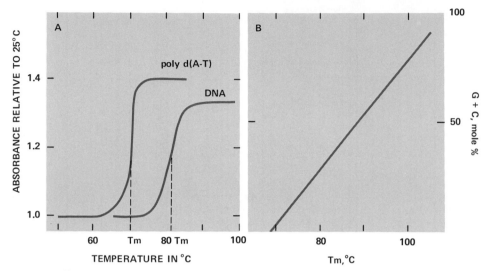

FIGURE 1-12
DNA melting curves (left) and dependence of T_m on G–C content (right).

tored by an increase in absorbance (hyperchromic effect) that results from the disruption of base stacking. The midpoint melting temperature (T_m) of the DNA is determined by its G–C content.

Renaturation

Denaturation is reversible even after the two chains have been completely separated. When complementary chains are incubated at a temperature about 25° below the T_m they begin to reassociate and eventually reform the original helix. Renaturation is usually measured by a decrease in absorbance (hypochromic effect), by sedimentation behavior, or by an insusceptibility to single-strand-specific nucleases.

An example of the precision of renaturation is seen in the annealing of one chain of a wild-type phage λ DNA with the complementary chain from a mutant λ carrying a deletion of 100 or more residues. The product looks normal in the electron microscope, except that in the location of the deletion the unmatched wild-type single strand forms a loop bridging the duplex regions[17]. Such

15. Marmur, J., Rownd, R. and Schildkraut, C. L. (1963) in *Prog. N. A. Res. and Mol. Biol.* **1**, 231.
16. Inman, R. B. (1967) *JMB* **28**, 103.
17. Davis, R. W. and Davidson, N. (1968) *PNAS* **60**, 243. Westmoreland, B. C., Szybalski, W. and Ris, H. (1969) *Science* **163**, 1343.

heteroduplex molecules make it possible to carry out refined physical mapping of the location of genes in chromosomes.

The kinetics of renaturation provide an accurate and sensitive measure of complementarity (or diversity) among a population of DNA molecules. Renaturation is a two-step process beginning with a slow nucleation event in which short sequences of matching bases form pairs. This is followed by a rapid "zippering" of the neighboring sequences to form the duplex. The rate of the overall process is governed by the second-order kinetics of the nucleation reaction. The rate constant, k_2, should be inversely proportional to the number of base pairs, N, in the DNA when the genome has no repetitive regions in its sequence[18]:

$$k_2 \propto \frac{1}{N}$$

Rates of renaturation can be presented in a plot (Fig. 1-13)[19] of C/C_0 versus log C_0t according to the second-order equation:

$$\frac{C}{C_0} = \frac{1}{1 + k_2 C_0 t}.$$

C is the concentration of single strands of one complementary type at time t, and C_0 is the concentration at t equals zero. When the fraction reassociated, C/C_0, is 50 percent, $C_0t = 1/k_2$. This value, called $(C_0t)_{1/2}$, is proportional to N and a direct measure of the complexity of the DNA.

The "C_0t curves" for viruses and bacteria (Fig. 1-13) approximate the second-order kinetics expected, and N values that are equal to the total number of bases in the chromosome. There are few if

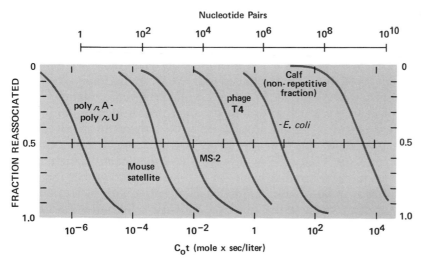

FIGURE 1-13
Reassociation of melted DNAs as a function of time and concentration. The sheared DNA was approximately 400 nucleotides long. (Britten, R. J. and Kohne, D. E. (1968) *Science* **161**, 529.) Copyright 1968 by American Association for the Advancement of Science.

18. Wetmur, J. G. and Davidson, N. (1968) *JMB* **31**, 349.
19. Britten, R. J. and Kohne, D. E. (1968) *Science* **161**, 529.

any repeating sequences. By contrast, the renaturation kinetics of eukaryotic DNA usually shows a complex curve reflecting the presence of several classes of sequences. Part of the DNA, characterized by sedimentation as a satellite of the main band (Chapter 1, Section 7), has a simple sequence repeated 10^7 times or more and renatures very rapidly. Most of the DNA is unique and renatures slowly; the N value is very large.

5. Size[20]

DNA is long and unbranched; its length has been measured directly in the electron microscope and by autoradiography. An accurate estimate of the size of duplexes containing as many as 50,000 base pairs has been obtained by ^{32}P-labeling of each of the 5′ ends[21]. Size has also been calculated from sedimentation rate, viscosity, and light-scattering measurements.

The more complex organisms require more DNA (Tables 1-2, 1-3). The simplest chromosome, such as that of the polyoma or SV40 viruses, carries information for only five to ten genes in its 1.7-micron length. Larger viruses, λ and vaccinia for example, have DNA that is ten and fifty times as long, respectively. The lengths of DNA in more complex organisms such as bacteria or animals are correspondingly greater. The chromosome of E. coli is more than one mm long (1000 times the length of the bacterium) and contains 4 million base pairs. A gross estimate of 4000 genes in E. coli can be made if one assumes that 1000 base pairs (coding for a protein of about 40,000 daltons) is the average length of a gene and there is only one copy of each.

It is difficult to obtain and measure very long DNA strands because of their sensitivity to shearing stresses during isolation. About 1960, the sizes of DNA isolated from phage, bacteria, and animal cells all measured about 15,000 base pairs (10×10^6 daltons). The size of the DNA was delimited by experimental technique, namely, by the uniform pressure of the thumb on a syringe delivering (and in the process shearing) DNA through a needle. When the shearing effect became known, and care was taken to avoid it, the measured size of DNA shot up. Phage DNA 50 μ long was then seen in the electron microscope and circular bacterial chromosomes one mm long were visualized by autoradiography. The longest stretch of animal DNA measured was also about one mm. However, human chromosomes contain from 48 to 240 million base pairs (molecular weights of 32×10^9 to 160×10^9 daltons) and their DNA, if continuous, should measure from 1.6 to 8.2 cm (Fig. 1-14).

20. Sober, H. A. (ed.) (1970) Handbook of Biochemistry, Selected Data for Molecular Biology, 2nd edition, The Chemical Rubber Co., Cleveland (Section H).
21. Weiss, B. and Richardson, C. C. (1967) JMB **23**, 405.

TABLE 1-2
Approximate mass-length equalities of DNA

	Base pairs[a]	Length		Molecular weight
		cm	μ	
Duplex DNA[b] (Na^+, B form)	3×10^3	1.0×10^{-4}	1.0	2×10^6

[a]The average molecular weight of one base pair is 660.
[b]1×10^{-12} g (one picogram) of duplex DNA contains 9.1×10^8 base pairs and is 30.9 cm long.

TABLE 1-3
Sizes of DNA molecules and genomes

Organism	Size of Genome		Shape
	Number of base pairs (thousands) (kb)	Total length[a] (mm)	
VIRUSES			
Polyoma, SV40	5.1	0.0017	Circular duplex
ϕX174	5.4	0.0018	Circular single strand; duplex replicative form
M13 (fd, f1)	5.74	0.0019	Circular single strand; duplex replicative form
P4	15.0	0.0051	Linear
T7	35.4	0.0120	Linear
P2, P22	40.5	0.0138	Linear
λ	49	0.0166	Linear
T2, T4, T6, P1	180	0.061	Linear
Vaccinia	183	0.062	Linear
BACTERIA			
Mycoplasma hominis	760	0.26	Circular
Escherichia coli	4000	1.36	Circular
EUKARYOTES			No. chromosomes (haploid)
Yeast	13,500	4.6	17
Drosophila (fruit fly)	165,000	56	4
Man	2,900,000	990	23
South American lungfish	102,000,000	34,700	19

[a]length = (kb)(3.4×10^{-4}) mm

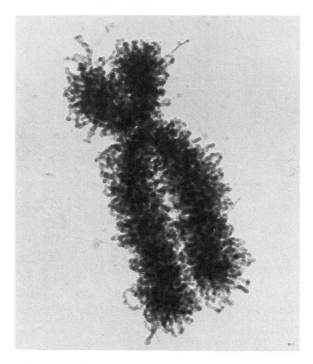

FIGURE 1-14
An electron micrograph of a human chromosome.
Chromosome XII from a HeLa cell culture. (Courtesy
of Dr. E. Du Praw.)

Analyses that exploit the viscoelastic properties of DNA have
now revealed the intact lengths of DNA in chromosomes of many
animal and bacterial cells[22]. The DNA is chromosome-sized. Chromosomal DNA up to four cm long, corresponding to molecular weights
up to 80×10^9, have been measured. There is remarkably good
agreement between the length of the DNA in the largest *Drosophila*
chromosome measured by this technique, and the DNA mass determined cytologically (Fig. 1-15). The molecules run the full length
of the chromosome without discontinuity at the centromere. The
continuity is conserved throughout the cell cycle. These data are
the most compelling that *an animal chromosome, like a bacterial
and viral chromosome, is composed of a single, threadlike molecule of DNA.*

6. Shape

DNA stretched out on an electron microscope grid appears as a
duplex circle (closed loop) and as a rod (Fig. 1-16). In nature, however, DNA is condensed in the chromosome of a cell, or in the head
of a virus[23], by several thousand-fold.

22. Kavenoff, R. and Zimm, B. H. (1973) *Chromosoma* **41**, 1; (1973) *CSHS* **38**, 1.
23. Richards, K. E., Williams, R. C. and Calendar, R. (1973) *JMB* **78**, 255.

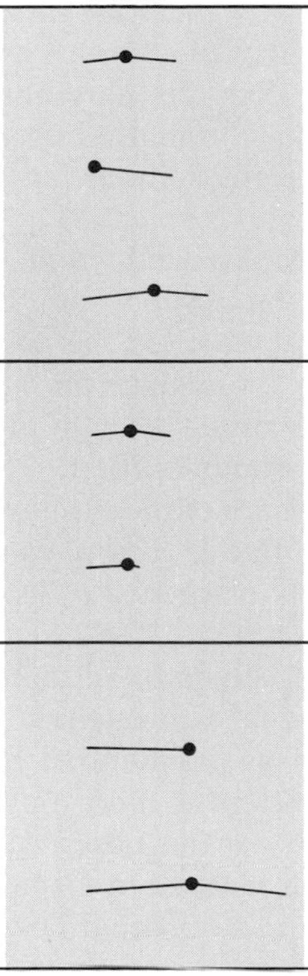

Species	Strain	Largest chromosome to scale	Molecular weight of largest DNA
D. melanogaster	wild type		41×10^9
	inversion		42×10^9
	translocation		58×10^9
D. hydei	wild type		40×10^9
	deletion		24×10^9
D. virilis			47×10^9
D. americana			79×10^9

FIGURE 1-15

Lengths of DNA in the largest *Drosophila* chromosome of several, species, determined by viscometry. (Kavenoff, R. and Zimm, B. H. (1973) *Chromosoma* **41**, 1; (1973) *CSHS* **38**, 1.)

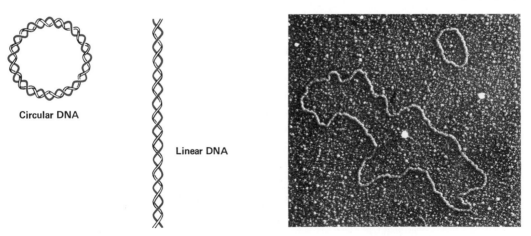

Circular DNA

Linear DNA

FIGURE 1-16

Diagrams of circular and linear forms of DNA and electron micrograph of *E. coli* plasmids (λdv, the defective phage λ plasmid, below, and the 15T⁻ miniplasmid, above). (Micrograph, courtesy of Dr. Tom Broker.)

It is unfortunate that so little is known about the arrangement of DNA in the chromosome[24]. In the structure of the eukaryotic chromosome, histone proteins have an important influence; non-histone proteins and RNA may play a part as well. The single molecule of duplex DNA in a eukaryotic chromosome is complexed with these proteins into a single chromatin fiber coiled upon itself. As a result, the length of the chromosome is far less than that of the DNA contained within it.

Circular duplexes isolated from viruses, bacteria, and mitochondria appear to be twisted upon themselves[25] (Fig. 1-17). These forms are called supercoiled (supertwisted, or superhelical) because they contain more than the standard number of base pairs appropriate to their duplex length and consequently possess extra negative turns in compensation. The supercoils are thought to originate during closure of the duplex circle, when part of its length is not base-paired but is complexed with proteins. Subsequent removal of the protein when the DNA is isolated leaves the circle underwound.

The underwound, supercoiled molecule may have the extra base pairs distributed evenly throughout its length, according to one view[26], or localized as a denatured bubble, according to another[27]. The latter model offers a more obvious, though not necessarily correct, explanation for the observed capacity of supercoiled viral DNA to be bound or cleaved at a specific region by proteins with a great affinity for single-stranded DNA[28]. By either view, it is clear that positive free energy is stored in the covalently closed, negatively supercoiled DNA and released when the DNA is uncoiled or *relaxed*.

The extra twist of the supercoil can be relaxed in several ways. Nicking of one strand, as by an endonucleolytic scission, permits the energetically favored untwisting; enzymatic ligation of the broken ends would restore the covalent circle. Alkali and heat relax the supercoil. Still another way is by intercalation of a planar molecule, such as ethidium bromide, between the stacked base pairs[29]. It has been calculated that each ethidium molecule untwists the supercoil by 12°, thus requiring 30 ethidium molecules per turn. After fully discharging all the negative supercoiling, further addition of ethidium would cause the DNA to become overwound and supercoil again, this time with positive turns.

Prokaryotic DNA is organized as a chromosome but without histones. Instead, there is a reliance on RNA in a unique fashion. Insights into the folding of DNA in the *E. coli* chromosome have come from recent success in isolating the intact structure[30,31].

24. The Organization of Genetic Material in Eukaryotes. (1973) *CSHS* **38**.
25. Vinograd, J., Lebowitz, J. and Watson, R. (1968) *JMB* **33**, 173.
26. Bauer, W. and Vinograd, J. (1970) *JMB* **47**, 419; Davidson, N. (1972) *JMB* **66**, 307.
27. Kato, A. C., Bartok, K., Fraser, M. J. and Denhardt, D. T. (1973) *BBA* **308**, 68.
28. Beard, P., Morrow, J. F. and Berg, P. (1973) *J. Virol.* **12**, 1303; Morrow, J. F. and Berg, P. (1973) *J. Virol.* **12**, 653.
29. Crawford, L. V. and Waring. M. J. (1967) *JMB* **25**, 23.
30. Stonington, G. O. and Pettijohn, D. E. (1971) *PNAS* **68**, 6; Pettijohn, D. E., Miles, E. and Hecht, R. (1973) *CSHS* **38**, 31.
31. Worcel, A. and Burgi, E. (1972) *JMB* **71**, 127; Worcel, A., Burgi, E., Carlson, L. and Robinton, J. (1973) *CSHS* **38**, 43; Delius, H. and Worcel, A. (1974) *JMB* **82**, 107.

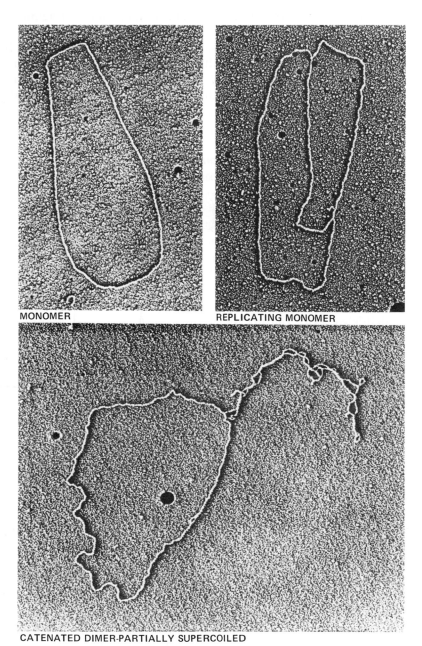

MONOMER

REPLICATING MONOMER

CATENATED DIMER-PARTIALLY SUPERCOILED

FIGURE 1-17
Supercoiled and relaxed forms of mitochondrial DNA from polyoma virus-transformed baby hamster kidney cells. The contour length of the monomer is about five microns. (Courtesy of Professor J. Vinograd.)

smaller peaks separated from the main peak (Fig. 1-19)[34]. The DNA in the smaller peak, or in the shoulder of the main peak, is called satellite DNA and may have a G–C content considerably different from that of the main DNA.

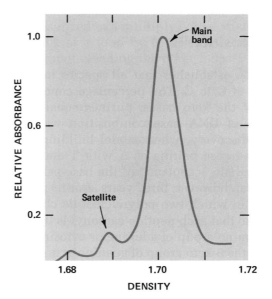

FIGURE 1-19
Embryonic nuclear DNA of *Drosophila melanogaster*.
(Courtesy of Professor D. S. Hogness.)

As many as 10^6 copies of certain sequences are reiterated in a satellite of mouse DNA[35]. An unusual satellite found in several crab species is essentially a copolymer of alternating A and T, with only three percent of the residues G and C interspersed[36]. The poly d(A–T)-like satellite (see Chapter 4, Section 15) may constitute as much as a fourth of the total DNA. In one species of *Drosophila*, the sequences of three satellite bands have been determined to be a particular heptanucleotide repeated some 10^7 times[37]. The sequences, lined up as shown below, are seen to be closely related:

5'-ACAAATT-3'
5'-ACAAACT-3'
5'-ATAAACT-3'.

A major satellite band, representing 11 percent of the DNA of the kangaroo rat, has the repeating decanucleotide sequence[38]:

5'-GGACACAGCG-3'.

34. Schachat, F. H. and Hogness, D. S. (1973) *CSHS* **38**, 371.
35. Pyeritz, R. E., Lee, C. S. and Thomas, C. A. Jr. (1971) *Chromosoma* **33**, 284.
36. Swartz, M. N., Trautner, T. A. and Kornberg, A. (1962) *JBC* **237**, 1961.
37. Gall, J. (1973) **in** *Symp. Mol. Cytogen.* (Hamkalo, B. and Papaconstantino, J., eds.) Plenum Press.
38. Fry, K., Poon, R., Whitcome, P., Idriss, J., Salser, W., Mazrimas, J. and Hatch, F. (1973) *PNAS* **70**, 2642.

A quantitative measure of the fraction of total DNA that is highly reiterated can be readily obtained by determining the speed with which fragmented, _denatured_ DNA reanneals.

Although the genetic function of satellite DNAs is for the most part undetermined, the genes for ribosomal RNA have been found in a discrete satellite band and are often reiterated there many times. In a satellite DNA band of the South African clawed toad (_Xenopus laevis_), the genes for ribosomal RNA are repeated hundreds of times along the length of the DNA; the 18S and 28S RNA genes are separated by segments (spacers) of relatively high G–C content[39].

An interesting aspect of DNA sequence, which may impart important secondary structural features to DNA, is a 180°-rotational symmetry residing in short sequences. Examples of such sequences are found at the sites of restriction and modification of DNA (Chapter 9, Section 5) and in a potentially regulatory site near a tRNA gene (Chapter 11, Section 4).

A poorly understood yet almost invariable aspect of base composition is the modification of certain bases, usually C, by methylation (Chapter 2, Section 10)[40]. In animal and plant DNAs, as many as 10 percent of C residues bear a 5-methyl group; and a small percentage of purine residues have N-methyl groups. There are also modifications of bases or substitutions by analogs in bacterial and phage DNAs that are thought to protect the DNA against intracellular nucleases (Chapter 9).

8. Functions

Ultimately, the structure of DNA must be understood in terms of all its functions, and the reverse is equally true. We must resolve and reconstitute each function outside the cell, and understand its location and regulation inside the cell. Yet, as we recognize an increasing number and variety of DNA functions, and as the known number of enzymes needed to catalyze them multiplies, this goal might seem to be receding. A further barrier to solving the problem lies in artifacts of academic organization more than in the nature of cellular organization. DNA structure is studied by chemists, function by biologists, and enzymes by biochemists.

It may be useful in this first chapter to enumerate the current ideas of DNA functions and the enzymes that act on DNA (Fig. 1-20, Table 1-4). In the succeeding chapters, as functions and enzymes are considered in more detail, physiologic functions for each of the enzymes will be suggested. But at the outset some limitations in attempting to assign enzyme functions should be emphasized.

(i) Enzyme activity measured in vitro is only a gross qualitative estimate of activity in the cell, and need not be related to its in vivo

39. Brown, D. D., Wensink, P. C. and Jordan, E. (1972) _JMB_ **63**, 57; Brown, D. D. (1973) _Sci. Amer._ **229**, No. 2, 21.
40. Evans, H. H., Evans, T. E. and Littman, S. (1973) _JMB_ **74**, 563.

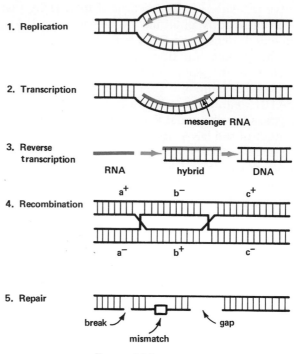

1. Replication

2. Transcription

messenger RNA

3. Reverse transcription

RNA   hybrid   DNA

a^+ b^- c^+

4. Recombination

a^- b^+ c^-

5. Repair

break   mismatch   gap

FIGURE 1-20
Some DNA functions.

TABLE 1-4
DNA functions and enzymes

Functions	Scheme	Enzymes involved*
Replication	$n(dNTP) + DNA \rightarrow (dNMP)_n \cdot DNA$	1, 3, 4, 5, 6
Transcription	$n(rNTP) \xrightarrow{DNA} (rNMP)_n$	3
Reverse transcription	$n(dNTP) \xrightarrow{RNA} (dNMP)_n$	2
Recombination†	$DNA + DNA \rightarrow DNAs$ $a^+b^-c^+ + a^-b^+c^- \rightarrow a^+b^+c^+ + a^-b^-c^-$ $+ a^+b^-c^+ + a^-b^+c^-.$	1, 4, 5, 6
Repair	DNA (with defect) $\rightarrow$ DNA (repaired)	1, 4, 5
Modification	DNA $\rightarrow$ DNA (methylated or glycosylated)	7
Restriction	DNA $\rightarrow$ DNA fragments	5

*Enzymes

1. DNA polymerases (DNA-directed).
2. Reverse transcriptase (RNA-directed DNA polymerase).
3. RNA polymerases (DNA-directed RNA polymerase).
4. Ligases.
5. Exonucleases and endonucleases.
6. Unwinding proteins.
7. Methyl and glycosyl transferases.

† $a^+b^-c^+$ represents arrangements of these genes in a segment of DNA.

functions. Finding a high or a low enzyme level, or none at all, does not prove its state of activity in the cell. The specific nature of the enzyme action and its control may be different in vivo.

(ii) Genetic studies, by providing missense, nonsense, and deletion mutants, may supply crucial information for evaluating roles of enzyme activities in vivo. However, missense and nonsense mutants may be leaky, and low levels of an enzyme activity in the cell may be sufficient to sustain growth rate and gross physiologic behavior. Enzyme activity completely lost by gene deletion may also be replaced by an auxiliary enzyme system.

(iii) Functions, grossly defined, include many individual steps, and different functions may have individual steps in common. Thus, different functions may involve many of the same enzymes. On the other hand, enzyme activities that appear to be similar in vitro may have entirely different locations and roles in the cell.

2

Biosynthesis of DNA Precursors

1. De Novo and Salvage Pathways of Nucleotide Synthesis[1]

The nature of DNA biosynthesis depends on the properties of its building blocks, their origins, structures, and functions. It is not often fully appreciated that details in the biosynthesis of DNA precursors, still not entirely clear to us today, have a profound influence on overall DNA synthesis by controlling rates and other characteristics as well[2].

The building blocks of DNA synthesis are deoxyribonucleoside monophosphates (deoxynucleotides), which with an added pyrophosphoryl group become activated precursors, deoxyribonucleoside triphosphates. Deoxyribonucleotides arise from direct reduction

---

1. Kornberg, A. (1957) **in** *The Chemical Basis of Heredity* (W. D. McElroy and B. Glass, eds.) Johns Hopkins Press, Baltimore, p. 579.
2. Bjursell, G. and Reichard, P. (1973) *JBC* **248**, 3904.

of ribonucleotides at the level of the di- or triphosphate (Fig. 2-1). (Also see Section 6 of this chapter.)

In virtually all cells, two fundamentally different kinds of pathways are used for the synthesis of nucleotides. One is a de novo pathway, in which ribose phosphate, certain amino acids, CO_2, and NH_3 are combined in successive reactions to form the nucleotides. Neither the free purine bases (adenine and guanine), nor the pyrimidine bases (cytosine, uracil, and thymine), nor the corresponding nucleosides (or deoxyribonucleosides) are intermediates in the de novo pathway. In other words, _nucleotides may be synthesized_ _without passing through a stage or pool containing free bases or_ _nucleosides._

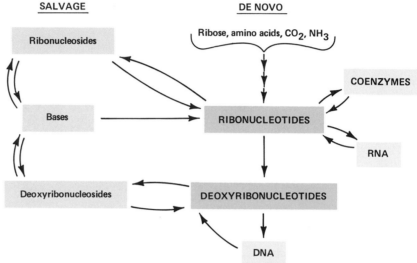

FIGURE 2-1
Schemes for salvage and de novo pathways.

By contrast, cells also have various mechanisms for making use of the free bases and nucleosides ultimately produced in the breakdown of nucleic acids (Fig. 2-1). There is a variety of routes whereby these components of nucleotides are converted back to the nucleotides themselves. Thus, we recognize here a second kind of pathway to which we apply the term _salvage pathway_. The use of the word "salvage" should not be taken in a sense that minimizes the importance of this pathway.

A cell lacking one of the enzymes required to operate the de novo pathway at an adequate rate for growth and maintenance absolutely requires help from a salvage pathway. The lack of an intact de novo pathway may be due to the inherent genetic makeup of the cell or to a deficiency resulting from disease, drugs or poisons, or physiological stresses. In any case, when the rate of production of nucleotides is inadequate, salvage of preformed purine and pyrimidine bases and nucleosides produced from breakdown of nucleic acids becomes vital.

Bases and nucleosides are salvaged from intracellular and extra-cellular sources. Breakdown of nucleic acids within bacterial, plant, and animal cells is a major source. This salvage pathway usually goes unnoticed because its efficiency often exceeds the capacity of extracellular components to enter the cell and compete with the pools of these endogenously generated precursors.

Extracellular sources from which bases and nucleosides are salvaged are the growth medium for bacteria and the extracellular fluids for animals. The latter fluids are supplied by digestion of foods in the alimentary tract and by the breakdown of cells elsewhere in the organism.

2. Purine Nucleotide Synthesis de Novo[3,4]

One of the first indications of how the purine skeleton is synthesized came from experiments in which various labeled nutrients were fed to pigeons. These creatures excrete their waste nitrogen as the purine uric acid in solid deposits. Chemical degradation of the uric acid revealed the ring positions of the labeled precursors. By this technique, glycine was shown to be the source of carbons 4 and 5 and nitrogen 7, formate the source of carbons 2 and 8, and bicarbonate, or CO_2, the source of carbon 6 (Fig. 2-2). Because of mixing of their

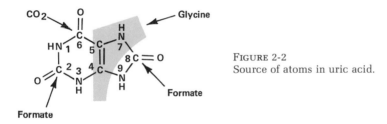

FIGURE 2-2
Source of atoms in uric acid.

nitrogen with intracellular pools of NH_3, aspartate as the source of nitrogen 1 and glutamine of nitrogens 3 and 9 could not be defined by nutritional experiments with intact cells and organisms.

Cell-free extracts of animal and bacterial cells are usually active in catalyzing the whole sequence of reactions of purine biosynthesis (Fig. 2-3). This fortunate circumstance has made it possible to separate and characterize the enzymes involved and thus elucidate the details of every step in the pathway. Purine biosynthesis is now an exceedingly instructive chapter in the fundamentals of general biochemistry and of biosynthetic metabolism specifically.

The initial reactant in purine assembly is phosphoribosylpyrophosphate, PRPP. This activated form of ribose 5-phosphate is formed by direct transfer of the β,γ-pyrophosphoryl group of ATP[5] to carbon 1 of the sugar. The transfer forms the α-ribosyl pyrophos-

3. Henderson, J. F. and Paterson, A. R. P. (1973) "Nucleotide Metabolism" Academic Press, N.Y.
4. Henderson, J. F. (1972) "Regulation of Purine Biosynthesis" Am. Chem. Soc. Monograph 170. Washington, D.C.
5. Khorana, H. G., Fernandes, J. F. and Kornberg, A. (1958) JBC 230, 941.

phate stereoisomer of ribose 5-phosphate. PRPP, which had actually

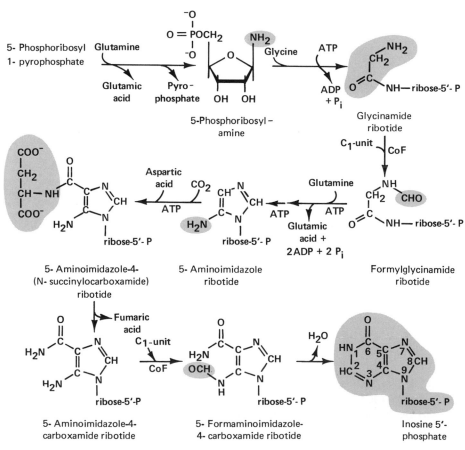

been discovered as an intermediate in pyrimidine biosynthesis[6], is a key reactant not only in purine and pyrimidine biosynthesis de novo, but also in the biosynthesis of the pyridine nucleotide coenzymes DPN and TPN, the amino acids histidine and tryptophan, and in the utilization, or salvage, of preformed bases and nucleosides (see Section 8 of this chapter).

Condensation of PRPP with glutamine, followed by removal of glutamic acid from the product, fixes the first atom, nitrogen 9, of the purine skeleton to the ribose phosphate. It is noteworthy that

6. Kornberg, A., Lieberman, I. and Simms, E. (1955) *JBC* **215**, 389.

FIGURE 2-3
Pathway of purine biosynthesis.

the condensation of glutamine with PRPP involves an inversion of configuration of the substituent on carbon 1 of the sugar and thereby establishes the β configuration of the future nucleotide.

The succession of condensations and ring closures shown in Fig. 2-3 produces _the primary product of purine biosynthesis, the hypoxanthine nucleotide inosinate_, from which adenylate and guanylate are then derived (Fig. 2-4). (See Table 1-1 in Chapter 1 for a summary of the names of bases, nucleosides, and nucleotides).

The individual steps in the sequences illustrated in Figures 2-3 and 2-4 are of great interest. They are the enzymatic loci for metabolic errors found in mutants or in other disordered states. Mutants that lack adenylosuccinase, required for synthesis of adenylate (Fig. 2-4), also fail to convert the succinyl intermediate in purine biosynthesis (Fig. 2-3); purification of the enzymes responsible for these

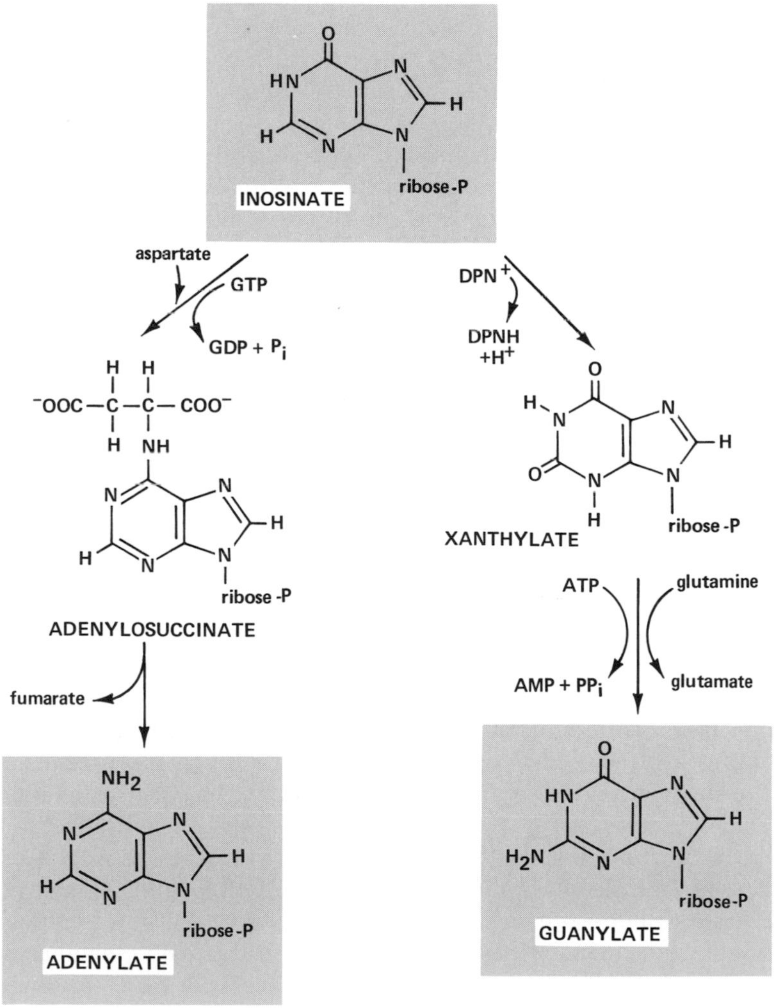

FIGURE 2-4
Conversion of inosinate to adenylate and guanylate.

two reactions indicates that the same one is used for both. The enzymes in these pathways are involved in the fine regulation of rates of reactions that are important in growth and cellular functions, and are targets for inhibitors (Table 2-1) used in the treatment of proliferative diseases, when blocking of DNA synthesis is desired.

There are two crucial stages in purine biosynthesis in E. coli and both appear to be finely regulated. One is the initial commitment to purine synthesis, which is under feedback regulation. The end product inhibits the pathway at the very first step[7]. The amino transferase that condenses PRPP with glutamine is inhibited by AMP, ADP, and ATP at one control site on the enzyme and by GMP, GDP, and GTP at another. A second stage of regulation involves a nicely tuned recognition of which of the purine nucleotides is in excess or in short supply. This regulation occurs at the divergence of the pathway from inosinate to adenylate and to guanylate. Adenylate synthesis requires GTP as the activating cofactor. Thus an abundance of GTP, signifying a relative lack of ATP, accelerates production of adenylate. Conversely, the rate of inosinate conversion to guanylate depends on the level of ATP, the specific cofactor for this reaction.

There is now available an assortment of inhibitors that can interrupt the pathway of purine biosynthesis at many points (Table 2-1). The antibiotic azaserine[8], which is isosteric with glutamine, competes with glutamine to block the reaction sequence at each of the two steps where glutamine participates. Amethopterin, aminopterin, and sulfonamides interrupt the purine pathway indirectly by reducing the supply of the folate coenzyme. Amethopterin and aminopterin prevent the regeneration of tetrahydrofolate from dihydrofolate produced in thymidylate synthesis (see Section 7 of this chapter). Sulfonamides prevent the initial synthesis of the folate coenzyme through competition with p-aminobenzoate, a precursor in the biosynthesis of folate.

The folate coenzyme is essential for the insertion of the one-carbon units at the carbon 8 and carbon 2 positions. A stringent need for the folate coenzyme at the latter step leads to the accumulation of 5-aminoimidazole-4-carboxamide ribose phosphate when E. coli is inhibited by sulfonamides, and in mutant strains of other microorganisms unable to form purines. Breakdown of this intermediate to the free base 5-aminoimidazole-4-carboxamide and its excretion into the culture medium by inhibited and mutant cells presented one of the earliest clues to the pattern of assembly of the purine skeleton.

The schemes in Figs. 2-3 and 2-4 illustrate the close interrelationships of purine biosynthesis and amino acid metabolism. Not only are certain amino acids such as glutamine, glycine, and aspartic acid precursors of the purines, but the amino acid histidine is actually derived from an adenine nucleotide, ATP (Fig. 2-5). Furthermore, the major byproduct of histidine biosynthesis is the 5'-phospho-

7. Henderson, J. F. (1972) "Regulation of Purine Biosynthesis" Am. Chem. Soc. Monograph 170. Washington, D.C.; Roth, D. G. and Deuel, T. F. (1974) JBC **249**, 297.
8. Mizobuchi, K. and Buchanan, J. M. (1968) JBC **243**, 4853; also following paper.

TABLE 2-1
Inhibitors of nucleotide biosynthesis

Name	Formula	Inhibitory action
PURINE INHIBITORS		
Azaserine	$\overset{-}{N}=\overset{+}{N}=CH-\underset{\underset{O}{\|\|}}{C}-O-CH_2-\underset{\underset{NH_2}{\|}}{CH}-COOH$	Competes with glutamine
2,6-Diaminopurine[a]		Incorporated into DNA as analog of adenine
8-Azaguanine[a]		Incorporated into DNA as analog of guanine
PYRIMIDINE INHIBITORS		
5-Bromouracil[a]		Incorporated into DNA as analog of thymine
5-Azaorotate		Synthesis of orotate
6-Azauridine		Synthesis of uridylate
6-Azacytidine		Synthesis of hydroxymethylcytosine
FOLATE INHIBITORS		
Amethopterin (Methotrexate)[b]		Dihydrofolate reductase
Aminopterin[c]		Dihydrofolate reductase
Sulfonamides		Synthesis of folate

[a]Formula shown in Fig. 2-13.
[b]4-amino-10-methylfolic acid.
[c]4-aminofolic acid.

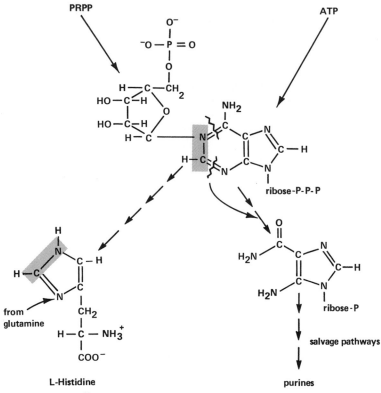

FIGURE 2-5
Histidine biosynthesis from ATP and PRPP.

ribosyl derivative of 5-amino-4-imidazole carboxamide, which is salvaged by insertion of carbon 2 to form inosinate, thus restoring the purine nucleotide.

The purine biosynthetic pathway is essentially the same sequence of reactions in a wide range of species such as *E. coli*, yeast, pigeon, and man. The unity of biochemistry was dramatically illustrated in the first half of this century by the discovery of the virtual identity of alcoholic fermentation in yeast with glycolysis in muscle, and was reiterated by the delineation of purine pathways during the decade of the 1950s. The validity of this unity of biochemical mechanism in nature is what justifies the confidence that studies of basic metabolic sequences in *E. coli* can be generally extended to the cells of more complicated and highly evolved organisms.

3. Pyrimidine Nucleotide Synthesis de Novo[9]

Biosynthesis of pyrimidine nucleotides differs in pattern from that of purine nucleotides in that the pyrimidine skeleton is assembled

9. O'Donovan, G. A. and Neuhard, J. (1970) *Bact. Rev.* **34,** 278; Jones, M. E. (1971) *Adv. Enzym. Regul.* **9,** 19; Blakeley, R. L. and Vitols, E. (1968) *ARB* **37,** 201; Koerner, J. F. (1970) *ARB* **39,** 291; Henderson, J. F. and Paterson, A. R. P. (1973) "*Nucleotide Metabolism*", Academic Press, N.Y.

first and is subsequently attached to ribose phosphate. The key pyrimidine intermediate is orotic acid.

The pathway of orotic acid synthesis (Fig. 2-6) starts with the transfer of a carbamyl group from carbamyl phosphate to aspartate by the enzyme aspartate transcarbamylase[10]. This first step to

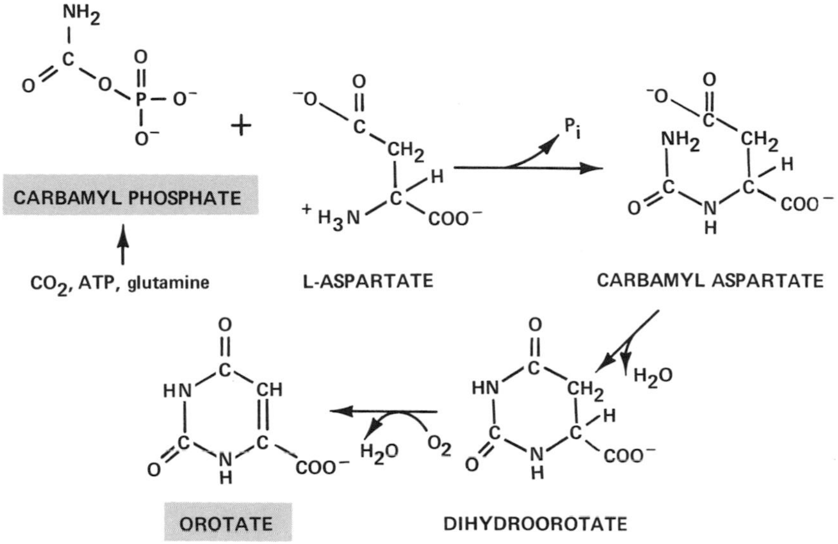

FIGURE 2-6
Pathway of orotate biosynthesis.

pyrimidine synthesis is a key point for regulating the rate of commitment of carbamyl phosphate, and aspartate, to the pathway. The control of the enzyme in *E. coli* is through feedback inhibition by CTP, a terminal product of the pathway. There is also feedback activation by ATP, a nucleic acid precursor whose accumulation signals the need for pyrimidine nucleotides. In fact, studies of the control of aspartate transcarbamylase have pioneered and consistently advanced our general understanding of the regulation of enzyme activity.

Cyclization of carbamyl aspartate, by dehydration, produces dihydroorotate, the reduced form of orotate. Oxidation of the latter yields orotate. The first clue that dihydroorotate serves as an intermediate in this pathway came from recognizing that microorganisms compelled to subsist on orotic acid as a source of carbon first reduced it to dihydroorotate[11]. The enzyme isolated from such cells proved to be a flavoprotein, with only a feeble capacity to oxidize the dihydroorotate back to orotate. By contrast, another enzyme, isolated subsequently, catalyzed an extremely facile oxidation of

10. Jacobson, G. R. and Stark, G. R. (1973) in *The Enzymes* (P. D. Boyer, ed.) 3rd edition **9**, 226, Academic Press, N.Y.
11. Lieberman, I. and Kornberg, A. (1953) *BBA* **12**, 223.

dihydroorotate to orotate[12]. From genetic and biochemical studies it has become clear that the flavoprotein is involved in the metabolism of orotate for energy needs and the other enzyme is used for biosynthesis of orotate. Thus biosynthesis is managed by a set of enzymes largely or entirely distinct from that which degrades orotate.

An appreciation in this instance of the duality of pathways of biosynthesis and degradation strengthens our belief that, in all cells, biochemical operations of energy production and biosynthesis are distinct from one another. This point needs emphasis because the early history of biochemistry focused on pathways of energy metabolism. And, when interest in biosynthesis began to develop, it was tacitly assumed that because individual degradative steps were catalyzed in both directions by the isolated enzymes, so must the entire pathway of degradation be reversible in a biosynthetic direction.

Another point of general interest relating to the physical organization of the pathway emerges from recent studies of pyrimidine biosynthesis[13]. The enzymes catalyzing each of the steps to orotate were known to be soluble and molecularly dispersed in broken cell extracts, as are the enzymes of purine biosynthesis. The components of such "soluble" enzyme systems were commonly regarded to be randomly dispersed in the cell sap. However, when due precaution was taken in handling cell extracts, it became clear that the enzymes catalyzing the first two or three steps in the pyrimidine pathway were organized as a complex. In fact, the extracted complex was tight enough to prevent mixing of exogenous carbamyl phosphate with carbamyl phosphate being generated from CO_2, glutamine, and ATP. It seems plausible that the entire pathways of pyrimidine and purine nucleotide synthesis are organized as multienzyme complexes and that such complexes may even be fixed to membranes or localized to compartments in the cell, where their rates and integration with other pathways can be finely regulated.

The second stage of pyrimidine biosynthesis involves condensation of orotate with PRPP and elimination of inorganic pyrophosphate (Fig. 2-7). This reaction, analogous to the initial step of purine biosynthesis, fixes the pyrimidine in the β-configuration. Subsequent decarboxylation produces uridylate. Cytidylate originates as CTP from UTP via an amination reaction, also under refined allosteric control[14].

Analogs of orotate and uridine (Table 2-1) inhibit at stages indicated in Fig. 2-7. 5-Azaorotate competes with orotate in the condensation with PRPP, and 6-azauridine presumably as a nucleotide, blocks the decarboxylation of orotidylate, the condensation product. 5-Azacytidine inhibits pyrimidine nucleotide synthesis specifically in phage T4 DNA replication and will be best described later (Section 10 of this chapter).

12. Taylor, W. H., Taylor, M. L. and Eames, D. F. (1968) *J. Bact.* **91**, 2251.
13. Davis, R. H. (1972) *Science* **178**, 835; Shoaf, W. T. and Jones, M. E. (1973) B. **12**, 4039.
14. Long, C. W., Levitzki, A. and Koshland, D. E., Jr. (1970) *JBC* **245**, 80.

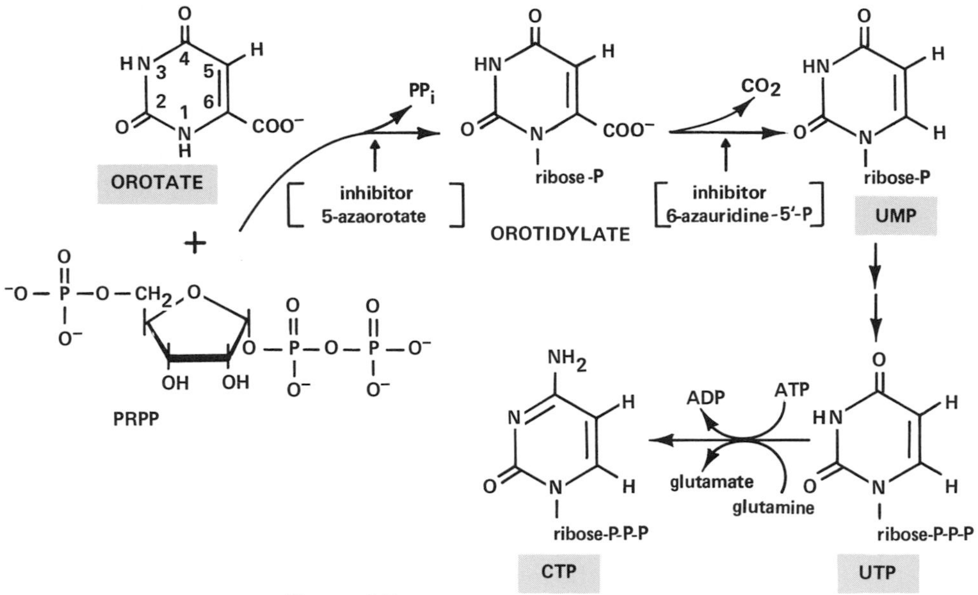

FIGURE 2-7
Conversion of orotate to UTP and CTP.

The de novo route of synthesis of thymidylate, from deoxyuridy-late, will be discussed in Section 7 of this chapter, following the description of conversion of ribonucleotides to deoxynucleotides.

4. Nucleoside Monophosphate Conversion to Triphosphate[15,16]

Nucleoside monophosphates do not participate directly in nucleic acid biosynthesis, but are first converted to diphosphates, and then to triphosphates, which are incorporated into RNA and DNA. The only other metabolic fate for the nucleoside monophosphates is, as already mentioned, interconversion, by amination and deamination reactions and by oxidations and reductions of their purine or pyrimidine moieties (Figure 2-4).

Conversion of the nucleoside monophosphates to diphosphates is catalyzed by kinases that are specific for each base, but nonspecific with regard to ribose or deoxyribose sugar[17]. As an example:

$$\text{(deoxy)adenylate}$$
$$\text{(d)AMP} + \text{ATP} \xrightleftharpoons{\text{kinase}} \text{(d)ADP} + \text{ADP}$$

15. Oeschger, M. P. and Bessman, M. J. (1966) JBC **241**, 5452; Nelson, D. J. and Carter, C. E. (1969) JBC **244**, 5254.
16. Parks, R. E. and Agarwal, R. P. (1973) **in** The Enzymes (P. D. Boyer, ed.) 3rd edition **8**, 307 Academic Press, N.Y.
17. Op. cit. in footnote 15.

The general reaction is:

$$(d)NMP + ATP \rightleftharpoons (d)NDP + ADP$$

Synthesis of the diphosphate is favored by the regeneration of ATP, through oxidative or substrate-level phosphorylations, and by the further phosphorylation of the diphosphate. ATP is the usual donor but may be replaced in some instances by dATP or other triphosphates.

The fate of nucleoside diphosphates is to become triphosphates. In only one instance, their reduction to deoxyribonucleotides, do they participate directly in a proved biosynthetic sequence. Conversion of the di- to triphosphates is achieved through the action of a powerful, ubiquitous, and *nonspecific* nucleoside diphosphate kinase[18]. The enzyme shows no preference either for purines or pyrimidines (including a large variety of synthetic analogs), or for ribose or deoxyribose. This lack of specificity applies to the donor triphosphate as well as the recipient diphosphate.

The reaction may be formulated:

$$(d)\text{nucleoside } X\text{-PP} + (d)\text{nucleoside } Y\text{-PPP} \rightleftharpoons$$
$$(d)\text{nucleoside } X\text{-PPP} + (d) \text{ nucleoside } Y\text{-PP}.$$

The donor, as may be expected, is almost invariably ATP and the reaction is favored by the same factors as in the previous reaction.

5. Significance of Pyrophosphate-Releasing Reactions[19]

In the synthesis of purine and pyrimidine nucleotides, condensations with PRPP are key reactions and are accompanied in each instance by the release of pyrophosphate, PP_i. The free energy change in these and other PP_i-releasing reactions is relatively small, and the synthetic direction of the reaction is demonstrably reversible upon addition of PP_i. However, an extremely potent inorganic pyrophosphatase is found in all cell extracts examined. PP_i is thus rarely detectable. Assuming that the PP_i released is promptly hydrolyzed, reaction sequences, as for example nucleotide synthesis, can be pulled far in the synthetic direction as follows:

$$\text{orotate} + \text{PRPP} \rightleftharpoons \text{orotidylate} + PP_i$$
$$\underline{PP_i + H_2O \longrightarrow 2\ P_i}$$
$$\text{Sum: orotate} + \text{PRPP} \longrightarrow \text{orotidylate} + 2\ P_i$$

As mentioned already, and as will be discussed in Chapter 3, the active forms of the nucleotides in nucleic acid synthesis are

18. Parks, R. E. and Agarwal, R. P. (1973) in The Enzymes (P. Boyer, ed.) 3rd edition **8**, 307 Academic Press, N.Y.
19. Kornberg, A. (1962) in *Horizons in Biochemistry* (M. Kasha and B. Pullman, eds.) Academic Press, N.Y. p. 251.

nucleoside triphosphates. Polymerization involves repeated nucleotidyl additions with the release of PP_i:

$$nNTP \rightleftarrows (NMP)_n + nPP_i$$

Removal of PP_i by hydrolysis drives the synthesis to completion. The same pattern has been observed in the biosynthesis of coenzymes, proteins, lipids, and polysaccharides:

Coenzymes: Nicotinamide mononucleotide $+$ ATP $\rightleftarrows$ DPN
$+ PP_i$ (see Fig. 3-1).

Proteins: Amino acid $+$ tRNA $+$ ATP $\rightleftarrows$ amino acyl tRNA
$+$ AMP $+ PP_i$

Lipids: Fatty acid $+$ coenzyme A $+$ ATP $\rightleftarrows$ fatty acyl CoA
$+$ AMP $+ PP_i$

Polysaccharides: Sugar-P $+$ UTP $\rightleftarrows$ UDP-sugar $+ PP_i$

Because of the reversibility of the PP_i-releasing reaction, an enzyme catalyzing it has often been called a pyrophosphorylase, by analogy with the phosphorylases. However, these enzymes are better understood as synthetases because the inevitable hydrolysis of PP_i makes pyrophosphorolysis unlikely. Phosphorylases by contrast usually serve in degradative pathways. They catalyze phosphorolytic cleavages producing phosphorylated intermediates which conserve the energy of glycosidic and phosphodiester-linked compounds. But the main direction of the pathways involving phosphorylases is catabolic and toward energy production.

A PP_i-releasing synthetase, if inevitably coupled with inorganic pyrophosphatase, may seem an energetically wasteful process, but for the end achieved, the expense is trivial. The de novo synthesis of DNA consumes the equivalent of at least 60 ATPs per nucleotide. At the cost of only 2 additional ATP molecules (resulting from the hydrolysis of PP_i), the integrity of DNA is assured against the vicissitudes of PP_i concentration.

6. Ribonucleotide Reduction to Deoxyribonucleotide[20, 21]

The production of deoxyribonucleotides directly from ribonucleotides was observed first in studies with whole animals. Cytidine labeled with an isotope of a particular specific activity in cytosine and another in ribose had the same ratio of specific activities after incorporation into DNA. How a ribonucleoside is reduced directly to the corresponding deoxyribonucleoside in animal tissue did not become clear until the enzyme responsible for the reaction was

20. Reichard, P. (1967) "The Biosynthesis of Deoxyribose" Ciba Lectures in Biochemistry, John Wiley & Sons, N.Y.
21. Reichard, P. (1972) Adv. Enzym. Regul. 10, 3.

discovered in extracts of *E. coli*. This enzyme, ribonucleoside diphosphate reductase[22], catalyzes the reaction: ribonucleoside diphosphate $\rightarrow$ deoxyribonucleoside diphosphate. The mechanism involves a nonheme iron atom bound to the enzyme and hydride ion (H⁻) attack (Fig. 2-8). The newly introduced hydrogen atom, as verified by ³H-label and NMR analyses, is added exclusively to the 2′ carbon and in the position previously occupied by the hydroxyl group. The source of reducing power is thioredoxin, a small protein (12,000 daltons) with two sulfhydryl groups in the cysteine residues, which are oxidized to cystine in the process. The thioredoxin is later restored to the reduced form by TPNH and a specific thioredoxin reductase[23].

Ribonucleoside diphosphate reductase (245,000 daltons) consists of two polypeptide subunits, B1 (160,000 daltons) and B2 (78,000 daltons) (Fig. 2-8). B2 is made up of two identical polypeptide chains and contains the iron. B1, composed of two nonidentical chains, binds the ribonucleotide substrate as well as a variety of

22. Thelander, L. (1973) *JBC* **248**, 4591.
23. Brown, N. C., Canellakis, Z. N., Lundin, B., Reichard, P. and Thelander, L. (1969) *EJB* **9**, 561.

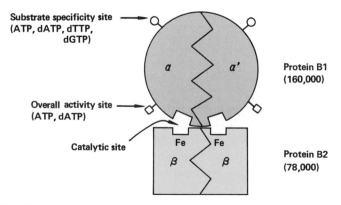

FIGURE 2-8
Schemes for the ribonucleoside diphosphate reductase reaction (top) and organization of the enzyme subunits (bottom). (Courtesy of Professor P. Reichard.)

nucleotides which regulate the reduction of ribonucleotides in highly specific ways, according to need for the deoxynucleotides. For example, ATP in *E. coli* (Fig. 2-9) is a stimulator (effector) for pyrimidine nucleotide reduction. As small amounts of dTTP accumulate, it also is a stimulator, in this case, for reduction of purine nucleotides. It is believed that various ratios of effectors bound at the substrate specificity site induce different states of the enzyme.

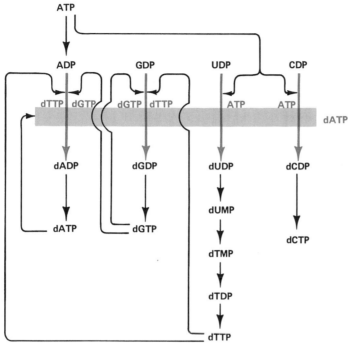

FIGURE 2-9
Scheme for allosteric control of ribonucleoside diphosphate reductase. (Courtesy of Professor A. L. Lehninger.)

Since the enzyme has four allosteric sites and can bind four different effectors, it is evident that a large number of enzyme states are possible. As deoxynucleotides build up in high concentrations, they become inhibitors. Among these, dATP is the strongest inhibitor of the overall activity in reduction of <u>all</u> ribonucleotides. This pattern of regulation has been deduced from observations with the pure enzyme and may, however, be different under operating conditions in the cell.

An illustration of the complexity and refinement of these regulatory patterns is the control of the ribonucleoside diphosphate reductase that bacteriophage T4 induces in *E. coli* upon infection[24]. This phage enzyme, which must maintain the more rapid production of deoxyribonucleotides for phage DNA synthesis, appears identical with the host enzyme except that dATP is the principal effector for

24. Berglund, O. (1972) *JBC* **247**, 7270, 7276.

reduction of pyrimidine nucleotides, not an inhibitor. Evidently, the nucleotide economy of the phage-infected cell requires a regulatory pattern different from that in the normal cell.

A puzzling feature of E. coli ribonucleoside diphosphate reductase is that its level of activity in vitro under optimal conditions is only about one percent of that required to account for the rates of deoxyribonucleotide synthesis in vivo. Nevertheless, the defect in a temperature-sensitive mutant unable to synthesize DNA (dna F, see Chapter 7, Section 10) has been identified in the structural gene for the B1 reductase polypeptide[25]. Furthermore, permeable cells (see Chapter 7, Section 10) can achieve reductase rates nearer the level in vivo [26]. Evidently some important association affecting the rate is disrupted in obtaining the soluble enzyme. The disturbingly low reductase activities measured in cell extracts illustrate, as emphasized earlier (Chapter 1, Section 7), how approximate and crude the estimate of enzyme activity in vivo usually is.

Nucleotide reduction in lactobacilli differs from the E. coli system[27]. Vitamin B_{12} coenzyme is the cofactor for the reductase rather than nonheme iron, and the substrate is ribonucleoside triphosphate, not diphosphate. Other features of the reaction including allosteric controls are quite similar to those in E. coli.

It seems likely that the reductase system operating in various animal cells has properties resembling those of the E. coli mechanism[28]. In animal cells the strong inhibition of DNA synthesis by hydroxyurea has also been correlated with the demonstration in vitro that its target is the interruption of reductase[28a].

7. Origin of Thymine[29]

Methylation of deoxyuridylate, dUMP, is the de novo route of synthesis of thymidylate. (See Chapter 1, Section 1-2 for the comment on nomenclature). The enzyme thymidylate synthetase catalyzes transfer of a hydroxymethyl ($-CH_2OH$) group from methylene tetrahydrofolate to deoxyuridylate; this is accompanied by reduction of the group to $-CH_3$ (Fig. 2-10), at the expense of tetrahydrofolate, to form deoxythymidylate[30]. Regeneration of the tetrahydrofolate coenzyme from the dihydrofolate product is essential not only for sustained thymidylate synthesis but also for other biosynthetic events that depend on the coenzyme, such as the synthesis of purines, histidine, and methionine.

25. Fuchs, J. A., Karlstrom, H. O., Warner, H. R. and Reichard, P. (1972) NNB **238**, 69.
26. Reichard, P. personal communication.
27. Goulian, M. and Beck, W. S. (1966) JBC **241**, 4233; Beck, W. S. (1967) JBC **242**, 3148.
28. Gleason, F. K. and Hogenkamp, H. P. C. (1972) BBA **227**, 466.
28a. Reichard, P. (1972) Adv. Enzym. Regul. **10**, 3.
29. O'Donovan, G. A. and Neuhard, J. (1970) Bact. Rev. **34**, 278; Jones, M. E. (1971) Adv. Enzym. Regul. **9**, 19; Blakeley, R. L. and Vitols, E. (1968) ARB **37**, 201; Koerner, J. F. (1970) ARB **39**, 291; Henderson, J. F. and Patterson, A. R. P. (1973) "Nucleotide Metabolism", Academic Press, N.Y.
30. Blakeley, R. L. (1969) in The Biochemistry of Folic Acids and Related Pteridines, John Wiley & Sons, N.Y. p. 253; Crusberg, T. C., Leary, R. and Kisliuk, R. L. (1970) JBC **245**, 5292.

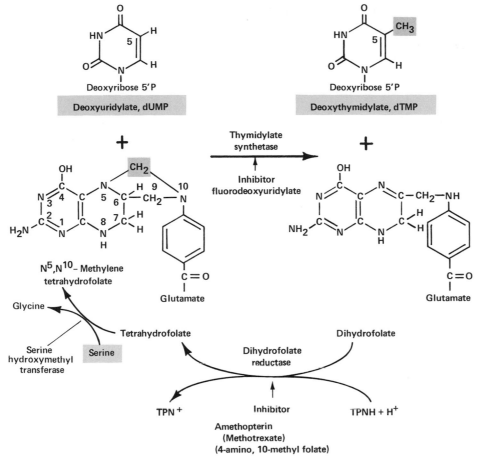

FIGURE 2-10
Conversion of deoxyuridylate to thymidylate
by thymidylate synthetase.

The thymidylate synthesis step affords opportunities for interrupting synthesis of DNA and the growth of cells, especially rapidly proliferating cancer cells. Clinically, inhibitors of either thymidylate synthetase or dihydrofolate reductase have been used. Thymidylate synthetase can be blocked by fluorodeoxyuridylate. The latter, derived by salvage reactions (see below) from administered fluorouracil or fluorodeoxyuridine, forms a covalent compound with the enzyme and immobilizes it[31]. Dihydrofolate reductase also can be virtually totally inhibited with suitable levels of the folate antagonist, amethopterin (Methotrexate; 4-amino-10-methylfolate) or aminopterin (4-aminofolate).

Thymidylate synthesis through salvage of thymidine is discussed in Section 9 of this chapter.

31. Santi, D. V. and McHenry, C. S. (1972) *PNAS* **69**, 1855; Santi, D. V., McHenry, C. S. and Sommer, H. B. (1974) *B.* **13**, 471.

8. Salvage Pathways of Nucleotide Synthesis[32-34]

Whereas de novo routes of purine and pyrimidine nucleotide synthesis are present and are virtually the same in all normal cells, the salvage routes are far more diverse in their nature and distribution. Available details about the salvage enzymes, especially in animal cells, are rather fragmentary and inadequate. Attempts to select animal cell mutants defective in nucleic acid metabolism or to understand changes in cancer cells are sometimes thwarted by the lack of such detail.

The principal salvage reactions are outlined in Fig. 2-11 and in Table 2-2.

The individual reactions are as follows.

(i) Conversion of Base to Ribonucleotide[35]

Because this reaction is theoretically reversible by pyrophosphorolysis, the enzyme was named nucleotide (inosinate or guanylate) pyrophosphorylase. Practically, however, the reaction is not reversible and provides a direct route for salvage of purine bases. In some bacterial cells uracil is salvaged in a similar way.

The importance of this pathway in man is illustrated by the Lesch-Nyhan syndrome[36], a severe disease of infants who lack the hypoxanthine-guanine nucleotide pyrophosphorylase. These free purines generated as breakdown products are not converted to nucleotides, with a consequent failure to regulate purine biosynthesis. The purines are diverted by oxidation to uric acid in excessive amounts, and gout-like tissue damage results.

Mutant cells defective in this particular enzyme can be selected in media containing azaguanine. Normal cells incorporate azagua-

---

32. Kornberg, A. (1957) in *The Chemical Basis of Heredity* (W. D. McElroy and B. Glass, eds.) Johns Hopkins Press, Baltimore p. 579; O'Donovan, G. A. and Neuhard, J. (1970) *Bact. Rev.* **34**, 278. Jones, M. E. (1971) *Adv. Enzym. Regul.* **9**, 19; Blakeley, R. L. and Vitols, E. (1968) *ARB* **37**, 201; Koerner, J. F. (1970) *ARB* **40**, 291.
33. Murray, A. W., Elliott, D. C. and Atkinson, M. R. (1970) *Prog. N. A. Res.* **10**, 87.
34. Murray, A. W. (1971) *ARB* **40**, 811.
35. Op. cit. in footnote 34.
36. Stanbury, J. B. Wyngaarden, J. B. and Fredrickson, D. S., eds. (1972) "*The Metabolic Basis of Inherited Disease*", 3rd edition, McGraw-Hill, N.Y. (Part VI).

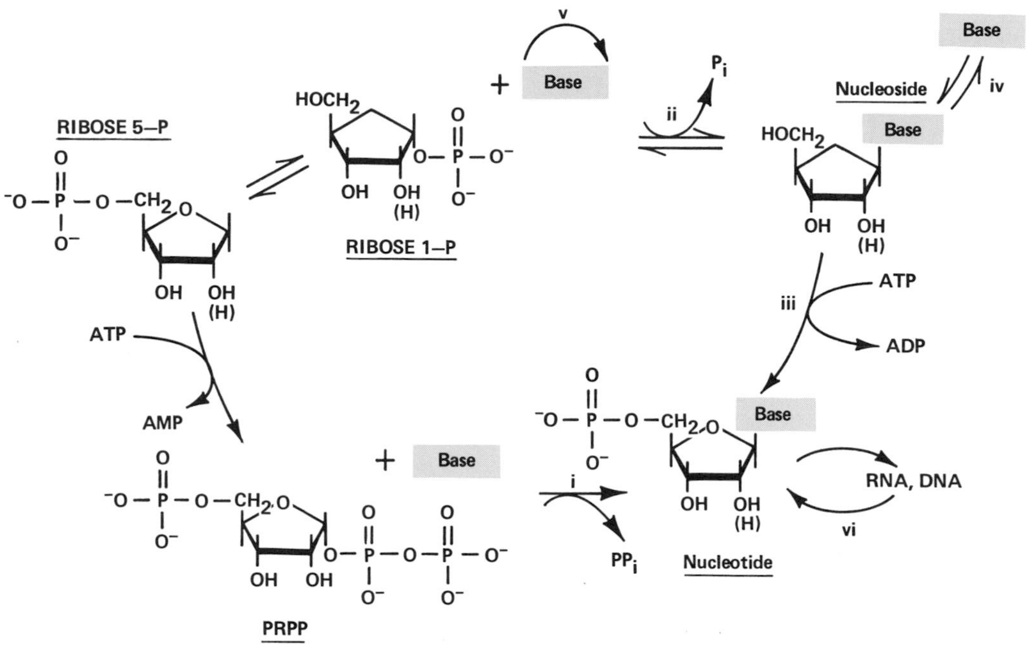

FIGURE 2-11
Principal routes for salvage of bases and nucleosides.

TABLE 2-2
Salvage pathways of nucleotide synthesis

Pathway	Reaction	Enzymes
(i) Conversion of base to ribonucleotide	PRPP + base $X \rightarrow$ ribonucleotide X	Nucleotide pyrophosphorylase
(ii) Interconversion of bases and nucleosides	Ribose 1-P + base $X \rightleftarrows$ nucleoside X + P_i	Nucleoside phosphorylase
(iii) Conversion of nucleoside to nucleotide	Nucleoside + ATP $\rightarrow$ nucleotide + ADP	Nucleoside kinase
(iv) Base exchange into deoxynucleosides	Base X + deoxyribonucleoside $Y \rightleftarrows$ Base Y + deoxyribonucleoside X	Nucleoside transglycosylases
(v) Interconversions by base alterations	Adenine $\rightarrow$ hypoxanthine Cytosine $\rightarrow$ uracil	Deaminases
(vi) Reutilization of nucleotides	rNMP $\rightarrow$ rNDP; dNMP $\rightarrow$ dNDP	Nucleoside monophosphate kinases
	rNDP $\rightleftarrows$ rNTP; dNDP $\rightleftarrows$ dMTP	Nucleoside diphosphate kinase

nine via hypoxanthine-guanine nucleotide pyrophosphorylase into DNA and the resulting toxicity to DNA destroys them. Those mutants that lack nucleotide pyrophosphorylase are spared.

(ii) Interconversion of Bases and Nucleosides

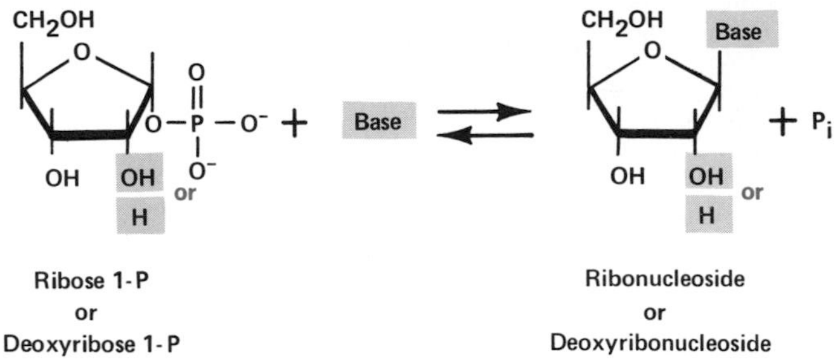

Ribose 1-P
or
Deoxyribose 1-P

Ribonucleoside
or
Deoxyribonucleoside

This equilibrium catalyzed by nucleoside phosphorylase, as with other phosphorolytic reactions, is not displaced in either direction. The reaction can therefore be used to salvage a preformed base if ribose 1-phosphate or deoxyribose 1-phosphate is provided, or to salvage sugar-phosphate and/or base from one available nucleoside to make another. For example, a generous supply of deoxyadenosine in the medium provides deoxyribose 1-phosphate and enables a thymidylate synthetase mutant to use thymine (see Section 9 of this chapter).

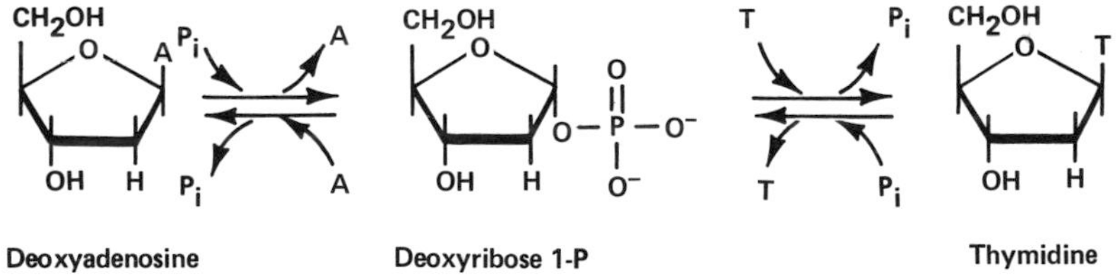

Deoxyadenosine

Deoxyribose 1-P

Thymidine

Three nucleoside phosphorylases are known: two for pyrimidine (uridine and thymidine) ribo- and deoxyribonucleosides, and one for purine (hypoxanthine and guanine) ribo- and deoxyribonucleosides.

Cells defective in de novo inosinate synthesis are not known to possess an inosine kinase but can still utilize the hypoxanthine from supplies of inosine. They do so by using the sequence of phosphorylase and pyrophosphorylase reactions. Ribose 1-phosphate generated in the phosphorylase reaction is readily converted to PRPP by a sequence of phosphoribomutase and PRPP synthetase reactions.

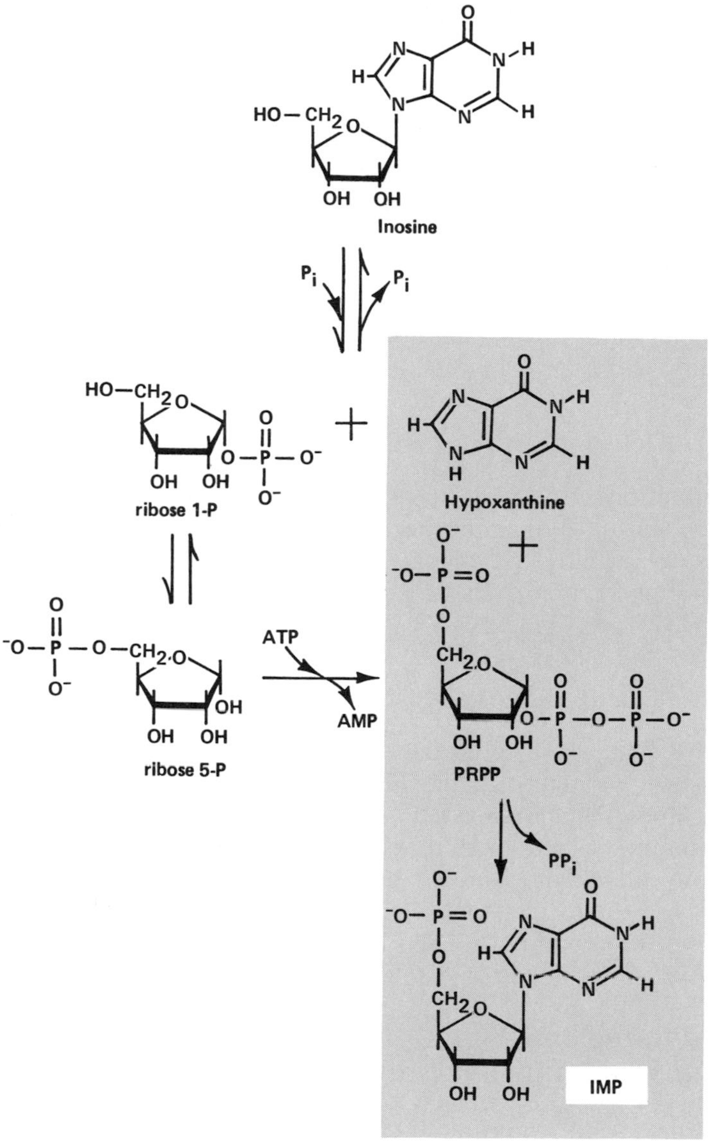
Inosine
Pi
Pi
ribose 1-P
Hypoxanthine
ATP
AMP
ribose 5-P
PRPP
PPi
IMP

(iii) Direct Conversion of a Nucleoside to a Nucleotide

$$\text{Nucleoside} + \text{ATP} \longrightarrow \text{Nucleotide} + \text{ADP}$$

Thymidine kinase, unique among the nucleoside kinases in its ubiquity and importance, will be described in the next section. Certain other specific kinases (e.g. adenosine, uridine) have been found in some cells. On the other hand, a kinase for inosine is uncommon, as has been mentioned, and this nucleoside is salvaged by the less direct routes described above.

(iv) Base Exchange Into Deoxynucleosides

Bases, including a large variety of chemically synthesized purines and pyrimidines, can be exchanged by enzymes found in some bacteria. In one case[37] the exchange is direct and does not involve phosphate or a phosphorylated intermediate:

$$\text{Base } X + \text{deoxyribonucleoside } Y \rightleftarrows \text{base } Y$$
$$+ \text{ deoxyribonucleoside } X$$

In another case[38], transfer of the deoxyribosyl group by thymidine phosphorylase takes place with the intermediate bound to the enzyme, and requires nonstoichiometric amounts of phosphate.

(v) Interconversions by Base Alterations

Deamination of adenine (base, nucleoside, and nucleotide) to the corresponding hypoxanthine forms, and of cytosine to the uracil forms, by specific enzymes is relatively common. Amination of the keto forms occurs in the de novo pathway (IMP $\rightarrow$ AMP; XMP $\rightarrow$ GMP; UTP $\rightarrow$ CTP).

(vi) Reutilization of Nucleotides

Perhaps the most important and certainly the least recognized salvage of nucleic acid breakdown products is of the nucleotides themselves. Nucleotides, with rare, or no exceptions, do not traverse cell membranes and must therefore be dephosphorylated to be taken into the cell. By the same token, nucleotides do not leave intact cells and may be reutilized for nucleic acid synthesis with relatively little dilution from external nucleotides, nucleosides, or free bases.

9. Thymine and Thymidine Conversion to Thymidylate[39, 40]

The salvage route by which thymidylate is synthesized from thymine and thymidine (Fig. 2-12) appears to be common to all cells. These pathways are also frequently exploited experimentally for labeling DNA, and to influence the rate of DNA synthesis[41].

    Thymine is converted to the nucleotide in two steps. The first is the conversion to thymidine by thymidine phosphorylase:

$$\text{thymine} + \text{deoxyribose 1-P} \rightleftarrows \text{thymidine} + P_i.$$

37. Beck, W. S. and Levin, M. (1963) *JBC* **238**, 702.
38. Krenitsky, T. A. (1968) *JBC* **243**, 2871; Gallo, R. C. and Breitman, T. R. (1968) *JBC* **243**, 4936.
39. O'Donovan, G. A. and Neuhard, J. (1970) *Bact. Rev.* **34**, 278; Jones, M. E. (1971) *Adv. Enzym. Regul.* **9**, 19; Blakeley, R. L. and Vitols, E. (1968) *ARB* **37**, 201; Koerner, J. F. (1970) *ARB* **40**, 291.
40. Cleaver, J. E. (1967) *"Thymidine Metabolism and Cell Kinetics"* Elsevier, N.Y.
41. Bjursell, G. and Reichard, P. (1973) *JBC* **248**, 3904.

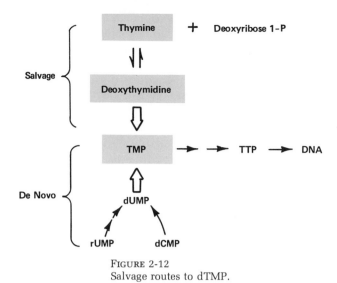

FIGURE 2-12
Salvage routes to dTMP.

The second step is the phosphorylation of thymidine by thymidine kinase:

$$\text{thymidine} + \text{ATP} \longrightarrow \text{dTMP} + \text{ADP}.$$

Deoxyribose 1-P is usually made available through the phosphorolysis of another deoxynucleoside:

$$\text{deoxyribonucleoside} + P_i \rightleftharpoons \text{deoxyribose 1-P} + \text{base}.$$

Accordingly, formation of thymidine has been sometimes found to be facilitated by deoxyadenosine.

Thymine is also salvaged in some cells by a nonphosphorolytic reaction which exchanges purine and pyrimidine bases directly into deoxyribonucleoside:

$$\text{thymine} + \text{deoxyadenosine} \rightleftharpoons \text{thymidine} + \text{adenine}.$$

Mutants defective in thymidylate synthetase require thymine or thymidine to make DNA and grow. Mutants of *E. coli* that lack thymidine phosphorylase fail to utilize thymine; those that lack thymidine kinase fail to utilize either thymine or thymidine.

A successful technique for selection of mutants defective in thymidylate synthetase exploits the salvage pathway and the pivotal role of dihydrofolate reductase. Exposure of normal cells to aminopterin inhibits the reductase and thereby regeneration of the tetrahydrofolate coenzymes. Even incomplete . inhibition of the reductase piles up the dihydrofolate form of the coenzyme as a result of repeated cycling through thymidylate synthetase. The cell dies for lack of purines and amino acids, the biosynthesis of which depends on the folate coenzymes. However, mutants defective in thymidylate synthetase do not suffer damage from aminopterin since no dihydrofolate is produced and no reductase is required. Thymine or thymidine supplied in the medium is converted to dTMP, permitting the mutants to multiply. It must be remembered that dihydrofolate is generated only in the thymidylate synthetase re-action and that tetrahydrofolate is the only active form of the coenzyme.

Thymidine kinase is found in all growing cells and is induced, or raised to much higher levels, during certain virus infections, or under conditions of increased growth rate, such as those existing during liver regeneration. The enzyme in E. coli has been purified and was shown to phosphorylate deoxyuridine and its 5-substituted derivatives, including the fluoro, chloro, bromo, and iododeoxy-uridine nucleosides[42]. Therefore, when various halogen analogs are administered as free pyrimidines or nucleosides, they are readily converted in the cell to nucleotides. Inhibition of thymidylate syn-thetase by fluorouracil and fluorodeoxyuridine occurs via fluoro-deoxyuridylate (Fig. 2-10) and therefore requires preliminary con-version to the nucleotide by thymidine kinase. Bromo analogs follow the same route, and are incorporated into DNA with great facility, but they do not inhibit thymidylate synthetase.

Although allosteric controls are generally found at branch points in de novo biosynthetic pathways, it is indicative of the importance of thymidine salvage that thymidine kinase is also under strict regulation[42]. A variety of deoxynucleoside diphosphates serve as stimulatory effectors whereas dTTP is a powerful inhibitor. Dimeri-zation of the enzyme[43] induced by stimulatory and inhibitory nucle-otides produces active or inactive conformations.

10. Uncommon Nucleotides

In addition to the four purine and pyrimidine nucleotides already discussed, DNA normally contains other nucleotides, usually in very small amounts. Certain phage DNAs include uncommon nucleotides in large proportion, even to the exclusion of a common one. Analogs of the natural nucleotides may also be incorporated into DNA experimentally. The most varied assortment of uncommon nucleotides is found in transfer RNA. However, their nature and their origin through modification of the common nucleotides in the RNA precursor is beyond the scope of this discussion.

There are three routes by which the uncommon nucleotides (Fig. 2-13) appear in DNA.

42. Okazaki, R. and Kornberg, A. (1964) JBC **239**, 269, 275.
43. Iwatsuki, N. and Okazaki, R. (1967) JMB **29**, 155.

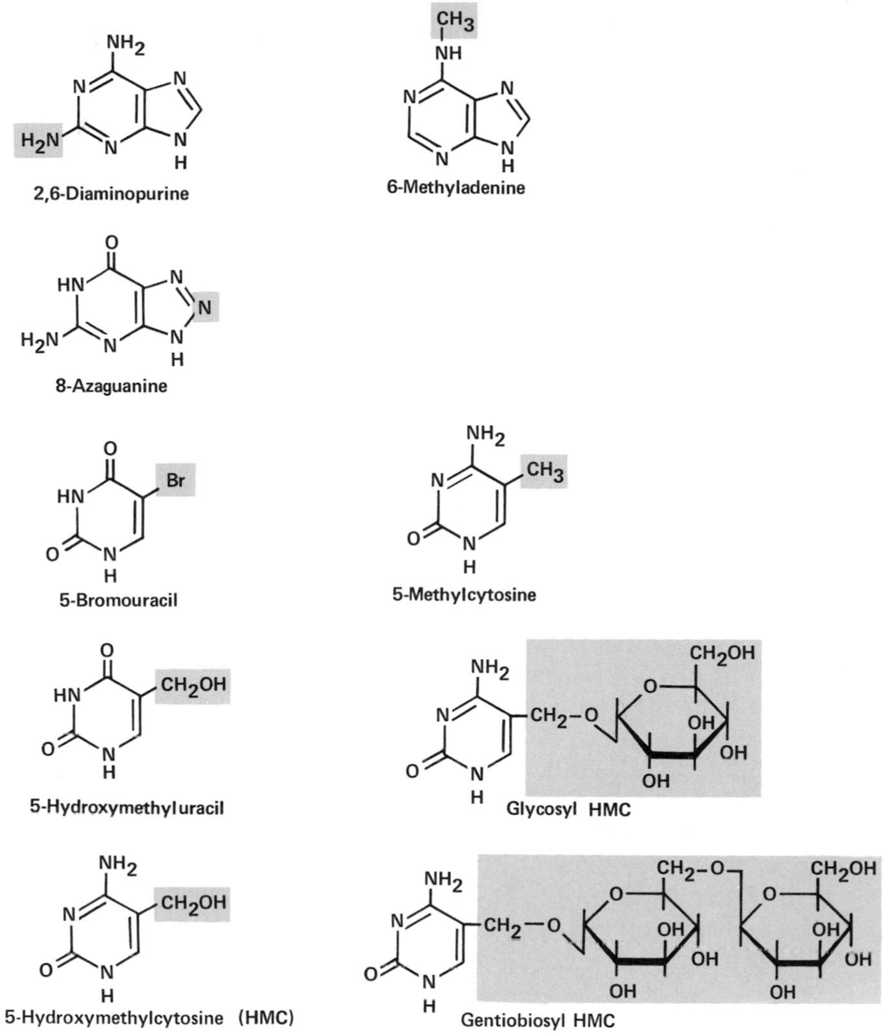

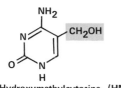

2,6-Diaminopurine

6-Methyladenine

8-Azaguanine

5-Bromouracil

5-Methylcytosine

5-Hydroxymethyluracil

Glycosyl HMC

5-Hydroxymethylcytosine (HMC)

Gentiobiosyl HMC

FIGURE 2-13
Uncommon nucleotides.

(i) Incorporation of a synthetic base analog. The analog is converted by a salvage pathway to the deoxynucleoside-5′-triphosphate and can compete with the naturally occurring nucleotide for base pairing with the template. 5-Bromouracil deoxynucleotide competes effectively enough to replace a large fraction of thymidylate residues, whereas incorporation of 8-azaguanine deoxynucleotide is very small.

(ii) Incorporation of an enzymatically altered nucleotide. Examples are found in certain phage DNAs. Hydroxymethylation of deoxycytidylate or of deoxyuridylate by enzymes induced by the phage leads to complete replacement of the C or T residues, respectively, with the 5-hydroxymethyl analogs in the DNA of the viral progeny (Chapter 8, Section 6).

(iii) Modification of DNA. The natural occurrence of 5-methylcytosine in animal and higher plant DNAs in place of a certain number of the cytosine residues (5-25 percent), and in unique patterns, could not at first be explained by polymerase action governed by base-pairing rules. Nor could the varied and distinctive patterns of glycosyl hydroxymethylcytosine (HMC) residues observed in the DNAs of phages T2, T4, and T6. The paradox was resolved by the discovery of specific glycosyl transferase enzymes which act on the polymerized DNA chain itself to produce the particular pattern of glucosylation (Chapter 8, Section 6). Similarly, specific methyl transferases were recognized which modify certain of the cytosine residues of an assembled DNA chain; other transferases are responsible for methylation of certain adenine residues. S-adenosyl methionine is the cofactor for these enzymes.

An understanding of the biosynthetic origin of HMC residues in phage T4 explains the specific inhibitory action of 5-azacytidine (Table 2-1) in T4 DNA replication. This analog by preventing the synthesis of the 5-HMC deoxynucleotide leads to incorporation of either dCMP or 5-aza dCMP residues into replicating T4 DNA. As a result, this DNA is rendered susceptible to a phage-induced nuclease which attacks DNA lacking HMC residues.

Although all the functions served by the uncommon nucleotides are probably not known, it is clear that glucosylation and certain methylations protect the DNA against destruction by nucleases in the cell. This topic will be dealt with further in Chapter 9, Sections 4, 5, and 6.

3

DNA Polymerases

1. Early Attempts at Enzymatic Synthesis of Nucleic Acids[1-4]

The presentation of the double-helix model of DNA in 1953[5] explained clearly the replication of DNA in terms of the complementarity of A to T and of G to C. Its simplicity dispelled at once a variety of earlier speculations as to how the DNA molecule might be reproduced and assembled, including the possibility of direct origin from RNA by reduction.

The suggestion was made in 1953 that A-, T-, G-, and C-containing precursors might orient themselves as base pairs with a DNA template and then be "zippered" together without any enzyme action[5]. However to the biochemist it is implicit that all biosynthetic and degradative events are catalyzed by enzymes, making possible refinements of control and specificity, as well as rapid rates of reaction.

1. Kornberg, A. (1957) in *"The Chemical Basis of Heredity"* (W. D. McElroy and B. Glass, eds.) Johns Hopkins Press, Baltimore, p. 579; (1959) *Revs. Mod. Phys.* **31**, 200.
2. Kornberg, A. (1960) *Science* **131**, 1503.
3. Kornberg, A. (1962) in *"Enzymatic Synthesis of DNA"* John Wiley & Sons, New York.
4. Kornberg, A. (1971) in *"Nucleic Acid-Protein Interactions"* (D. W. Ribbons, J. F. Woessner, and J. Schultz, eds.) North-Holland Publishing Co., Amsterdam, p. 3.
5. Watson, J. D. and Crick, F. H. C. (1953) *Nat.* **171**, 737; (1953) *CSHS* **18**, 123.

The first attempts at understanding the mechanism of nucleic acid synthesis were directed by what had been learned from studies of nucleotide and coenzyme biosynthesis. The primary products of nucleotide biosynthesis were found to be nucleoside 5'-monophosphates, and not the familiar 2' and 3' nucleotide isomers obtained from the breakdown of RNA by alkali or ribonucleases. Synthesis of the coenzyme diphosphopyridine nucleotide (DPN), a relatively simple nucleotide condensation product, illustrated the basic mechanism of nucleotidyl transfer: a nucleophilic attack by nicotinamide mononucleotide (NMN) on the α-phosphate of ATP with the elimination of PP_i (Fig. 3-1)[6]. This nucleotidyl transfer mech-

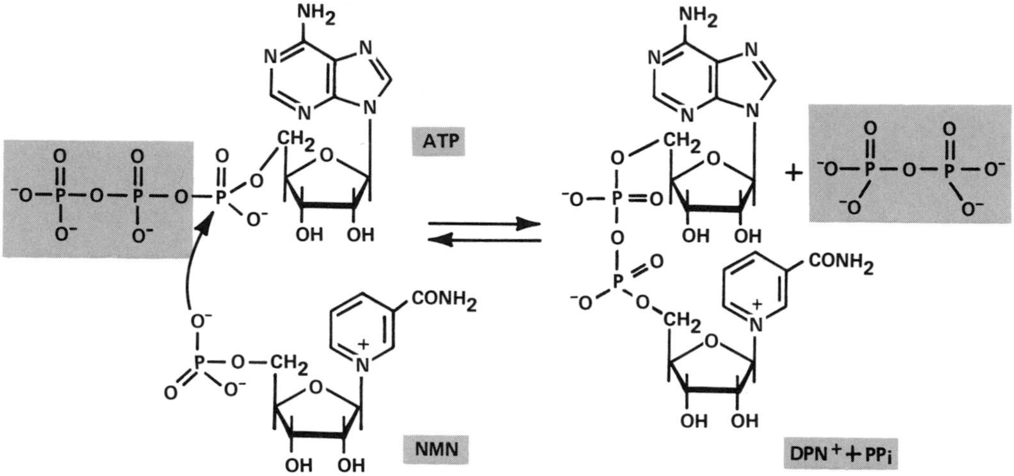

FIGURE 3-1
Coenzyme biosynthesis.

anism, an activation of the nucleotide as a nucleoside 5'-triphosphate, is the prototype for transfers of nucleotides to form all the other nucleotide coenzymes and also for the mixed acid anhydrides, the form in which activated fatty acids, amino acids, and sulfate are found[7].

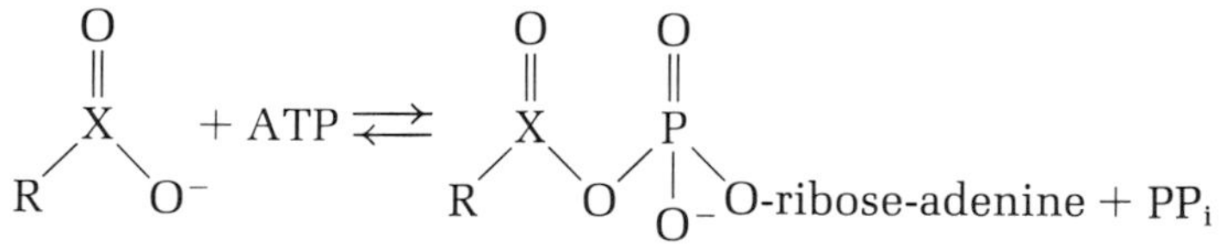

It might seem unreasonable today to have considered as nucleic acid precursors any other than 5' isomers. Yet, as mentioned before (Chapter 2, Section 3), biochemistry before 1950 had emphasized energy production, and relatively little work had been done on

6. Kornberg, A. (1971) in "Nucleic Acid-Protein Interactions" (D. W. Ribbons, J. F. Woessner, and J. Schultz, eds.) North-Holland Publishing Co., Amsterdam, p. 3.
7. Kornberg, A. (1962) in "Horizons in Biochemistry" (M. Kasha and B. Pullman, eds.) Academic Press, Inc., New York, p. 251.

biosynthetic pathways. It was assumed that, since individual enzymatic reactions could be pulled and pushed back and forth, so could entire pathways. Glycogen could be synthesized by glycogen phosphorylase. Why not protein synthesis by reversal of peptidases? Lipids by lipases? And RNA by ribonuclease? Inasmuch as the 2',3' cyclic nucleotide produced by ribonuclease degradation of RNA was also utilized by it to reverse the reaction, these nucleotides were commonly regarded as likely to be the building blocks in nucleic acid synthesis. During the fifties, however, it became clear that pathways of energy production and biosynthesis are distinct. They are dual and divided highways. How rational it is that the physical integration and control of a sequence of reactions be sharply separated from another sequence whose purpose is unrelated or even conflicting. It may be hard to imagine now why it was not obvious to biochemists from the beginning that there are separate routes of biosynthesis and degradation.

Efforts to reconstruct in cell-free systems what we now regard as the biosynthetic route of nucleic acids were started with the observation that (^{14}C)-labeled ATP was incorporated by E. coli extracts into an acid-insoluble product presumed to be RNA. Similar observations were made of incorporation of (^{14}C)-thymidine by such extracts into an acid-insoluble compound that was DNase-sensitive and presumed to be DNA[8].

Attempts to characterize the RNA-synthesizing action were overshadowed by the simultaneous discovery, in extracts of the bacterium *Azotobacter vinelandii*, of the enzyme polynucleotide phosphorylase, which polymerizes ADP and other nucleoside diphosphates into a polyribonucleotide polymer[9]. At physiological concentrations of phosphate, however, this enzyme rapidly phosphorolyzes RNA, and with the discovery of RNA polymerase a few years later the function of polynucleotide phosphorylase was associated more with the salvage of nucleotides from RNA than with their polymerization to RNA.

The incorporation of (^{14}C)-thymidine was found to occur by way of the 5'-triphosphate, dTTP. With the purification of the enzyme responsible for the conversion of dTTP to DNA[10], it became apparent that preformed DNA and the other three commonly occurring deoxyribonucleotides were required for the reaction. Since the purified enzyme, called DNA polymerase, used the preformed DNA as a template for directing the assembly of a new DNA chain, it was obvious at once why all four nucleotides were absolutely required. Directing enzyme function with a template was unique and unanticipated by any evidence in enzymology, and was for some biochemists, even after a number of years, very hard to believe.

8. Kornberg, A. (1957) in "The Chemical Basis of Heredity" (W. D. McElroy and B. Glass, eds.) Johns Hopkins Press, Baltimore, p. 579; (1959) Revs. Mod. Phys. 31, 200.
9. Grunberg-Manago, M. and Ochoa, S. (1955) JACS 77, 3165; Ochoa, S. and Heppel, L. A. (1957) in "The Chemical Basis of Heredity" (W. D. McElroy and B. Glass, eds.) Johns Hopkins Press, Baltimore p. 615.
10. Lehman, I. R., Bessman, M. J., Simms, E. S. and Kornberg, A. (1958) JBC 233, 163.

2. Basic Features of Polymerase Action: The Template-Primer[11]

DNA polymerases, catalysts for DNA chain growth, have been found in extracts of all cells, bacterial, plant, and animal, where DNA synthesis has been measured. These enzymes may be called DNA-directed DNA polymerases. By analogy, enzymes that transcribe DNA chains into RNA are called DNA-directed RNA polymerases, or simply, RNA polymerases (Chapter 10). Enzymes that transcribe RNA into DNA have been found in animal viruses and animal cells, and are called RNA-directed DNA polymerases, or reverse transcriptases (Chapter 6, Section 7).

The DNA and RNA polymerases are unique among enzymes in that the choice of substrate is determined by a template as well as by the enzyme. Bacterial DNA polymerases can make animal DNA, and an animal polymerase can make bacterial DNA, given the respective DNA template chain. Thus, there is no requirement for recognition of a specific sequence of bases in the chain by DNA polymerases.

Biosynthesis of DNA has two basic features:

(i) A phosphodiester bridge is formed between the 3' hydroxyl group at the growing end of a DNA chain, called *primer* DNA, and the 5' phosphate group of the incoming deoxynucleotide. The over-all chain growth is 5' → 3' (Fig. 3-2).

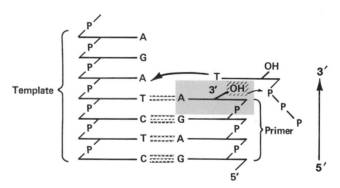

FIGURE 3-2
Template-primer in action.

(ii.) Each deoxynucleotide added to the chain is selected by base-pair matching to a DNA chain called *template* DNA (Fig. 3-2). No known DNA polymerase is able to *initiate* chains. For this and other functions related to chromosome replication, additional proteins are required.

Distinguishing between the terms *primer* and *template* is important in discussing polymerase action.

The term primer was used earlier in biochemistry to refer to the glycogen needed to start, or *prime*, glycogen synthesis. In DNA

11. Kornberg, A. (1969) Science **163**, 1410.

synthesis, however, DNA chains required in the reaction serve two distinct functions. *Primer* should be restricted to the DNA chain from which growth occurs at its 3′-hydroxyl terminus, and *priming* to designate this function. The *primer terminus* is the terminal nucleotide of the primer bearing the 3′-hydroxyl group (Fig. 3-2).

The term *template* applies to the DNA chain that furnishes directions for the sequence of nucleotides, and the term *directing*, not priming, designates this second function. Thus the term *template-primer* embraces two different functions and meanings.

The enormous variety of secondary structures of DNA can be roughly divided into five categories of template-primer (Fig. 3-3). These include synthetic polymers as well as natural DNA. Polymerases are often distinguished on the basis of their capacity to act on one or another kind of DNA template-primer structure. This will be considered in detail for each of the enzymes discussed; the properties of *E. coli* DNA polymerase I (Chapter 4) illustrated in Fig. 3-3 are given simply as an example.

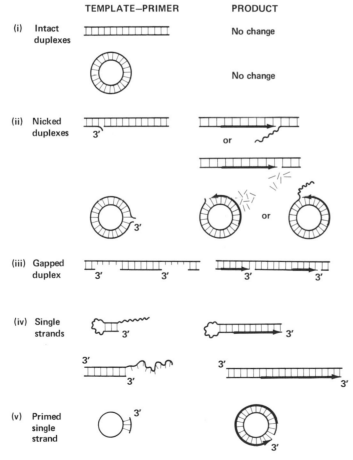

FIGURE 3-3
Structures of various template-primers and their use by *E. coli* DNA polymerase I.

The five categories are:

(i) *Intact duplex*. Intact double-stranded DNA, such as linear phage T4 DNA, and circular duplex viral or mitochondrial DNA, is inert as a substrate for all known DNA polymerases. The circular duplex has no ends. The linear duplex, in which the strands are the same length, cannot be extended by DNA polymerase, nor can either strand be replicated. RNA polymerase, however, (Chapter 10) can initiate transcription on such a structure. Part of its multi-subunit structure is devoted to the recognition of signals for the initiation of an RNA chain.

(ii) *Nicked duplex*. Duplex DNA may be "activated" as a template-primer for certain DNA polymerases by single-strand scissions introduced by an endonuclease such as pancreatic DNase (Chapter 9, Section 1). A *nick* is defined as a break introduced in a DNA strand that can be sealed by the enzyme DNA ligase, which requires contiguous 3' hydroxyl and 5' phosphate groups.

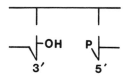

Extensive nicking may lead to fragmentation of the duplex when scissions on opposite strands are within a few nucleotides of each other. Such fragments may have protruding 5' ends which serve as templates for all DNA polymerases.

(iii) *Gapped duplex*. DNA is made more active as a template when short gaps are introduced at nicks by the action of exonuclease III, an enzyme which degrades one strand of a duplex structure in the 3' → 5' direction at a nick.

(iv) *Single strand*. Such DNA molecules are not active as template-primers unless they have duplex regions. They may self-anneal to produce such regions, or they may anneal partially with other molecules. Duplex DNA may become effectively single-stranded when large sections are removed from its ends, or gaps in it are greatly enlarged.

(v) A *primed single strand* can be prepared by annealing an oligonucleotide to a synthetic homopolymer chain or to a natural, circular single strand such as some viral DNA's (Chapter 4, Section 17). This priming renders the single strands active as templates for some DNA polymerases.

3. Basic Features of Polymerase Action: Polymerization Step[12]

The overall reaction catalyzed by DNA polymerase is:

$$(\text{dNMP})_n + \text{dNTP} \rightleftarrows (\text{dNMP})_{n+1} + \text{PP}_i$$

12. Kornberg, A. (1969) Science **163**, 1410.

The enzyme has two substrates. One is the base-paired primer terminus of a template-primer containing a free 3'-hydroxyl group. The other substrate is a deoxynucleoside 5'-triphosphate.

The reaction consists of a nucleophilic attack of the 3'-hydroxyl group of the primer terminus on the α-phosphorus of the incoming nucleoside triphosphate. The products are a new primer-terminus, i.e. a nucleotide linked to the primer chain by a 3', 5'-phosphodiester bond, and inorganic pyrophosphate. Subsequent hydrolysis of pyrophosphate by inorganic pyrophosphatase (Chapter 2, Section 5) provides a driving force for polymerization and renders the reaction highly irreversible.

Direction of Chain Growth

The overall direction of chain growth is $5' \rightarrow 3'$; addition of the triphosphate occurs at the 3'-hydroxyl end of the chain.

Mechanisms of polymer biosynthesis may be grouped in two classes called _tail growth_ and _head growth_. Chain growth of DNA, and also of RNA, are examples of tail growth (Fig. 3-4).

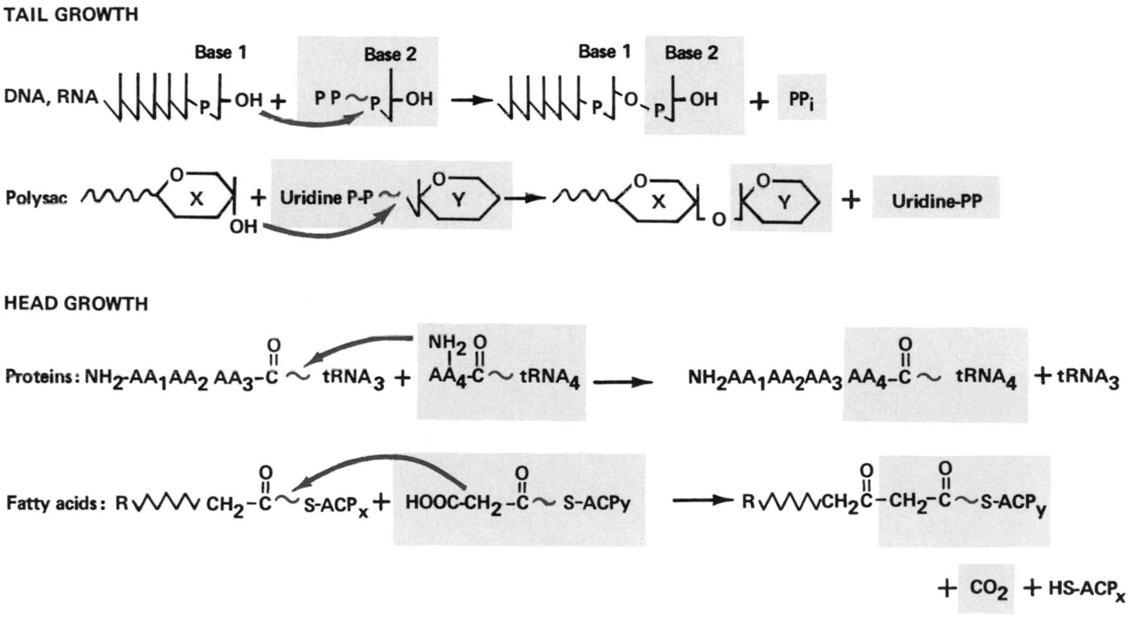

FIGURE 3-4
Mechanism of tail growth and head growth of biopolymers.

Tail growth involves the attack of a nonactivated end of the polymer on the activated end of the monomer. This growth mechanism is that used in the polymerization of many polysaccharides where the nonactivated hydroxyl end of the sugar polymer attacks a phosphoryl-activated monomer.

Head growth involves the attack of a monomer on the activated end of the polymer. Since the monomer itself contains an activated end not altered in the reaction, the lengthened chain again contains an activated head, supplied by the monomer.

The head growth mechanisms, exemplified in three different processes — protein synthesis on a ribosome, fatty acid synthesis in a synthetase complex, and coupling of the trisaccharide units on the inner cell membrane in O antigen synthesis — have one feature in common. The activated monomer, as well as the activated head, is held and oriented on the surface of a particle. Tail growth synthesis seems, on the other hand, to be used when the geometry of assembly is not so rigidly defined and either component, usually the monomer, is brought in from solution.

Chain Growth in Replication

The fact that the strands in a DNA double helix run in opposite directions posed a problem to biochemists trying to understand how DNA is replicated. How could *both* strands of the helix be replicated at the same time if growth by the DNA polymerase mechanism is strictly $5' \longrightarrow 3'$? An adequate answer to this question is now in hand and will be presented in Chapter 7. However, for a number of years this dilemma led to a search for a mechanism involving $3' \longrightarrow 5'$ chain growth.

A possible tail growth mechanism for simultaneous replication of the antiparallel strands of the DNA duplex invoked the $3'$ isomers of the nucleoside triphosphates. It postulated attack by the $5'$-hydroxyl tail end of the chain on the activated monomer, with a consequent growth $3' \longrightarrow 5'$ (Fig. 3-5). Thus far there is no evidence for the biosynthesis of nucleoside $3'$-mono- or triphosphates; nor any indication, after intensive efforts to find it, of an enzyme that can polymerize chemically synthesized deoxyribonucleoside $3'$-triphosphates (Fig. 3-5).

Another possible mechanism for achieving $3' \longrightarrow 5'$ growth involves an attack of the $3'$-hydroxyl group of the incoming nucleoside

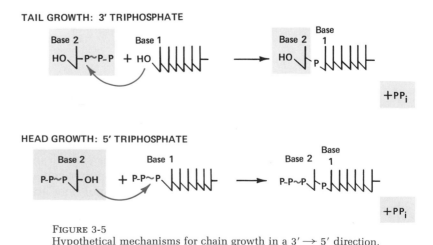

FIGURE 3-5
Hypothetical mechanisms for chain growth in a $3' \longrightarrow 5'$ direction.

triphosphate on the 5'-α-phosphate of the primer terminus. This describes a head growth mechanism, an attack of an activated monomer on the activated polymer. No evidence for this kind of DNA polymerization has been found either. The reason such a mechanism would be disadvantageous in checking the fidelity of base-pairing at the primer terminus in a "proofreading" operation will be considered now.

Base-Pair Matching

Fidelity of replication is assured by accurate base-pairing. Before the polymerization step can occur, the incoming nucleoside triphosphate must be paired with its template partner. If the correct base pair has not been formed, the enzyme rejects the mispaired nucleotide (Chapter 4, Section 7).

In addition to matching the incoming nucleotide to its template partner, the enzyme can also check the accuracy of base-pairing between the primer terminus and the template. If these bases are unpaired, subsequent polymerization will not occur. Rather, the mispaired primer terminus will be excised by a function of polymerase called the 3' $\longrightarrow$ 5' exonuclease. The activity degrades unpaired nucleotides from the 3' end of the chain. When a correctly paired duplex structure is reached, polymerization can proceed (Chapter 4, Section 9).

This proofreading function, a feature of several *E. coli* DNA polymerases, provides a rationale for the tail growth, rather than the head growth, mechanism of polymerization. Were a mismatch present at the primer terminus, and the head growth mechanism employed for polymerization, removal of the activated triphosphate end of the polymer would prevent further polymerization.

Generalizations About DNA Polymerase Action

From what we know thus far of the DNA polymerases, these generalizations apply:

(i) The nucleotide added is a monomer activated as a 5'-triphosphate. Claims that the diphosphate[13], or an activated form of the monophosphate[14], is the DNA precursor in the cell are likely in error due to compartmentalization (metabolic channeling) of the triphosphate in different cellular pools[15].

(ii) The phosphodiester bridge is made by a nucleophilic attack of the 3'-hydroxyl group of a primer terminus on the α-phosphorus atom of the activated monomer with the elimination of its terminal pyrophosphate.

(iii) A primer with a 3'-hydroxyl group is required; new chains cannot be initiated.

13. Rubinow, S. I. and Yen, A., (1972) *NNB* **239**, 73.
14. Werner, R. (1971) *NNB* **233**, 99.
15. Fridland, A. (1973) *NNB* **243**, 105.

(iv) Overall chain growth is $5' \rightarrow 3'$.

(v) Correct base-pairing is required at the primer terminus, as well as in the matching of the incoming nucleotide to the template. In the case of *E. coli* DNA Polymerase I, a base at the primer terminus that is unpaired because of mismatching or fraying is removed by the $3' \rightarrow 5'$ exonuclease activity of the polymerase; a high fidelity of replication is thereby assured.

(vi) The polarity of the newly synthesized chain is the opposite of that of the template; the growing chain and template are antiparallel.

(vii) Any base sequence in the template will be copied with no particular preference or specificity.

4. How to Measure Polymerase Action[16,17]

The Measurement

Activity of a polymerase is quantitatively determined by measuring conversion of the monomer to polymeric form. There are at least four ways in which this change has been measured.

(i) Conversion of a radioactive mononucleotide (^{32}P in the α-phosphate; ^{14}C or ^{3}H in the nucleoside) into an acid-insoluble form[16,17]. Polynucleotides of ten or more residues are generally insoluble in three to five percent perchloric or trichloracetic acid. The precipitate is collected by centrifugation or on a glass filter, and washed. The blank values, due to residual substrate, may be reduced to <0.01 percent of the starting radioactivity. More discriminating methods for characterizing the macromolecular product and separating it from substrate depend on methods such as gel filtration and sedimentation gradients.

(ii) Conversion of a radioactive mononucleotide into a polyelectrolyte form, which is then tightly adsorbed to an ion exchanger[18]. A small square of DEAE-cellulose paper, which binds highly negatively charged compounds, can be washed free of the substrate while retaining polynucleotides down to the size of a trinucleotide. Advantages of this method are that many samples can be processed at once and that acid-soluble polynucleotides are detected.

(iii) Hypochromic shift as mononucleotides are polymerized[19]. The decrease in absorbance at 260 nm, due to stacking interactions of the parallel bases in the helix, provides a direct measure of the extent of reaction, and is most suitable for kinetic analysis. However, the method is relatively insensitive. It is useful only when a substantial fraction of the substrate is converted to polymer and the template concentration is not high.

16. Lehman, I. R., Bessman, M. J., Simms, E. S. and Kornberg, A. (1958) *JBC* **233**, 163.
17. Jovin, T. M., Englund, P. T. and Bertsch, L. L. (1969) *JBC* **244**, 2996.
18. Brutlag, D. and Kornberg, A. (1972) *JBC* **247**, 241.
19. Radding, C. M. and Kornberg, A. (1962) *JBC* **237**, 2877.

(iv) Increase in viscosity when mononucleotides are polymerized[20]. The advantages and limitations are similar to those of the optical method.

Polymerase activity can be measured, in some special instances, by observing a change in the macromolecular properties of the template-primer. One example is the conversion of a radioactive, single-stranded template, such as phage ϕX174 or M13 DNA, to a particular duplex form separable by sedimentation or gel electrophoresis. A change or increase in biological activity of such DNA molecules may also be observed.

Substrates

(i) The template-primers used by DNA polymerases vary in secondary structure, and these have been discussed in Section 2 of this chapter. Since the polymerases differ in their capacity to utilize various template-primers, the nature of this substrate is an important parameter of the assay.

Concentrations of template-primer used in the assays are generally between 10^{-5} and 10^{-4} M in nucleotide residues; depending on the size of the DNA, molecular concentrations are three to six orders of magnitude lower.

The effects that nucleases have on the measurement and analysis of polymerase action cannot be overemphasized[21]. A trace of endonuclease may render an inert duplex active for a polymerase. Alternatively, an endonuclease that creates 3'-phosphoryl termini may convert an active template-primer into an inhibitor that binds the enzyme in an unproductive complex. Finally an exonuclease such as exonuclease III, which can act as a phosphomonoesterase on 3'-phosphoryl termini and can enlarge nicks into gaps, can easily increase the reaction rates of certain polymerases by factors of a hundred or a thousand. Obviously, an excess of nuclease would lead to net loss of DNA, and without adequate DNA pools to trap and protect incorporated radioactive substrates, high levels of DNase would obscure polymerase activity. Chapter 9 gives a more detailed discussion of nucleases.

(ii) The deoxynucleoside triphosphates, or precursors convertible to them, are generally required in concentrations near 10^{-5} M. The purine and pyrimidine nucleotides required are those needed for pairing with the template.

Cofactors

Mg^{2+} is invariably required at concentrations near 10^{-3} M although it may be partially replaced by Mn^{2+} at lower concentrations. Re-

20. Schachman, H. K., Adler, J., Radding, C. M., Lehman, I. R. and Kornberg, A. (1960) JBC **235**, 3242.
21. Lehman, I. R. (1967) ARB **36**, 645.

sponses to temperature, ionic strength, pH, buffer ions, detergents, sulfhydryl, and other chemical agents vary among the polymerases. These properties have served to distinguish *E. coli* polymerases I, II, and III (see Chapters 4 and 5).

Other Factors

Agents known to influence the secondary structure of the template-primer have profound effects on the action of some polymerases. Proteins that bind specifically to single-stranded DNA and effect unwinding of duplexes stimulate the replication of single-stranded template-primers by certain polymerases (Chapter 7, Sections 7 and 8). Spermidine, which stabilizes the duplex structure, stimulates the replicative action of another polymerase system (Chapter 7, Section 11).

Protein factors may also prove to be essential in other ways. In one instance, a protein devoid of polymerase activity is required for the action of a polymerase in forming a proper complex with the template-primer (Chapter 5, Section 3).

4

DNA Polymerase I of *E. coli*

1. Isolation and Physicochemical Properties[1-5]

DNA polymerase I of *E. coli* is so designated because it was the first polymerase recognized. We will consider this enzyme in detail because it has been studied more extensively than the others, and because it appears to embody the salient features of DNA polymerase action.

1. Kornberg, A. (1969) *Science* **163**, 1410.
2. Lehman, I. R., Bessman, M. J., Simms, E. S. and Kornberg, A. (1958) *JBC* **233**, 163.
3. Richardson, C. C., Schildkraut, C. L., Aposhian, H. V. and Kornberg, A. (1964) *JBC* **239**, 222.
4. Jovin, T. M., Englund, P. T. and Bertsch, L. L. (1969) *JBC* **244**, 2996.
5. Jovin, T. M., Englund, P. T. and Kornberg, A. (1969) *JBC* **244**, 3009.

The enzyme was isolated by conventional steps of protein fractionation (Table 4-1)[6-8]. Extracts were prepared by mechanical rupture of cells followed by removal of nucleic acids using streptomycin precipitation and autolysis (involving nuclease and possibly some protease action). Fractionation by ammonium sulfate, ion exchange on DEAE- and phosphocellulose, and Sephadex gel filtration yielded a pure enzyme. (Phosphocellulose, resembling DNA as it does, has proved to be especially effective for affinity chromatography not only of this and other DNA polymerases, but also of proteins that bind to DNA, such as the *gal* and *lac* repressors.) From 100 kilograms of *E. coli* paste, 0.6 to 1.0 gram of pure enzyme was obtained, with an overall yield of about twenty percent from the extract.

TABLE 4-1
Procedure for purification of DNA polymerase I

Fraction	Protein mg/ml	Specific activity units/mg protein	Yield %
1. Extract	20	3.5	[100]
2. Streptomycin	12	23	47
3. Autolysis	1.3	170	40
4. Ammonium sulfate	23	250[a]	36
		650[b]	
5. DEAE-cellulose	11	750	36
6. Phosphocellulose	3.7	10,000	28
7. Sephadex G-100	4.0	18,000	22

The homogeneous enzyme has approximately 2 units of polymerase activity per 0.1 μg or 1 pmole. This represents 20,000 pmoles of nucleotide incorporated into DNA in a 30-minute assay at 37° per mole, or a turnover number of 667 units, with an optimal template-primer.
[a]DNA was the template-primer for assay of this and previous fractions.
[b]Poly d (A-T) was the template-primer for assay of this and subsequent fractions.

The molecular weight of the homogeneous enzyme is 109,000 daltons. The turnover number is 667 nucleotides polymerized per molecule of enzyme per minute at 37° (Table 4-1). It is a single polypeptide containing approximately 1000 amino acid residues[9]. (Table 4-2). The one remarkable feature of the amino acid composition is the presence of only a single disulfide bond and one sulfhydryl group. The latter is probably not part of the active site because it can be modified either with iodoacetate or mercuric ion to give derivatives with full activity. The reaction with mercuric ion will give either a polymerase monomer, with one mercury atom per protein molecule, or, in the presence of a molar excess of enzyme, a dimer, also fully active, with two protein molecules linked by a mercury atom[10].

6. Lehman, I. R., Bessman, M. J., Simms, E. S. and Kornberg, A. (1958) *JBC* **233**, 163.
7. Richardson, C. C., Schildkraut, C. L., Aposhian, H. V. and Kornberg, A. (1964) *JBC* **239**, 222.
8. Jovin, T. M., Englund, P. T. and Bertsch, L. L. (1969) *JBC* **244**, 2996.
9. Op. cit. in footnote 8.
10. Jovin, T. M., Englund, P. T. and Kornberg, A. (1969) *JBC* **244**, 3009.

TABLE 4-2
Amino acid composition of DNA polymerase I

Amino acid	Content moles/109,000 g protein
Lysine	61
Histidine	19
Arginine	48
Aspartic acid, asparagine	88
Threonine	53
Serine	40
Glutamic acid, glutamine	126
Proline	53
Glycine	63
Alanine	102
Half-cystine	3
Valine	61
Methionine	24
Isoleucine	55
Leucine	112
Tyrosine	33
Phenylalanine	25
Tryptophan	9
(Ammonia)	(58)
Total	975

The reaction of polymerase with a single atom of ^{203}Hg provides a convenient way of incorporating a radioactive label of about 10,000 counts per minute per microgram of enzyme, without affecting enzymatic activity. This label has served as a marker in DNA binding studies which will be described below. Fluorescent derivatives of polymerase have been prepared with the acylating agent N-carboxymethylisatoic anhydride[11].

There is evidence that the enzyme contains zinc. One or more atoms per molecule have been found in the purified preparation[12], and o-phenanthroline markedly inhibits the polymerase activity, suggesting the presence of and requirement for a metal such as zinc.

2. Preview of Polymerase Functions[13]

DNA polymerase performs a multiplicity of functions, and these are manifested and measured in a variety of ways. But the enzyme is not unusually complex, as the detailed discussion to follow will point out. Although our lack of knowledge of even the amino acid

11. Jovin, T. M., Englund, P. T. and Kornberg, A. (1969) *JBC* **244**, 3009.
12. Springgate, C. F., Mildvan, A. S., Abramson, R., Engle, J. L. and Loeb, L. A. (1973) *JBC* **248**, 5987.
13. Kornberg, A. (1969) *Science* **163**, 1410.

sequence, let alone the folded structure of the enzyme, precludes adequate interpretations of its action, some simplifying generalizations are nevertheless useful at this point.

The principal, discrete activities of the enzyme may be described as: polymerization, pyrophosphorolysis, pyrophosphate exchange, $3' \rightarrow 5'$ exonucleolytic degradation and $5' \rightarrow 3'$ exonucleolytic degradation.

Polymerization, pyrophosphorolysis, and pyrophosphate exchange are all features of the polymerization activity; pyrophosphorolysis and pyrophosphate exchange are simply demonstrations of the reversal of the polymerization reaction. In view of the large concentrations of pyrophosphate required, and the potency of inorganic pyrophosphatase in all cells, the physiological significance of these reactions is likely to be only minor.

Chain growth in a $5' \rightarrow 3'$ direction is dictated by the specificity for two substrates: the $3'$–hydroxyl primer terminus and the deoxynucleoside $5'$–triphosphate. Synthesis of chains antiparallel to the template is a direct consequence of the initial antiparallel orientations of the primer strand to the template strand.

The $3' \rightarrow 5'$ exonuclease is a component of the polymerase machinery essential for recognition of a correctly base-paired primer terminus. It supplements the capacity of polymerase to recognize base-pair matching of triphosphate substrate with template. A mismatched terminal nucleotide on the primer chain permits it to occupy a site that enables the enzyme to hydrolyze and thereby remove it.

The $5' \rightarrow 3'$ exonuclease activity has been found, to date, in polymerase I but in no other polymerases. It degrades duplex DNA from a $5'$ end releasing mono- or oligonucleotides. Its capacity to excise distorted sections of DNA that lie in the path of the chain extension by polymerase seems well suited for removing a DNA lesion such as a thymine dimer (Chapter 9) or RNA from a DNA–RNA hybrid duplex (Chapter 7, Section 5). This activity will be described in greater detail in Sections 12 and 13 of this chapter.

Results from proteolytic cleavage of the enzyme into a large and a small fragment show that the polymerization, pyrophosphorolysis, pyrophosphate exchange, and $3' \rightarrow 5'$ exonuclease activities, all features of the polymerization operation, reside in the large fragment. The $5' \rightarrow 3'$ exonuclease activity resides in the small fragment, distinct from the other activities. Thus, polymerase I contains at least two enzymes in one polypeptide chain.

In the cell, DNA polymerase may be associated with other proteins which are related in function. However, it is difficult to determine whether the persistence of a particular activity in an enzyme preparation represents a significant association or merely an adventitious one. A trace of nucleoside diphosphate kinase supports the utilization of deoxynucleoside diphosphates as substrates[14]. But pure polymerase preparations have little or none of this kinase activity and are specific for triphosphates.

14. Miller, L. K. and Wells, R. D. (1971) *PNAS* **68**, 2298.

3. Multiple Sites in an Active Center[15]

DNA polymerase is roughly spherical in shape with a diameter near 65 Å. It can therefore contact a DNA helix, which is about 20 Å across, for a length of nearly two turns (twenty base pairs).

We picture the active center of the enzyme as a polypeptide surface specifically adapted to recognize and accommodate several nucleotide structures (Fig. 4-1). Evidence will be presented in succeeding sections that, within the active center, there are at least five major sites.

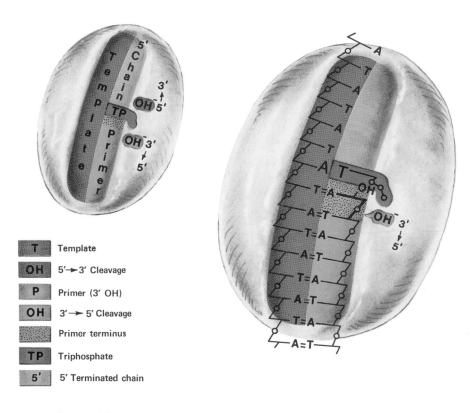

T	Template
OH	5′ → 3′ Cleavage
P	Primer (3′ OH)
OH	3′ → 5′ Cleavage
	Primer terminus
TP	Triphosphate
5′	5′ Terminated chain

FIGURE 4-1
Scheme for multiple sites in the active center of DNA polymerase I.

(i) A site for a portion of the _template_ chain. This area binds the chain where a base pair is to be formed and for a distance of several nucleotides on either side. It seems likely that this is the site where an interior region of single-stranded DNA is bound, but it is uncertain whether it is bound in this site in an extended, or a tightly stacked, helical conformation.

(ii) A site for the growing chain, the _primer._ The polarity of the primer chain must be opposite to that of the template.

15. Kornberg, A. (1969) _Science_ **163**, 1410.

(iii) A site with special recognition for the 3'-hydroxyl group of the terminal nucleotide of the primer, the <u>primer terminus</u>. This region will be discussed later as a site that is also involved in the hydrolytic cleavage, in the $3' \rightarrow 5'$ direction, of an unpaired 3'-hydroxyl-terminated primer chain.

(iv) A site for a deoxynucleoside triphosphate.

(v) A site for the $5' \rightarrow 3'$ exonucleolytic cleavage of a DNA chain positioned beyond the triphosphate site and in the path of the growing chain.

4. DNA Binding Site[16,17]

Binding of DNA to polymerase has been studied by measuring the sedimentation velocity, in sucrose gradients, of mixtures containing a variety of DNA structures (Fig. 4-2) labeled with ^{3}H or ^{32}P, and polymerase labeled with ^{203}Hg. The mixtures were layered on top

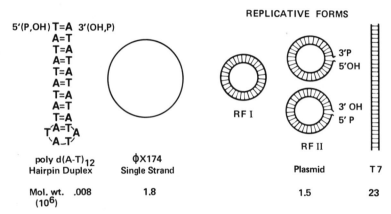

FIGURE 4-2
DNA structures tested for binding to DNA polymerase I. The replicative forms are those of the mini plasmid of E. coli 15T⁻.

of the gradients and the enzyme-DNA complexes were identified after sedimentation (Table 4-3). Binding occurred along single-stranded chains and especially at nicks and ends of duplexes, but not to any measurable extent at helical regions.

Double-stranded, circular DNA was not bound at all by polymerase. However, when these duplex forms were denatured to make and maintain them in single-strand form, they were bound by polymerase in proportion to their length and to the same extent as single-stranded ϕX174 DNA—about one molecule per 300 nucleotide residues. Nicks containing 3'-hydroxyl and 5'-phosphate termini, introduced into a circular, duplex DNA by pancreatic deoxyribonuclease, were active points for replication. Nicks with 3'-phosphate

16. Englund, P. T., Kelly, R. B. and Kornberg, A. (1969) *JBC* **244**, 3045.
17. Griffith, J., Huberman, J. A. and Kornberg, A. (1971) *JMB* **55**, 209.

and 5'-hydroxyl termini introduced by micrococcal nuclease (Chapter 9, Section 1) were not active points for replication; they inhibited replication by nonproductive binding of the enzyme. Regardless of the kind of nick, polymerase molecules formed complexes in numbers exactly equivalent to the number of nicks (Table 4-3 and Fig. 4-3).

TABLE 4-3
Influence of DNA structure on binding
of DNA to DNA polymerase I

Conformation	Per DNA molecule	
	Nicks or ends	Polymerase molecules bound
$d(A—T)_{12}$ oligomer		
Hairpin	1	1
ϕX174 DNA		
Circular, single-stranded	0	20
Circular, duplex	0	<0.1
Plasmid DNA (RF)		
Irreversibly denatured	0	21
3'-Hydroxyl nick	1	1
3'-Phosphate nick	5	6
T7 DNA		
Linear, duplex	2.5	2.6
Single-stranded	2	240

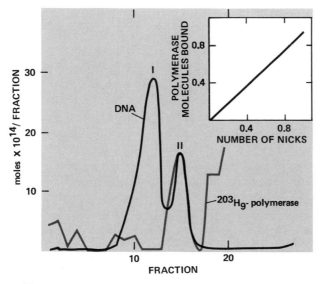

FIGURE 4-3
Binding of DNA polymerase I to nicks in duplex DNA. The enzyme fails to bind to RF I, and binds to RF II in equimolar amounts.

An alternating copolymer of deoxyadenylate and deoxythymidy-late, poly d(A–T), was partially digested by deoxyribonuclease to yield oligomers of varying sizes. Those of approximately 24 and 40 nucleotide residues, d(A–T)$_{12}$ and d(A–T)$_{20}$ respectively, were induced to assume *hairpin* or *nicked* conformations by melting and quick cooling at low ionic strength and low d(A–T) concentration (Fig. 4-4). With excess enzyme, d(A–T)$_{12}$ or d(A–T)$_{20}$ sedi-

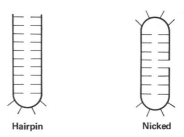

Hairpin   **Nicked**

FIGURE 4-4
Hairpin and nicked conformations of d(A–T)$_{12}$.

mented almost quantitatively with the enzyme, indicating very high binding affinity. With the d(A–T) oligomer present in excess, all the enzyme sedimented as a complex containing an equimolar amount of poly d(A–T), an indication that DNA polymerase binds only one d(A–T) oligomer per enzyme molecule. The same results were obtained with the oligomers terminated either by a 3′-hydroxyl or a 3′-phosphate group. (Although a 3′-phosphate terminus is not held in the primer terminus site, the chain does bind to the primer site of the enzyme.)

In experiments with linear, duplex DNA of bacteriophage T7, 2.6 molecules of the enzyme were found for every T7 DNA molecule in the complex. The two ends of the T7 molecule and the presence in the T7 preparation of one nick per two molecules, on the average, nicely accounts for this result. Even though the end of a duplex structure does not serve a template-primer role in polymerization, it binds the template, primer, and primer terminus sites on the enzyme.

Electron microscopy[18] provides an independent means of determining DNA binding to polymerase. With appropriate techniques, DNA may be spread on the microscope grid without complexing it to cytochrome c or other basic proteins, as is usually done in DNA microscopy (Fig. 4-5). Thus, the polymerase is the only protein present, and the nature of its interaction with DNA is clearly seen. Fig. 4-6a illustrates the interaction of the enzyme with a DNA duplex consisting of (dA)$_{4000}$ and shorter lengths of poly dT, (dT)$_{200}$. Since the (dT)$_{200}$ chains abut each other, this duplex contains nicks spaced at intervals of 200 base pairs, or near 600 Å on the average. The DNA polymerase molecules are seen in the figure as a string of beads

18. Griffith, J., Huberman, J. A. and Kornberg, A. (1971) *JMB* **55**, 209.

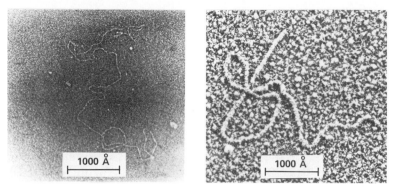

FIGURE 4-5
Electron micrographs of DNA (dA$_{4000}$·dT$_{200}$) visualized (left) by a high-resolution procedure (Griffith) and (right) by the protein film method (Kleinschmidt). (Courtesy of Dr. Jack Griffith.)

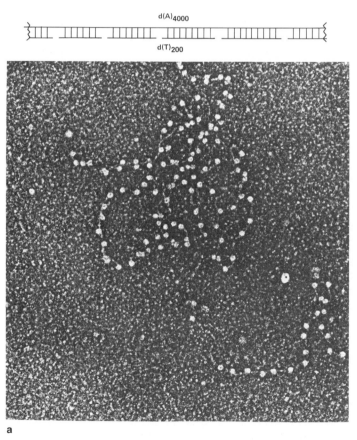

a

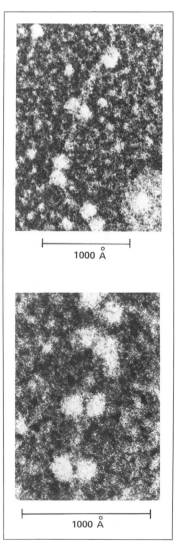

b

FIGURE 4-6
DNA polymerase bound to dA$_{4000}$·dT$_{200}$ (a). Molecules are bound where dT$_{200}$ oligomers aligned on dA$_{4000}$ abut each other. These nicks are at regular intervals of about 600 Å. (b) DNA polymerase dimers bound to dA$_{4000}$·dT$_{200}$ (Courtesy of Dr. Jack Griffith.)

on the poly dA·poly dT strand spaced at distances of approximately 600 Å. Enzyme molecules were never seen bound to fully helical, intact DNA, such as extensive lengths of phage λ DNA. When enzyme dimers, linked through their sulfhydryl groups by Hg, were bound to the nicked polymer (Fig. 4-6b), the binding involved only one enzyme molecule, while the other enzyme molecule in the dimer assumed a variety of positions on the grid. Evidently the Hg bridge is very flexible.

Binding of DNA duplexes at ends or nicks is probably not simply a consequence of fraying and single-strandedness in these regions, but also includes specific binding directed to the 3′ and 5′ nucleotide termini at these points.

5. Primer Terminus Binding Site[19]

Binding of DNA to the enzyme must be regarded as involving at least three distinguishable sites.

(i) The area that orients the template chain — single-stranded beyond the primer terminus and base-paired behind it.

(ii) The area occupied by the primer chain — single-stranded (as for 3′ → 5′ exonuclease action, see Section 9 of this chapter) — or base-paired to the template chain.

(iii) The special 3′ end of the primer chain, the primer terminus.

Evidence for a specific primer terminus site is based largely on the presence and the nature of a single and highly specific nucleoside monophosphate binding site. As determined by equilibrium dialysis measurements of a large variety of nucleotides, polymerase binds a ribonucleotide that bears a *free 3′-hydroxyl* group and is *not* a triphosphate (Fig. 4-7). Each of the four common deoxynucleoside monophosphates was bound to and competed for a single site on the enzyme, a site entirely distinct from the triphosphate site (see Section 6 of this chapter).

Nucleotides with un-ionized base analogs or with substitutions of sulfur or carbon for oxygen in the furanose ring were bound in the site. However, absence of the 5′-phosphate group, its replacement by a phosphinate or phosphonate, or addition to it of additional phosphates or another nucleotide made such molecules inactive for binding. Nucleotides in which the 3′-hydroxyl group was replaced, as shown by a 3′-deoxy, 3-O-methyl or a 3′-phosphoryl substitution, or in which arabinosyl, lyxosyl, or xylosyl configurations replaced the ribosyl were also unable to bind in the primer terminus site.

The interpretation that the nucleoside monophosphate binding site represents the primer terminus site rests on the close correspondence between binding and the primer terminus functions. The

19. Huberman, J. A. and Kornberg, A. (1970) *JBC* **245**, 5326.

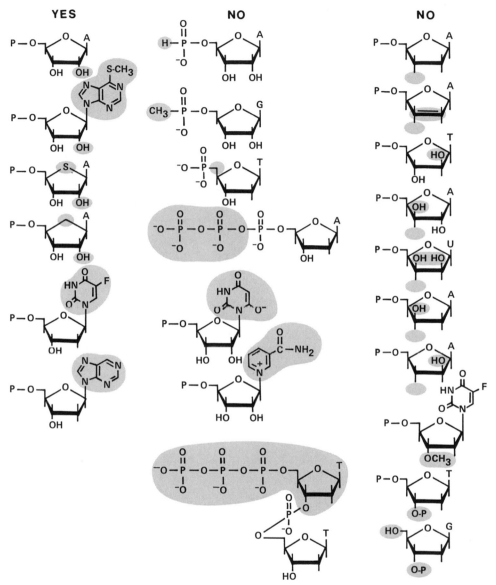

FIGURE 4-7
Structural requirements for binding to the primer terminus site determined by deoxyribonucleotide analogs that bind (left column) and those that do not (other columns). The deviant features of the analogs are shown in color.

3'-hydroxyl group is essential for binding, for extension by polymerase action, and for the $3' \rightarrow 5'$ exonuclease function. The most instructive example is the behavior of the 3'-deoxynucleotide analog. A 2',3'-dideoxynucleoside triphosphate serves as a polymerase substrate to elongate a DNA chain by one residue[20]. The chain terminated with such a dideoxynucleotide was inert to further elongation, and relatively inert to exonuclease action at this terminus, or (as described in Section 9 of this chapter), to attack by inorganic pyrophosphate. Without a 3'-hydroxyl group the chain terminus appears unable to bind properly in the primer site, and is therefore not an effective substrate for $3' \rightarrow 5'$ exonuclease action.

6. Deoxynucleoside Triphosphate Binding Site[21]

Triphosphate binding studied by equilibrium dialysis with an ultra micro apparatus showed that there is a single binding site for the four deoxynucleoside triphosphates. Although the four triphosphates differ in binding affinity, the interpretation of the particular values is necessarily limited. DNA template and primer were not present (see below), and their influence on binding, which is likely to be profound, has yet to be assessed.

Is the triphosphate site made up of separate sites for each of the four triphosphates, or does the enzyme have a single site for which all four triphosphates compete? Equilibrium dialyses were run with each of the six possible combinations of two triphosphates, and each of the pair was labeled distinctively. These competition experiments established that there is a single binding site on the enzyme for which all four triphosphates compete. Binding required Mg^{2+}, presumably because it is the magnesium chelate of the triphosphate that is recognized by the enzyme. Strength of binding decreased with increasing temperature and ionic strength.

Further explorations (Fig. 4-8) of the specificity of binding in this site emphasize the primary importance of the triphosphate moiety; the sugar and base components are only of secondary importance. For example, the 2',3'-dideoxynucleoside triphosphate, lacking the 3'-hydroxyl group vital for chain growth, nevertheless was bound with the same affinity as the natural triphosphate. A dinucleoside tetraphosphate, in which a nucleotide is linked to the 3'-hydroxyl group, was also bound in the triphosphate site. Even inorganic triphosphate was bound, although with very low affinity. However, neither deoxynucleoside mono- and diphosphates nor inorganic pyrophosphate showed any detectable binding in or competition for the triphosphate site.

The effects of DNA template and primer on the binding of triphosphates are difficult to study experimentally. An active template and

20. Atkinson, M. R., Deutscher, M. P., Kornberg, A., Russell, A. F. and Moffatt, J. G. (1969) B. **8**, 4897.
21. Englund, P. T., Huberman, J. A., Jovin, T. M. and Kornberg, A. (1969) JBC **244**, 3038.

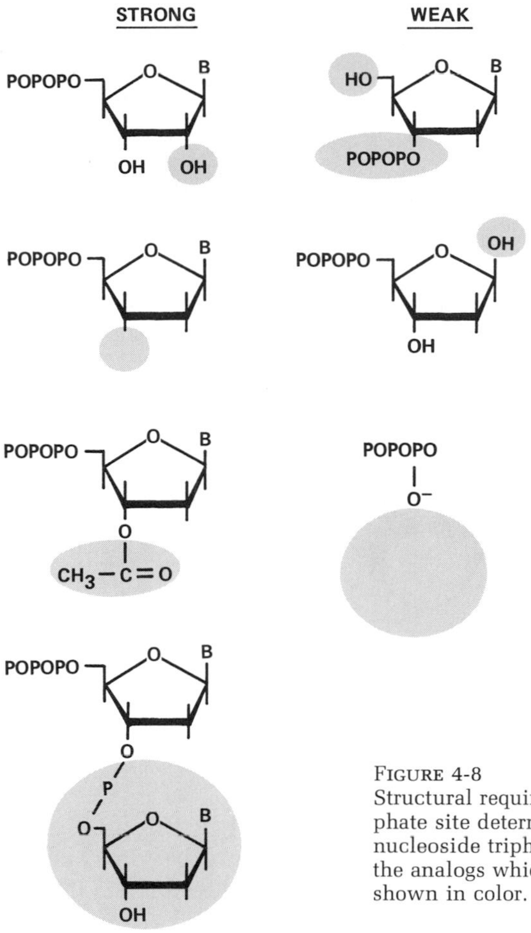

FIGURE 4-8
Structural requirements for binding to the triphosphate site determined by analogs of a deoxyribonucleoside triphosphate. The deviant features of the analogs which bind strongly or weakly are shown in color.

primer invariably promote polymerization or are degraded, and binding measurements are thus complicated. Among the questions still to be answered are whether a template confers specificity on triphosphate binding and what influence the primer terminus exerts on the entry and orientation of the triphosphate in the site.

7. Polymerization Step[22, 23]

Inasmuch as the essence of DNA polymerase action is to catalyze the condensation of the primer terminus with a triphosphate, the sites for these two components should be next to one another. Their proximity was determined by binding a spin-labeled ATP derivative[24] (Fig. 4-9) in the triphosphate site and binding simultaneously

22. Kornberg, A. (1969) *Science* **163,** 1410.
23. Kelly, R. B., Cozzarelli, N. R., Deutscher, M. P., Lehman, I. R. and Kornberg, A. (1970) *JBC* **245,** 39.
24. The 6-amino group contains tetramethylpiperidine nitroxide, which has a free electron, a spin label.

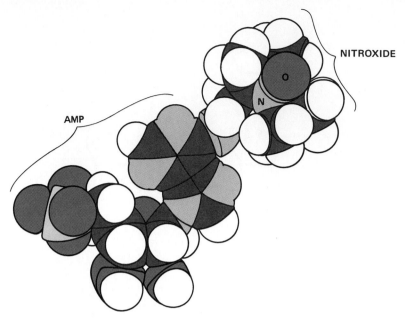

FIGURE 4-9
Spin-labeled analog of AMP. The tetramethylpiperidine nitroxide is attached
to the 6-amino group of AMP. (Compare with formula of dAMP in Figure
1-1.) The experiment described in the text employed ATP labeled in this
way; AMP was unlabeled.

AMP (a 3′-hydroxyl ribonucleotide) in the primer terminus site[25].
Based on measurements of the perturbation caused by the spin-
labeled ATP in the nuclear magnetic resonance (NMR) spectrum of
the AMP, the 2-proton of the AMP was 7.1 Å away from the free
electron of the triphosphate analog, indicating stacking of the AMP,
presumably under the ATP.

Linear, duplex DNA partially degraded from each 3′-hydroxyl end
is an excellent template-primer. The denuded portions, if not too
extensive, are replaced with great facility by all the DNA polymerases. DNA polymerase I is unusual in its capacity to use a nicked
duplex such as poly d(A–T) as a template-primer, with which it
achieves optimal rates (Table 4-1).

The mechanism of this polymerization may be postulated as follows: The triphosphate is bound on the enzyme adjacent to the
3′-hydroxyl group of the terminal nucleotide of the primer, and
oriented so that it can be brought into direct contact and form a
base pair with the template (Fig. 4-10). It is plausible (though not
substantiated) that, when the correct base pair is formed, movement
or translation of the chain relative to the enzyme is concurrent with
diester bond formation. As the primer terminus loses its 3′-hydroxyl
group during transformation to a diester bond, it is no longer held
in the primer terminus site. Through movement of the entire chain,

25. Krugh, T. R. (1971) B. **10**, 2594.

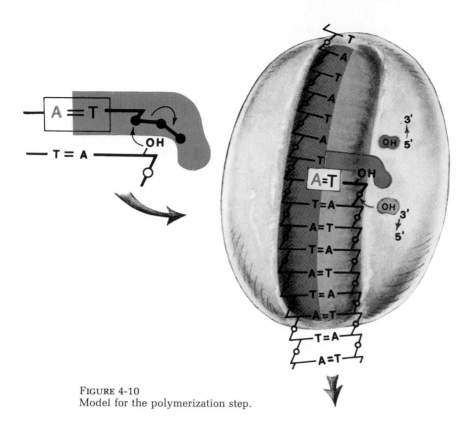

FIGURE 4-10
Model for the polymerization step.

the old primer terminus is replaced by the newly added nucleotide, which has a terminal 3′-hydroxyl group, and is therefore held in the primer terminus site. The new primer terminus is now ready to attack another triphosphate and add the next nucleotide. Inorganic pyrophosphate is displaced only as formation of the diester bond is being completed and the chain movement is translocating the newly added nucleotide into the primer terminus site.

The behavior of the dideoxy analog of the triphosphate is instructive regarding the mechanism of polymerization[26]. Despite the absence of the 3′-hydroxyl group, this analog retains affinity for the triphosphate site (see the preceding section) and forms a polymerization complex protecting the primer terminus from nucleolytic attack (see below). However, the rate of covalent linkage to the primer is reduced 1000-fold by lack of the 3′-hydroxyl group.

In view of the strict requirement for a 3′-hydroxyl group in the primer site[27], it seems likely that there is a similar requirement in the structure of one or more of the late intermediates in the sequence of polymerization steps. This may explain why the lack of the hydroxyl group in the dideoxy analog sharply reduced the rate of polymerization. This finding illustrates a degree of specificity dis-

26. Atkinson, M. R., Deutscher, M. P., Kornberg, A., Russell, A. F. and Moffatt, J. G. (1969) B. 8, 4897.
27. Huberman, J. A. and Kornberg, A. (1970) JBC 245, 5326.

played by polymerase in addition to its recognition of a correct base pair.

The specificity of DNA polymerase is very likely based on its demand for one of the four base pairs (A–T, T–A, G–C, C–G) rather than on the recognition of an appropriate incoming triphosphate. All the correct base pairs, as discussed in Chapter 1, contain regions of identical dimensions and geometry, and are symmetrical. When the correct base pair is within the active site, the enzyme may respond, possibly by change in conformation, so that the subsequent catalytic steps can then proceed. If a nonmatching triphosphate were to bind to the enzyme, the correct base pair could not be formed, there would be no conformational change, and the triphosphate would be rejected.

An early, and still impressive, demonstration of the crucial importance of base-pairing in polymerization is found in the experiments substituting the natural bases with analogs[28, 29]. As shown in Table 4-4, utilization of analogs by polymerase can be described simply by their base-pairing potentialities: uracil and 5-substituted uracils took the place of thymine; 5-substituted cytosines of cytosine; and hypoxanthine of guanine. Whereas hypoxanthine, capable of forming two hydrogen bonds with cytosine, was effective at a reduced rate of incorporation, xanthine, which can form only one hydrogen bond with cytosine, was inert. The behavior of analogs was essentially the same when tested with DNA polymerases from various sources.

Polymerization proceeds until the available template is filled in. If, as in Fig. 4-11, the duplex is completed, no template is available beyond the primer terminus and polymerization ceases.

TABLE 4-4
Replacement of natural bases by analogs in polymerase action

Analog, used in form of deoxynucleoside triphosphate	Deoxynucleoside triphosphate replaced by analog			
	dTTP	dATP	dCTP	dGTP
	Percent of control value			
Uracil	54	0	0	0
5-Bromouracil	97(100)	0(0)	0(0)	0(0)
5-Fluorouracil	32(9)	0(0)	0(0)	0(0)
5-Hydroxymethylcytosine	(0)	(0)	(98)	(0)
5-Methylcytosine	0	0	185	0
5-Bromocytosine	0(0)	0(0)	118(104)	0(0)
5-Fluorocytosine	0(0)	0(0)	63(67)	0(0)
N-methyl-5-fluorocytosine	0	0	0	0
Hypoxanthine	0	0	0	25
Xanthine	0	0	0	0

Values in parentheses were obtained with T2-phage polymerase (Chapter 5, Section 5), the others with DNA polymerase I.

28. Bessman, M. J., Lehman, I. R., Adler, J., Zimmerman, S. B., Simms, E. S. and Kornberg, A. (1958) *PNAS* **44**, 633.
29. Kornberg, A. (1962) **in** *Enzymatic Synthesis of DNA.* John Wiley & Sons, N.Y.

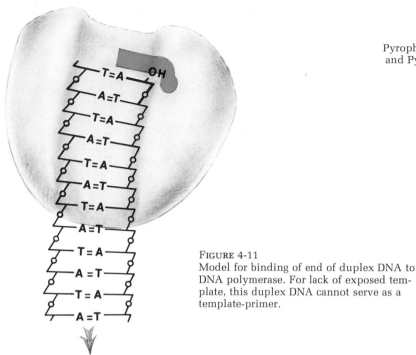

FIGURE 4-11
Model for binding of end of duplex DNA to
DNA polymerase. For lack of exposed template, this duplex DNA cannot serve as a
template-primer.

This hypothetical model for polymerization, even if correct in basic outline, does not account for all of the enzyme's features. For example, the functions of added metal ions (Mg^{2+}, Mn^{2+})[30] and endogenous Zn^{2+} [31] are not discussed. Mg^{2+} is assumed to chelate the triphosphate substrates but may serve in additional ways. Zn^{2+} may facilitate the nucleophilic attack of the 3'-hydroxyl group by proton abstraction, but there is no experimental evidence available. Unexplained, at present, is the capacity of the enzyme for fine discrimination of stacking interactions, helical forms, and other structural features of primer and template chains. A proper understanding will require a better three-dimensional recognition of the orientations within the active center. To accommodate the screwlike translation of an essentially helical structure through the active center demands a surface adapted to one or more turns of the helical duplex.

8. Pyrophosphorolysis and Pyrophosphate Exchange[32]

The pyrophosphorolysis reaction catalyzed by DNA polymerase is the degradation of a DNA chain by inorganic pyrophosphate, and

30. Slater, J. P., Tamir, I., Loeb, L. A. and Mildvan, A. S. (1972) *JBC* **247**, 6784.
31. Springgate, C. F., Mildvan, A. S., Abramson, R., Engle, J. L. and Loeb, L. A. (1973) *JBC* **248**, 5987.
32. Deutscher, M. P. and Kornberg, A. (1969) *JBC* **244**, 3019.

is the reversal of polymerization (Fig. 4-12):

$$\text{(dNMP)}_n + x\text{PP}_i \overset{\text{DNA}}{\rightleftarrows} \text{(dNMP)}_{n-x} + x\text{dNPPP}$$

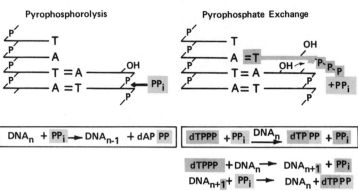

FIGURE 4-12
Equations for pyrophosphorolysis and pyrophosphate exchange.

Pyrophosphorolysis reaches a steady state when the accumulation of triphosphates supports synthesis at a rate that balances their removal. The reaction, like $3' \to 5'$ exonuclease hydrolysis (next section), requires a proper primer terminus. With a chain terminated in a dideoxynucleotide, the rate is reduced at least 1000-fold[33]. But unlike hydrolysis from the $3'$ end, pyrophosphorolysis absolutely requires the template strand of a duplex DNA. The reaction is therefore truly a reversal of polymerization and not competitive with hydrolysis, as had been thought at one time.

The enzyme also supports exchange of pyrophosphate into the β,γ-groups of a nucleoside triphosphate:

$$^{32}\text{PP}_i + \text{dNPPP} \overset{\text{DNA}}{\rightleftarrows} \text{PP}_i + \text{dNP}^{32}\text{P}^{32}\text{P}$$

This reaction can occur even when only a single triphosphate is present, but otherwise requires all the primer and template conditions demanded of replication. There is no evidence to suggest formation of a nucleotidyl-enzyme intermediate. Instead, inorganic pyrophosphate exchange appears to be the result of a sequence of a polymerization step and a pyrophosphorolytic step, repeated many times over. Because the rate of pyrophosphate exchange is considerably faster than that of pyrophosphorolysis, it seems reasonable to suggest that the attack by pyrophosphate occurs at an intermediate state short of completion of the polymerization step, and that this intermediate form is attained more readily from the direction of polymerization.

33. Atkinson, M. R., Deutscher, M. P., Kornberg, A., Russell, A. F. and Moffatt, J. G. (1969) B. **8**, 4897.

9. The $3' \rightarrow 5'$ Exonuclease: Proofreading[34]

A most important feature of DNA polymerase as catalyst for poly-merizing mononucleotides into long chains of DNA is that it can also degrade these chains. The pure enzyme can degrade DNA from the primer terminus in a $3' \rightarrow 5'$ direction ($3' \rightarrow 5'$ exonuclease). The $3' \rightarrow 5'$ exonuclease activity of DNA polymerase had been desig-nated exonuclease II[35] before it was realized that it is an integral part of DNA polymerase in both structure and function. The turnover number of this exonuclease activity is, under conditions optimal for it, only about two percent that of the polymerizing activity (see also Table 5-7).

The enzyme also has an entirely distinct capability to degrade DNA from the 5' end of the chain in the $5' \rightarrow 3'$ direction ($5' \rightarrow 3'$ exonuclease, see Sections 12 and 13 of this chapter). The nature of the two nucleolytic functions acting in opposite directions on a DNA chain, and their significance for DNA metabolism, became clear when with proteolytic cleavage of the polypeptide chain it became possible to separate them (see Section 14 of this chapter).

The essential function of the $3' \rightarrow 5'$ exonuclease is to recognize and cleave a non-base-paired terminus. This conclusion has emerged from several lines of evidence. The $3' \rightarrow 5'$ exonuclease activity could be shown to degrade both single- and double-stranded DNA, and, although temperature had no more than the usual influence on single-strand degradation, a rise in temperature from 30° to 37° enhanced the hydrolysis of double-stranded DNA more than tenfold. This indicates that a frayed end created by partial melting of the helix is the site of the enzyme action. The alkaline pH optimum, at which melting of the helix is favored, is additional evidence.

Studies with synthetic DNA polymers have confirmed that the $3' \rightarrow 5'$ exonuclease requires a frayed or unpaired 3'-hydroxyl termi-nus. When pairs of synthetic polymer chains containing correctly paired 3'-hydroxyl ends were used as substrates, in the absence of deoxynucleoside triphosphates required for polymerization, fraying of the primer terminus exposed it to repeated nuclease action (Fig. 4-13). However, this nuclease action was completely suppressed by a deoxynucleoside triphosphate, but only by the one that matched the template and could be incorporated into the polymer.

Polymers with mismatched ends always lost the nonmatching termini before polymerization was initiated (Fig. 4-14). The mis-paired termini tested included purine-purine, pyrimidine-pyrimi-dine, and purine-pyrimidine mismatches.

Deoxynucleoside triphosphates that could base-pair with the tem-plate but could not be incorporated into polymer failed to suppress nuclease action. For example, phosphonate analogs of dTTP could bind to the triphosphate site and base-pair with the poly (dA) tem-

---

34. Brutlag, D. and Kornberg, A. (1972) JBC **247**, 241.
35. Lehman, I. R. and Richardson, C. C. (1964) JBC **239**, 233.

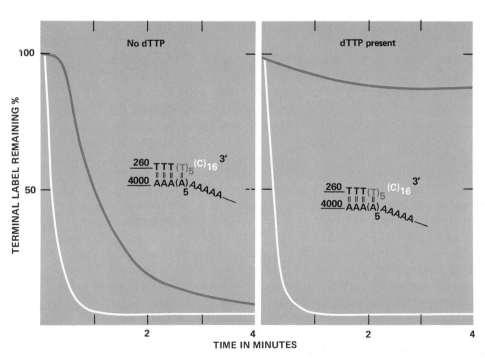

FIGURE 4-13
Retention of base-paired primer terminus. The
terminus is preserved under polymerization
conditions (+dTTP) and removed by $3' \rightarrow 5'$
exonuclease action in the absence of polymerization (−dTTP).

FIGURE 4-14
Removal of a mismatched primer terminus. The $3' \rightarrow 5'$ exonuclease activity
removes the mismatched C residues whether or not dTTP is present to support
polymerization. When the T residues are reached they are retained in the presence
of dTTP.

plate used, but did not serve as substrates and were correspondingly ineffective in suppressing nuclease.

Of special interest is the dideoxynucleotide analog ddTTP, which was incorporated at a rate 1000-fold slower than dTTP and to the extent of only one nucleotide per chain[36]. Nevertheless, ddTTP inhibited $3' \rightarrow 5'$ exonuclease activity as completely as did dTTP. These results indicate that the primer terminus, in its complex with template and triphosphate, is invulnerable to nuclease even when the rate-limiting step in polymerization is exceedingly slow.

Template-directed chain extension catalyzed by DNA polymerase results in a properly base-paired $3'$ terminus which is utilized for subsequent polymerization. Pyrophosphorolysis, as the reversal of polymerization, would be expected to show specificity for the product of polymerization, i.e., a properly base-paired $3'$ terminus. Experiments with pairs of synthetic polymer chains showed that only a properly base-paired terminus was removed by pyrophosphorolysis (Table 4-5). A mispaired primer terminus was not attacked

TABLE 4-5
Nature of cleavage of the primer terminus in the presence of inorganic pyrophosphate[a]

	Products released	
	Mono-phosphate (hydrolysis)	Triphosphate (pyrophos-phorolysis)
Structure at $3'$ primer terminus	Percentage of total	
Single-stranded chain $(d(T)_{260}[^3H]d(T)_1)$	100	< 0.5
Double-stranded, base-paired $(d(T)_{260}[^3H]d(T_1) \cdot d(A)_{4000})$	53	47
Single-stranded chain $(d(T)_{260}[^3H]d(C)_1)$	100	< 0.5
Double-stranded, mispaired $(d(T)_{260}[^3H]d(C)_1 \cdot d(A)_{4000})$	100	< 0.5

[a]Source: Brutlag, D. and Kornberg, A. (1972) *JBC* **247**, 241.

by pyrophosphate. This specificity clearly distinguishes pyrophosphorolysis, which attacks only base-paired termini, from hydrolysis, in which attack is only on unpaired termini. Thus, there may be two forms of enzyme-DNA complex. One, with a base-paired primer terminus, would allow incorporation of a triphosphate or attack by pyrophosphate. The other, in which the $3'$ terminus is frayed, would be available only for hydrolysis.

These are the characteristics we should expect of the $3' \rightarrow 5'$ nuclease if it acts as a mechanism to correct rare errors in the polymerization process. Correction of such errors cannot be postponed. Repair systems that excise a mismatched nucleotide in DNA (Chapter

36. Atkinson, M. R., Deutscher, M. P., Kornberg, A., Russell, A. F. and Moffatt, J. G. (1969) B. **8**, 4897.

9, Section 2) cannot determine which of the two strands contains the mismatched member of an incorrect pair.

Inasmuch as polymerase cannot extend a mispaired terminus generated by itself, its exonuclease would remove this mismatched nucleotide as it does in the synthetic polymers studied here. Fidelity of template copying is greater as a result of two base-pairing selection steps. The first step is the determination that the primer terminus is properly paired before the addition of the next nucleotide; the second is the selection of the correct nucleotide for polymerization. This base-pairing specificity of the $3' \rightarrow 5'$ exonuclease of DNA polymerases emphasizes the importance of this activity for accurate template-directed polymerization.

10. Polymerase as Mutator or Antimutator[37]

Incorrect base-pairing of a nucleotide with the template may come about if an unusual tautomeric base form should occur in the template or entering nucleotide. For example, the rare enol form of T and the normal keto form of G might form a base pair (Fig. 4-15) that DNA polymerase does not distinguish from the normal G–C pair[38,39].

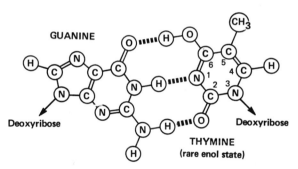

FIGURE 4-15
Base pairing of guanine with thymine in its rare enol state.

The probability of such a mismatch has been estimated at about one in 10^4 pairings, a frequency far greater than observed for spontaneous mutations.

Another possible source of mispairings is a defect in polymerase itself, as for example in polymerases induced by T4 phages with lesions in the gene for the polymerase, gene 43. Since several of these mutants in polymerase were observed to have greatly increased frequency of spontaneous mutations as well as a higher frequency of reversions, gene 43 may be regarded as a mutator gene[40]. It was once

37. Muzyczka, N., Poland, R. L. and Bessman, M. J. (1972) *JBC* **247**, 7116.
38. Watson, J. D. and Crick, F. H. C. (1953) *CSHS* **18**, 123.
39. Trautner, T. A., Swartz, M. N. and Kornberg, A. (1962) *PNAS* **48**, 449.
40. Speyer, J. F., Karam, J. D. and Lenny, A. B. (1966) *CSHS* **31**, 693.

reasoned that the acceptance of noncomplementary bases by the defective polymerase was responsible for the increased rate of mutations. In vitro studies with the purified mutant enzyme demonstrated misincorporation of dTTP with a poly (dC) template[41].

Mutants in the same gene were later found with markedly _reduced_ spontaneous mutation rate. Thus the same gene can behave as a mutator or an antimutator[42]. Is the antimutator property explained by increased base-pair selectivity during polymerization?

Recent studies suggest that spontaneous mutation rates reflect relative rates of polymerization and proofreading at the primer terminus[43,44]. Mutations altering the normal ratio of polymerase activity to 3′ → 5′ exonuclease activity might be expected to be mutators when the ratio is increased, and antimutators when the ratio is decreased. This interesting interpretation comes from studies with homopolymer pairs similar to those cited in the previous section. Purified DNA polymerases induced by mutator, wild-type, or antimutator phages exhibited strikingly different patterns of 3′ → 5′ exonuclease activity. As seen in Fig. 4-16a, removal of the primer terminus was slower with mutator polymerase and very rapid with antimutator polymerase when compared to wild-type enzyme, all

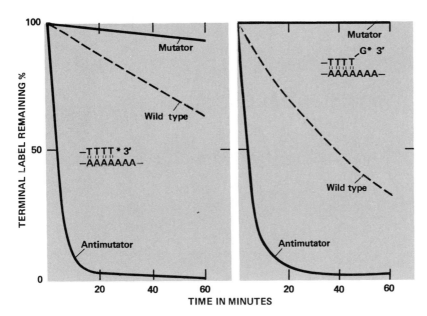

FIGURE 4-16a
The 3′ → 5′ exonuclease activities of mutant (mutator, antimutator, and neutral) and wild type (normal) polymerases compared. Removal of a base-paired (T) terminus (left) and a mismatched (G) terminus (right).

41. Hall, Z. W. and Lehman, J. R. (1968) _JMB_ **36**, 321.
42. Drake, J. W., Allen, E. F., Forsberg, S. A., Preparata, R. M. and Greening, E. O. (1969) _Nat._ **221**, 1128.
43. Muzyczka, N., Poland, R. L. and Bessman, M. J. (1972) _JBC_ **247**, 7116.
44. Hershfield, M. S. (1973) _JBC_ **248**, 1417.

measured at the same level of polymerase activity. In the presence of the deoxynucleoside triphosphate required for polymerization, the $3' \rightarrow 5'$ nuclease activity was very low for the normal and mutator enzymes, but rather vigorous for the antimutator (Fig. 4-16b). In fact, with the latter, the turnover of the deoxynucleoside triphosphate (dTTP), measured by its hydrolysis to dTMP (Fig. 4-16c), resulted in a loss of 93 percent of the triphosphate; this compared with a loss of only 0.5 percent with the mutator and 4 percent with the normal enzymes.

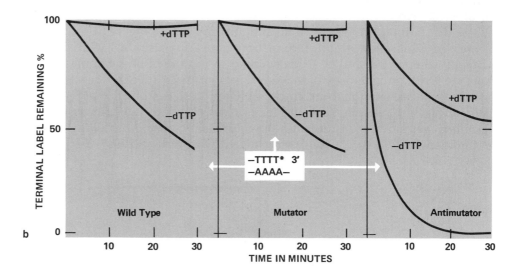

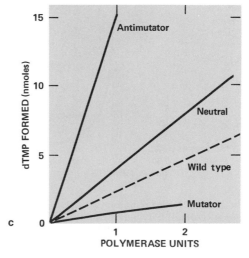

FIGURE 4-16b, c
The $3' \rightarrow 5'$ exonuclease activities of mutant (mutator, antimutator, and neutral) and wild type (normal) polymerases compared. (b) Influence of dTTP, which supports polymerization, in retention of a base-paired (T) terminus by wild type, mutator, and antimutator polymerases. (c) Turnover of dTTP (polymerization followed by exonucleolytic removal) by mutant and wild type polymerases compared. The polymer substrates were like those in (b).

11. Exonuclease Actions at the Primer Terminus[45, 46]

The activity of the $3' \rightarrow 5'$ exonuclease has been seen as an essential component of DNA polymerase action. Also active in shaping and modifying the primer terminus are two other $3' \rightarrow 5'$ exonucleases identified in *E. coli*. Their distinctive operations, summarized in Table 4-6, may at times be essential in the DNA synthesis that takes place in the processes of replication and of repair and recombination (see Chapter 9).

Exonuclease I is absolutely specific in degrading single-stranded DNA[46]. A displaced $3'$ end of a DNA strand (Table 4-6), incapable of extension by polymerase for lack of a matching template, can be

TABLE 4-6
Actions of several $3' \rightarrow 5'$ exonucleases on various $3'$ termini

	Action on 3′ primer terminus			
Structure	Chain extension by DNA polymerase	$3' \rightarrow 5'$ hydrolysis by DNA polymerase	$3' \rightarrow 5'$ hydrolysis by exonuclease I	$3' \rightarrow 5'$ hydrolysis by exonuclease III[a]
	Yes	No[b]	No	Yes
	No	Yes	No[c]	Yes
	No	Yes	Yes	No
	No	Yes	Yes	No

[a]Exonuclease III is unique among exonucleases in its phosphomonoesterase action on a $3'$-phosphate terminus.
[b]No, in the presence of triphosphates; yes, in their absence.
[c]Hydrolysis is very slow.

rapidly pared down to a point near where the duplex structure would support polymerization.

The specificity of exonuclease I was analyzed by use of terminally labeled polynucleotides, with base-paired or mismatched ends, as used in studies of $3' \rightarrow 5'$ exonuclease function of polymerase[47]. With enzyme present in excess over the available $3'$ termini, the terminal nucleotide was removed rapidly from $d(T)_{260}[^3H]d(T)_1$ but only

---

45. Brutlag, D. and Kornberg, A. (1972) *JBC* **247**, 241.
46. Lehman, I. R. (1971) **in** *The Enzymes*, Academic Press, N.Y. **4**, 251.
47. Op. cit. in footnote 45.

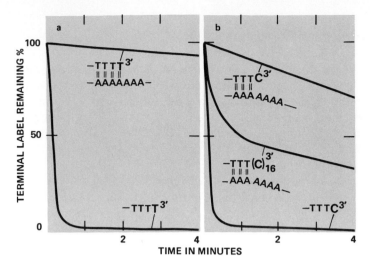

FIGURE 4-17
Exonuclease I action on termini of base-paired and single-stranded chains (left) and on mismatched (C) terminal regions, one and 16 residues long, in duplex chains (right). The polymer substrates were like those in Figures 4-13 and 4-14.

slowly, if at all, from the same polymer annealed to $d(A)_{4000}$ (Fig. 4-17a). Attack on the mispaired terminal nucleotide of $d(T)_{260}[^3H]d(C)_1$ was similar (Fig. 4-17b). Whereas, removal of the dCMP residues from the single strand was rapid, removal from $d(T)_{260}[^3H]d(C)_1 \cdot d(A)_{4000}$ was very slow. When the number of terminal dCMP residues was increased to 16, exonuclease I removed up to 60 percent of the dC residues of $d(T)_{260}[^3H]d(C)_{16} \cdot d(A)_{4000}$ rapidly, but the remaining nucleotides were removed very slowly. These data indicate that exonuclease I removed unpaired nucleotides that are within six to eight residues of a base-paired region very slowly compared with nucleotides in a single-stranded chain. Thus exonuclease I shows a general specificity for extended single-stranded structures.

The other important exonuclease operating at the primer terminus is exonuclease III[48]. It has two distinctive features. One is its capacity to act as a phosphatase on a chain terminated by a 3'-phosphate group. Such a 3'-phosphate terminus, generated by enzymatic or chemical cleavage, is not only inert as a primer, but by binding unproductively in the active center of DNA polymerase is a potent inhibitor of its action. Removal of the phosphate group restores the primer terminus as a substrate for DNA polymerase.

The other feature of exonuclease III action is its relative specificity for duplex DNA[49]. The enzyme hydrolyzed the terminal residues of $d(T)_{260}[^3H]d(T)_1$ very rapidly only when the polymer was annealed to $d(A)_{4000}$ (Fig. 4-18). Likewise, the residues of $d(T)_{260}[^3H]d(C)_1$ were sensitive to exonuclease III only when the polymer was in the double-stranded conformation (Fig. 4-18). In experiments with

48. Richardson, C. C. and Kornberg, A. (1964) *JBC* **239**, 242; Richardson, C. C., Lehman, I. R. and Kornberg, A. (1964) *JBC* **239**, 251.
49. Brutlag, D. and Kornberg, A. (1972) *JBC* **247**, 241.

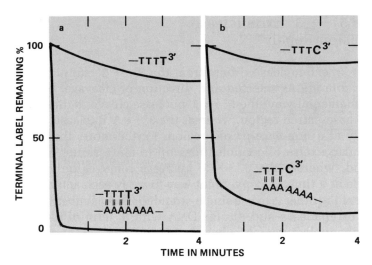

FIGURE 4-18
Exonuclease III action on termini of single-stranded chains (upper curves) and on base-paired or mismatched termini of duplex chains (lower curves). The polymer substrates were like those in Figures 4-13 and 4-14.

polymers containing blocks of dCMP residues, resistance to hydrolysis increased in proportion to the length of the sequence that did not anneal to the $d(A)_{4000}$. From these results it appears that exonuclease III hydrolyzes double-stranded polymers which contain one, two, or even three mispaired terminal nucleotides at a significant rate. About 90 percent of the terminal C residues were removed from $d(T)_{260}[^3H]d(C)_1 \cdot d(A)_{1000}$ in four minutes (Fig. 4-18). But those polymers in the population with four or more C residues were attacked very slowly, and so nearly ten percent of them remained.

These studies of exonuclease III show that the enzyme acting at a nick in a duplex enlarges it to a gap; acting at an end, a protruding 5'-ended chain is created. Such gaps and ends may be essential intermediates in recombinational events that later require DNA synthesis.

What is known of the properties of the purified $3' \rightarrow 5'$ exonucleases suggests that their action at the 3' end of a DNA chain may be complementary in some instances and redundant in others (Table 4-6.) The rate of exonuclease I cleavage on a protruding single-stranded chain decreases as the enzyme approaches a duplex region. It is at this point that exonuclease III action may set in. This switchover occurs at about five residues from the base-paired region. The exonuclease activity of DNA polymerase on displaced single strands appears to embrace the combined actions of exonucleases I and III. The redundant action of these enzymes and the multifunctional nature of some indicates that each may be involved in more than one physiological process, and may explain why mutants deficient in one or another of these enzymes appear to be healthy and to maintain DNA functions at a normal level. The overlapping of functions of these exonucleases thus insures the integrity of the essential operations in DNA metabolism.

become inactive for polymerization. An example of this is a primer terminus terminated in a dideoxynucleotide (Section 5 of this chapter) which cannot be extended nor readily removed by the $3' \rightarrow 5'$ exonuclease.

This ability to excise mismatches and distortions resulting from ultraviolet irradiation suggests that polymerase removes and repairs faulty regions produced in DNA in vivo[56]. The physiologic role of the $5' \rightarrow 3'$ exonuclease of DNA polymerase and the in vivo evidence for excision and repair of ultraviolet, X-ray, and related DNA lesions will be discussed in Chapter 9, Section 2. However it is worth pointing out here that coupling of polymerizing and excising functions in a single enzyme may be of significant advantage in repair of in vivo damage. As a thymine dimer is removed by the $5' \rightarrow 3'$ exonuclease excision function, concurrently the polymerase function would fill in the gap caused by the excision. Therefore, excision-repair by DNA polymerase would at no time leave a single-strand gap exposed to attack at the $3'$ end by exonucleases such as exonuclease III. Rather, only a nick with a $3'$-hydroxyl and a $5'$-phosphate would be left from which the polymerase can be displaced by DNA ligase (see Chapter 7, Section 6). Perhaps the extensive DNA degradation observed after ultraviolet irradiation of E. coli mutants deficient in DNA polymerase I (see Section 18 of this chapter) is the result of production of single-strand gaps during excision of thymine dimers by secondary repair systems. The lifetime of these gaps may be long enough to enable enzymes such as exonuclease III to enlarge them, and thereby cause extensive degradation of the DNA.

A possible physiological repair function served by the $5' \rightarrow 3'$ exonuclease is removal of the RNA fragment that primes initiation of DNA synthesis. The RNA in model RNA–DNA hybrid polymers was shown to be removed by the exonuclease, as was the RNA priming fragment of enzymatically synthesized DNA. A discussion of this follows in Chapter 7, Section 5. Possibly related to this excision of RNA from DNA is the preferential degradation of uracil-containing DNA by the $5' \rightarrow 3'$ exonuclease[57]. The A–U base pair may be distorted enough to invite excision, as is the case with a thymine dimer. This action may be a mechanism for excluding uracil from DNA.

13. The $5' \rightarrow 3'$ Exonuclease: Nick Translation[58]

DNA polymerase binds at a nick, and in the presence of the required deoxynucleoside triphosphates extends the primer terminus. Early studies showed that polymerization was accompanied by a burst of hydrolysis of the template-primer that matched the extent of polymerization (Fig. 4-21). With $d(A)_{4000}$ as template and sufficient

56. Cozzarelli, N. R., Kelly, R. B. and Kornberg, A. (1969) JMB **45**, 513.
57. Wovcha, M. G. and Warner, H. R. (1973) JBC **248**, 1746.
58. Kelly, R. B., Cozzarelli, N. R., Deutscher, M. P., Lehman, I. R. and Kornberg, A. (1970) JBC **245**, 39.

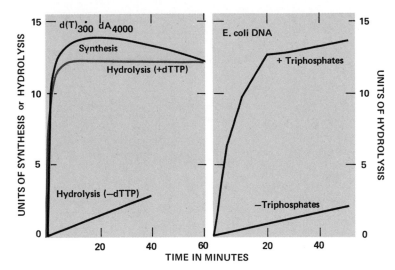

FIGURE 4-21
Nick translation by DNA polymerase I. When concurrent polymerization and 5′ → 3′ exonuclease actions balance each other (as on
the left), a nick in the DNA is linearly advanced (translated) along
the chain. Hydrolysis is stimulated manyfold by concurrent synthesis.

[³H]dT$_{300}$ to cover it, synthesis of a poly dT chain proceeded to the
extent of matching the input d(A)$_{4000}$ with a concurrent hydrolysis
of the d(T)$_{300}$. Before the 5′ → 3′ exonuclease of DNA polymerase
was recognized, investigators were puzzled as to how the known
3′ → 5′ exonuclease and polymerase activities could operate simultaneously and antagonistically on the same primer terminus. Even
more puzzling was the observation that hydrolysis of DNA was
enhanced ten-fold upon the addition of the deoxynucleoside triphosphates that made DNA synthesis possible[59] (see also Table 5-7).

With the discovery of the 5′ → 3′ exonuclease activity, it became
clear that polymerization at a nick was coordinated with nuclease
action, so as to move (translate) the nick linearly along the helix
without change in mass of the DNA. The nuclease action at the nick
was exclusively 5′ → 3′ and occurred at a rate ten times faster in
advance of a growing chain than in the absence of polymerization.
This concurrent polymerization and hydrolysis was found to proceed for a brief period and stop, presumably for one of three reasons
(Fig. 4-22): (i) replication reached the end of the template or a nick
in it and therefore terminated, (ii) polymerase was dislodged and
the nick sealed in the presence of ligase, or (iii) the 5′ oligonucleotide, sufficiently long to escape cleavage, was displaced and then
adopted as template by the polymerase, thereby producing a branch
or fork on the growing molecule.

Only when the latter occurs does net synthesis of DNA take place.
Some of these actions were nicely illustrated in the early studies on

---

59. Lehman, I. R. (1967) *ARB* **36**, 645.

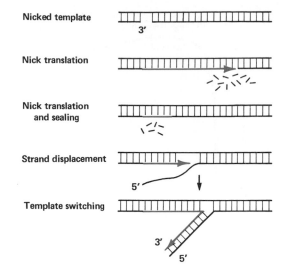

Nicked template

Nick translation

Nick translation
and sealing

Strand displacement

Template switching

FIGURE 4-22
Polymerase action at a nick may lead to nick
translation (to an extent determined by ligase
sealing), strand displacement, or template
switching.

replication of single-stranded circular DNA (see Section 17 of this
chapter) or of a nicked circular duplex.

14. Proteolytic Cleavage:
Two Enzymes in One Polypeptide[60-63]

Proteolytic surgery of DNA polymerase cleaves the polypeptide
chain into two active fragments[64, 65] and provides an unusual opportunity to examine the complex enzyme in two parts.

The products of proteolysis of the enzyme (109,000 daltons) are
a large fragment[66] (76,000 daltons), containing the polymerase and
$3' \rightarrow 5'$ exonuclease activities, and a smaller one[67] (36,000 daltons)
with only the $5' \rightarrow 3'$ exonuclease activity. The single cysteine
residue and the single disulfide group of the intact enzyme as
well as the binding sites for deoxyribonucleoside triphosphate and
3'-hydroxyl deoxyribonucleotide (primer terminus) are retained in
the large fragment.

The large fragment may be obtained in two ways: by cleavage of
intact enzyme with subtilisin, trypsin, or other proteases; or by isolation from E. coli extracts, presumably a product of proteolysis.
Extracts of one strain of E. coli have routinely yielded rather large
amounts of the large fragment. Whether generation of the fragment
occurred in the cell, or after extraction, has not been determined.

60. Klenow, H. and Henningsen, I. (1970) PNAS **65**, 168.
61. Brutlag, D., Atkinson, M. R., Setlow, P. and Kornberg, A. (1969) BBRC **37**, 982.
62. Setlow, P., Brutlag, D. and Kornberg, A. (1972) JBC **247**, 224.
63. Setlow, P. and Kornberg, A. (1972) JBC **247**, 232.
64. Op. cit. in footnote 60.
65. Op. cit. in footnote 61.
66. Op. cit. in footnote 62.
67. Op. cit. in footnote 63.

The large fragment carries out DNA synthesis on the 3'-hydroxyl side of a nick in double-stranded DNA. It catalyzes 3' → 5' exonuclease action on both single-stranded DNA and unpaired regions in double-stranded DNA with production of deoxynucleoside monophosphates (see Section 9). The large fragment also catalyzes primed poly d(A–T) synthesis at a relatively linear rate generating exceptionally long poly d(A–T) molecules, but is inactive in unprimed (de novo) poly d(A–T) synthesis (see next section of this chapter). Quantitative utilization of dATP and dTTP substrates, as well as the kinetics of synthesis and the size of the polymer product show that 5' → 3' exonuclease activity is absent. During synthesis at the 3'-hydroxyl side of a nick the 5'-phosphate end of the nick is not degraded.

The small fragment retains only the 5' → 3' exonuclease activity. This fragment resembles the 5' → 3' exonuclease of the intact enzyme in degrading DNA to mono- and oligonucleotides and in its capacity to excise mismatched regions such as thymine dimers. But it differs from the intact enzyme in that deoxynucleoside triphosphates, which support polymerization, fail either to stimulate the exonuclease, or to increase the proportion of oligonucleotides among the products.

No evidence of physical interaction has been found when the large and small fragments are brought together[68]. However, with a mixture of small fragment, large fragment, nicked DNA, and suitable deoxynucleoside triphosphates, evidence of interaction was striking. The same influences of polymerization on 5' → 3' exonucleases were seen with the mixtures as with the intact enzyme. This suggests that at the locus of a nick in DNA the two fragments are adjacent to one another, making possible coordinated polymerization and 5' → 3' exonuclease action.

Cleavage of DNA polymerase into two active fragments that show no measurable affinity for one another and are similar to the intact enzyme in catalytic activity indicates that within the single polypeptide chain of DNA polymerase are two distinct enzymes; one enzyme containing the polymerase and 3' → 5' exonuclease function, and the other enzyme containing the 5' → 3' excision function. The two enzymes are held together by a polypeptide link susceptible to a protease such as trypsin (Fig. 4-23). This polypeptide hinge ensures that both the polymerase and excision functions act simultaneously at the same nick in a DNA molecule. The consequences of this fact for DNA metabolism will be discussed in Section 18 of this chapter.

Separation of the enzyme activities has also been achieved by genetic means. A mutation has rendered the polymerase but not the 5' → 3' exonuclease activity temperature-sensitive (pol A12, see Table 4-11) whereas another mutation has destroyed the 5' → 3' exonuclease without affecting the polymerase (see Section 18).

68. Setlow, P. and Kornberg, A. (1972) *JBC* **247**, 232.

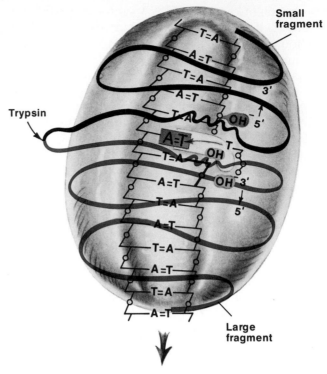

FIGURE 4-23
The polypeptide chain of DNA polymerase I is represented in a
simplified form to indicate regions corresponding to the large
and small fragments (and their functions) and the location
susceptible to proteolytic cleavage.

The sharp physical and functional distinctions of the two components of DNA polymerase suggest that they have evolved as separate
proteins, or subunits, coded by separate genes. Better insights into
this fascinating possibility will be one of the rewards of comparative
studies of the variety of polymerases within a cell and among the
cells of different species.

15. De Novo Synthesis of Repetitive DNA[69,70]

The synthesis in the absence of any template-primer of a huge polymer of deoxyadenylate and thymidylate in which the residues are
arranged in perfectly ordered alternation was one of the most remarkable and unanticipated events in the history of DNA polymerase
studies[71]. The event was remarkable because this synthesis of DNA

69. Kornberg, A. (1965) in Evolving Genes and Proteins (V. Bryson and H. J. Vogel, eds.) Aca-
    demic Press, Inc., N.Y. p. 403.
70. Burd, J. F. and Wells, R. D. (1970) JMB 53, 435.
71. Schachman, H. K., Adler, J., Radding, C. M., Lehman, I. R. and Kornberg, A. (1960) JBC
    235, 3242.

was observed to take place in the total absence of added template to dictate the reaction, or of primer to start it. Since the initial discovery, the de novo synthesis of a large variety of homopolymers and copolymers by DNA polymerase has been observed and studied[72, 73].

There are important reasons for examining this subject in some detail.

(i) The apparent violation of basic requirements by polymerase for template and primer demands explanation.

(ii) Studies of de novo synthesis have demonstrated the template-primer capacities of template-primers as small as six base-pairs.

(iii) Growth of large polymers directed by a small template-primer can be explained by a mechanism called "reiterative replication."

(iv) The particular kind of polymer produced, whether homopolymer or copolymer, whether composed of A and T or G and C, can be influenced by a variety of agents including, intercalating dyes, pH, ionic strength, and temperature, in addition to the particular polymerase employed and the nuclease activities present.

(v) "Satellite" DNAs that closely resemble the repetitive polymers in their simplicity constitute significant fractions of eukaryotic chromosomes, especially in the region of the centromere. The evolutionary origin of these unusual DNAs and even, perhaps, of the main body of DNA itself may be reflected in some ways in the polymerase-catalyzed de novo synthesis.

(vi) Finally, availability of an assortment of repetitive DNAs with sharply defined composition and sequence has been and will remain invaluable in the isolation and characterization of polymerases and nucleases, in physicochemical analyses of DNA structure, and in a variety of related studies.

The first DNA-like polymer discovered was the alternating copolymer of A and T, now designated poly d(A–T)[74] (Fig. 4-24). Since then, syntheses de novo have been observed of homopolymer pairs such as poly dG·poly dC, poly dI·poly dC, and poly dA·poly dT, and the alternating copolymer, poly d(I–C).

Studies of the kinetics of de novo polymer development have been assisted by the use of chemically prepared, _small_ template-primers. The de novo synthetic reaction can be regarded as occurring in four stages: (i) initiation, (ii) reiterative replication, (iii) polymer-directed replication, and (iv) degradation of the polymer by nuclease action when the deoxynucleoside triphosphate substrates have been exhausted. Progress of events in the first two stages and part of the third have not been detectable by available methods of measurements including spectrophotometry, viscometry, and tracer incor-

72. Burd, J. F. and Wells, R. D. (1970) *JMB* **53**, 435.
73. Radding, C. M., Josse, J. and Kornberg, A. (1962) *JBC* **237**, 2869.
74. Schachman, H. K., Adler, J., Radding, C. M., Lehman, I. R. and Kornberg, A. (1960) *JBC* **235**, 3242.

cations of this behavior come largely from studies of the priming action of chemically synthesized poly d(A–T) oligomers. A reiterative mechanism is also suggested by de novo poly d(A–T) synthesis stimulated by natural DNAs with a relatively high content of A and T residues.

Reduction of lag time for poly d(A–T) synthesis was observed in the presence of d(A–T) oligomers of six to fourteen residues in length (Fig. 4-25). Although the optimal temperature for polymer replication was near 40°, the optimum for priming by the hexamer d(A–T)$_3$ was near 0°, by d(A–T)$_4$ near 10°, and by d(A–T)$_5$ near 20° (Fig. 4-26). This behavior can be interpreted as an expression of the

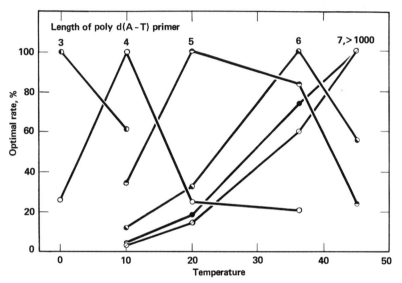

FIGURE 4-26
Influence of temperature on the priming action of d(A–T) oligomers.

condition under which the oligomer is stable as a duplex in solution and can function as a template-primer in a reiterative mechanism (Fig. 4-27).

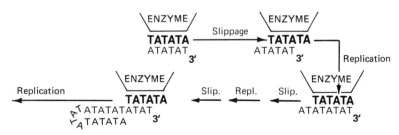

FIGURE 4-27
A scheme of reiteration and slippage to account for priming by d(A–T) oligomers. Note that the template strand of the oligomer is held fixed by the enzyme and remains constant in size after repeated stages of slippage and replication.

Assuming that one strand of the oligomeric duplex is bound and held fixed by the polymerase in the template site, the other strand may then serve as primer for extensive chain growth. Slippage of the primer strand in the $3' \longrightarrow 5'$ direction on the template can provide an additional template region for replication. Successive episodes of slippage and replication reiterate sequences of A–T residues until the primer chain has grown to a size sufficient to fold back upon itself as a hairpin-like duplex.

An important feature in the influence of temperature on priming by the series of oligomers is that, as shown in Fig. 4-26, *each size has and maintains a distinctive temperature optimum!* Were the duplex of $d(A-T)_3$ hexamers that serves as a nucleating element to grow immediately to the size of a $d(A-T)_6$ duplex and larger, the optimal temperature for polymer synthesis for this oligomer would be near 40° rather than 0°. As indicated in Fig. 4-27 this remarkable behavior of the oligomers can be interpreted by assuming that the hexamer serving as template is held fixed by the enzyme and remains constant in size while supporting the growth of the primer that can slip along it.

The suggestion that reiteration of A–T sequences in DNA may account for de novo synthesis was supported by the reduction in lag time when A–T-rich DNAs were added to a reaction mixture of dATP, dTTP, and polymerase. A positive correlation was observed with the A–T content of the DNA, the most profound effect being with a fraction of crab DNA very rich in A and T.

Polymer-directed Replication

A linear rate of synthesis by reiterative replication can become exponential only when new polymer molecules are generated to serve as template-primers for the huge excess of idle polymerase molecules. Clear indications that nuclease action is required for rapid and exponential rates of polymer synthesis came from two directions. One was the failure to obtain de novo poly dG·poly dC synthesis when polymerase preparations were freed of traces of endonuclease I, and the restoration of polymer synthesis when small amounts of endonuclease were added back to the pure polymerase reaction mixture. The other line of evidence was the behavior of the large proteolytic fragment of polymerase in poly d(A–T) synthesis[78]. This enzyme preparation, lacking the $5' \longrightarrow 3'$ exonuclease, is unable to carry out de novo synthesis. When primed with poly d(A–T), the large fragment sustains a *linear* rate of polymerization rather than the exponential rate seen with the intact enzyme. It seems plausible that the $5' \longrightarrow 3'$ exonuclease of the intact enzyme generates fragments with priming activity.

---

78. Setlow, P., Brutlag, D. and Kornberg, A. (1972) *JBC* **247**, 224.

A peak level of polymer synthesis is passed when the rate of degradation of the polymer by nuclease exceeds the rate of synthesis. The rate of degradation becomes overwhelming when synthesis ceases entirely, upon exhaustion of one of the required deoxynucleoside triphosphates. In primed poly d(A–T) synthesis, catalyzed by the large proteolytic fragment, there is 100 percent conversion of dATP and dTTP to polymer. This is in contrast to that observed with the intact enzyme, whose utilization of the substrates at the peak extent of synthesis never exceeds 70 percent. Presumably, degradation by $5' \rightarrow 3'$ exonucleolytic excisions proceeds throughout the period of synthesis as well as after exhaustion of the triphosphates.

Factors Affecting Polymer Synthesis

In view of the complex kinetics of de novo polymer synthesis, and the nice coordination of nuclease and polymerase action required, one might expect that a number of reaction parameters would influence the rates and extent of reaction. Nevertheless, some of the striking effects observed are surprising and largely unexplained. Proflavin[79] or nogalomycin[80], compounds that intercalate between bases in the DNA helix, brought about the synthesis of homopolymer chains of poly dA·poly dT rather than the copolymer, poly d(A–T). Polymerization of dITP and dCTP in 50 mM phosphate buffer yielded the homopolymer pair exclusively at pH 7.3, whereas only the copolymer was produced at pH 7.9. Simply changing the buffer at pH 7.3 from phosphate to an amine while maintaining the same ionic strength, also caused a complete shift from homopolymer to copolymer synthesis. These profound changes in kinetics and specificity of de novo polymer synthesis occurred in response to what appear as relatively slight perturbations in reaction conditions. Such results suggest important consequences of conformational changes that presumably occur in the polymerizing and nucleolytic enzymes and in the DNA molecules they act upon.

Natural Occurrence of Repetitive DNA

Upon analysis by equilibrium sedimentation in a buoyant density gradient the DNA of several crab species revealed, in addition to the major component characteristic of animal DNA, a satellite band of very light density[81]. In some species, the satellite band was as much as a fourth of the total DNA. The satellite was startlingly similar in composition and sequence to the alternating copolymer poly d(A–T)[82]. It was composed of approximately 97 percent A and T residues, of

79. McCarter, J. A., Kadohama, N. and Tsiapalis, C. (1969) Can. J.B. **47**, 391.
80. Olson, K., Luk, D. and Harvey, C. L. (1972) BBA **227**, 269.
81. Sueoka, N. (1961) JMB **3**, 31.
82. Swartz, M. N., Trautner, T. A. and Kornberg, A. (1962) JBC **237**, 1961.

which all but a few were alternating; the three percent of G and C residues were peppered throughout the DNA. Similar DNAs were also isolated from other animals (Chapter 1, Section 7). Protein coded by a DNA rich in an -ATAT- sequence would contain copolymeric runs of isoleucine and tyrosine, but no such proteins have been found. Furthermore, the amount of A–T satellite varies widely from one crab species to another. The satellite DNA has been located in the heterochromatin fraction in the centromere region of the chromosome. This DNA does not appear to be transcribed, and its function remains obscure.

High levels of A and T residues have also been found in the mitochondrial DNA of several yeast species[83]. These residues are organized into alternating copolymers and into homopolymeric regions with some G and C residues interspersed. Here too the origins and significance of these sequences is uncertain. Perhaps interruptions in the normal course of DNA replication have permitted these sequences, so susceptible to slippage, to be reiterated many times over.

16. Ribonucleotides as Substrate, Primer Terminus, and Template[84-87]

Having considered the unorthodox synthesis of DNA-like polymers without added template or primer, one other serious deviation from the basic rules of DNA polymerase behavior remains to be discussed. It is the capacity of DNA polymerase, under certain conditions, to substitute a _ribo_nucleoside triphosphate substrate, a _ribo_nucleotide primer terminus, or an _RNA_ template for the appropriate deoxyribo counterparts.

Ribonucleotide Substrates[84, 88]

Under certain conditions DNA polymerase incorporates a ribonucleotide into polymers. This reaction requires Mn^{2+} in place of Mg^{2+}, and the ribonucleoside triphosphate in place of the corresponding deoxyribo substrate. Although the reaction is not sustained, apparently limited by an especially feeble incorporation of UTP, the capacity of the enzyme to utilize ribonucleoside triphosphate is of interest from many standpoints. It provides: (i) an additional measure of the specificity of the polymerization event, (ii) a possible clue to the function of Mg^{2+}, (iii) a suggestion for the in

83. Borst, P. (1972) ARB **41**, 333.
84. Berg, P., Fancher, H. and Chamberlin, M. (1963) in _Informational Macromolecules_ (H. J. Vogel, V. Bryson and J. O. Lampen, eds.) Academic Press, Inc., N.Y. p. 467.
85. Van de Sande, J. H., Loewen, P. C. and Khorana, H. G. (1972) _JBC_ **247**, 6140.
86. Wells, R. D., Flugel, R. M., Larson, J. E., Schendel, P. F. and Sweet, R. W. (1972) _B._ **11**, 621.
87. Karkas, J. D., Stavrianopoulos, J. G. and Chargaff, E. (1972) _PNAS_ **69**, 398.
88. Op. cit. in footnote 85.

vivo mutagenic role of Mn^{2+}, and (iv) a useful technique for DNA sequence determination by affording distinctive points for alkaline or ribonuclease cleavage (Chapter 11).

The rate of rCTP incorporation in place of dCTP, in the presence of the other three deoxynucleoside triphosphates and Mn^{2+}, approached that of dCTP. Incorporation of rGTP as substitute for dGTP was also effective, but that of rATP for dATP was much less so. With rUTP in place of dTTP, incorporation was barely detected, a behavior also observed with dUTP as an analog for dTTP under standard assay conditions.

It would seem that the poor performance of uracil as an analog for thymine is due to the structural features of the A–U base pair. Its stacking properties and other features seem to be recognized by the polymerase as aberrant. DNA containing A–U base pairs is preferentially degraded by DNA polymerase I through the $5' \rightarrow 3'$ exonuclease excision activity[89]. The absence of uracil from DNA may depend on both a dUTPase to prevent incorporation in the first place, and its excision if incorporated.

Mn^{2+} distorts polymerization not only by permitting ribonucleotide incorporation, but also by permitting misincorporation of deoxynucleotides themselves[90-92]. When the repair of short regions of defined DNA sequences was observed with only dTTP present, Mn^{2+} permitted T to form a base pair with G in the template. Limited replication in which GTP or CTP were incorporated also led to a significant percentage of misincorporations. However under certain experimental conditions good fidelity of incorporation has been observed using CTP to replace dCTP[93]. Therefore, limited ribonucleotide insertions made possible by Mn^{2+} should, under appropriate incubation conditions, permit sequencing of long DNA segments[94, 95]. Whether Mn^{2+} exerts its permissive effect on ribonucleotide substitutions and misincorporations at the stage of base-pairing, proofreading, or binding of the template-primer is not yet clear.

Ribonucleotide Primer Terminus and Template

Incorporation of ribonucleotides into a DNA chain synthesized by DNA polymerase indicated that the enzyme was able to use them not only as substrates in place of deoxynucleotides but, further, that the added ribonucleotide could serve as a primer terminus. It was also observed that the ribohomopolymer pair poly rA·poly rU, synthesized by polynucleotide phosphorylase, could serve as a template-

89. Wovcha, M. G. and Warner, H. R. (1973) JBC **248**, 1746.
90. Van de Sande, J. H., Loewen, P. C. and Khorana, H. G. (1972) JBC **247**, 6140.
91. Klenow, H. and Henningsen, I. (1969) EJB **9**, 133.
92. Hall, Z. W. and Lehman, I. R. (1968) JMB **36**, 321.
93. Op. cit. in footnote 91.
94. Op. cit. in footnote 90.
95. Salser, W., Fry, K., Brunk, C. and Poon, R. (1972) PNAS **69**, 238.

primer[96]. Recently, intensive research on RNA-directed DNA polymerases (the reverse transcriptases) of animal viruses and animal cells has added new interest to the utilization of RNA as template-primer by DNA polymerases (see Chapter 6, Sections 6, 7).

When DNA polymerase I utilized poly rA·poly rU as template-primer, the corresponding poly dA·poly dT was produced. The stages in this replication are likely represented by these equations[97]:

Formation of hybrids:

$$\text{poly rA·poly rU} + \text{dATP} + \text{dTTP} \rightarrow \text{poly rA·poly dT} + \text{poly dA·poly rU}$$

Replication of hybrids:

$$\begin{cases} \text{poly rA·poly dT} + \text{dATP} \rightarrow \text{poly dA·poly dT} + \text{poly rA} \\ \text{poly dA·poly rU} + \text{dTTP} \rightarrow \text{poly dA·poly dT} + \text{poly rU} \end{cases}$$

Reannealing:

$$\text{poly rA} + \text{poly rU} \rightarrow \text{poly rA·poly rU}$$

Sum:

$$\text{poly rA·poly rU} + \text{dATP} + \text{dTTP} \rightarrow \text{poly rA·poly rU} + \text{poly dA·poly dT}$$

What in sum total had appeared to be conservative replication of ribohomopolymers could be accounted for by ribo-deoxyribo hybrid pairs as intermediates. Such hybrid pairs are excellent template-primers.

These results make it clear that ribohomopolymers do support deoxypolymer synthesis. As members of hybrid duplexes, such as poly rA·poly dT or poly rG·poly dC, the ribohomopolymers were as good templates as the unmixed deoxyhomopolymer pairs.

Formation of the hybrid intermediates would be expected to be brought about by covalent extension of a ribo primer terminus. Although the deoxyribo product appeared to be cleanly separated from the template-primer by equilibrium sedimentation in a density gradient, a result which was interpreted to rule out covalent linkage, the use of α-[^{32}P]-deoxynucleoside triphosphates in the reaction mixture would likely demonstrate covalent attachment of the DNA homopolymer to a trace amount of RNA primer[98].

The relative rates with which the enzyme uses ribopolymers as templates and primers vary considerably, often under apparently identical conditions[98a]. In the case of poly rA·poly rU, values reported for its template-priming range from one to 25 percent of the rates observed with poly dA·poly dT. Single-stranded homopolymers, ribo or deoxy, are generally inert, but in some instances appear to support significant synthesis. Since factors that profoundly alter the effectiveness of a template-primer, ribo or deoxyribo, are often ignored, they deserve to be enumerated. These include:

(i) state of the template-primer—the number of 3'-hydroxyl primer termini with a sufficient length of template to support extension; the

96. Lee-Huang, S. and Cavalieri, L. F. (1964) PNAS 51, 1022.
97. Chamberlin, M. (1965) Fed. Proc. 24, 1446.
98. Wells, R. D., Flugel, R. M., Larson, J. E., Schendel, P. F. and Sweet, R. W. (1972) B. 11, 621.
98a. Karkas, J. D. (1973) PNAS 70, 3834.

presence of annealed oligonucleotide primer fragments, even in homopolymer samples isolated by isopycnic sedimentation; the presence of inhibitory 3'-phosphoryl termini.

(ii) purity of the enzyme—the presence of nucleases that stimulate extension by creating additional primer termini, by removing inhibitory termini, and by producing oligonucleotide priming fragments; the presence of nucleases that inhibit by destroying template-primer or the product itself.

(iii) incubation conditions—the effects of even small changes in temperature, pH, ionic strength, buffer, and metal ions.

Natural RNA template-primers were utilized in vitro to direct DNA synthesis by DNA polymerase[99]. Ribosomal (28S) RNA from *Drosophila*, or L-cells, and tobacco mosaic virus RNA were active, the addition of poly$(dT)_{6-9}$ being essential for replication of the viral RNA, but not for ribosomal RNA. The rates of these DNA syntheses are only a few percent of the rates with natural DNA as template-primer. In evaluating these observations the admonitions about the state of the template-primer, the purity of the enzymes, and incubation conditions should be recalled. Nevertheless, the capacity of *E. coli* DNA polymerase I to be directed by RNA is irrefutable, although its physiologic significance is still not apparent.

17. The Product of Polymerase Synthesis[100-102]

The DNA produced after replication of a template-primer many times over resembles the template-primer in physical and chemical properties. The product has the sedimentation and viscoelastic characteristics of a DNA fiber several microns in length. But it is unusual in two ways[103]. One is the readiness with which the DNA renatures after thermal or alkaline melting; despite rapid cooling or neutralization, the product, unlike natural DNA, regains its original optical density and native properties. The other unusual feature is the branched appearance of the DNA in electron micrographs[104].

Both the _rapid renaturability_ and the _branching_ can be attributed to the presence of hairpin-like structures. This feature may come about by a template switch, some time during replication, at a nick in a duplex. At this point first one strand of the duplex could be used as a template and then later the complementary strand. Such _template-switching_ can be more easily visualized in a space-filling DNA model than from a diagram on paper. Free rotation of the phosphodiester bridge in the DNA backbone permits a DNA strand to

99. Loeb, L. A., Tartof, K. D. and Travaglini, E. C. (1973) *NNB* **242**, 66.
100. Schildkraut, C. L., Richardson, C. C. and Kornberg, A. (1964) *JMB* **9**, 24.
101. Inman, R. B., Schildkraut, C. L. and Kornberg, A. (1965) *JMB* **11**, 285.
102. Goulian, M. and Kornberg, A. (1967) *PNAS* **58**, 1723; Goulian, M., Kornberg, A. and Sinsheimer, R. L. (1967) *PNAS* **58**, 2321.
103. Op. cit. in footnote 100.
104. Op. cit. in footnote 101.

displace its complement without any looping or distortion of the helix, and would enable a polymerase molecule to continue replication of the new template without changing direction. The factors that might prompt template-switching are still unknown.

Subsequently, by a process called _branch migration_[105], the displaced strand of the template-primer may be restored to its original duplex state, permitting the product to anneal as hairpins (Fig. 4-28).

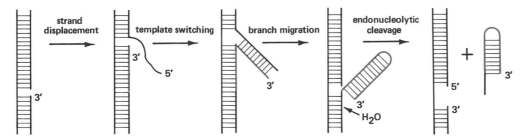

FIGURE 4-28
Model for synthesis of hairpin and branched structures by DNA polymerase I action.

Endonucleolytic cleavages are required to release the covalent attachment of the product to the template-primer.

The chemical composition of the product reflects the template-primer[106]; with a single-stranded DNA as template-primer the synthetic complementary strand has exactly the composition predicted by base-pairing[107]. After extensive replication of duplex DNA the A–T and G–C contents of the product are the same as in the template-primer. Inasmuch as the accuracy of analysis of base composition is only of the order of one percent, a more sensitive measure of the fidelity of replication may be gained by the use of template-primers lacking one of the bases. Thus, the misincorporation of guanine nucleotides with poly d(A–T) as template-primer was looked for, but not detected even at a level of one nucleotide per 10^5 A and T nucleotides polymerized[108].

Whereas base analysis of the product confirms that the overall composition matches the template-primer, it does not inform us of the fidelity with which each of the sixteen possible linkages of one nucleotide to another are made. The content or frequency of each of the sixteen dinucleotide arrangements in the product can be determined by a procedure that has come to be known as nearest-neighbor analysis[109]. The procedure relies simply on the incorporation of α-(^{32}P)-deoxynucleoside triphosphate and enzymatic hydrolysis of the product to 3′-deoxynucleotides (Fig. 4-29). Thus, from a product in which α-(^{32}P)-dATP was incorporated, one obtains the labeled

105. Lee, C. S., Davis, R. W. and Davidson, N. (1970) _JMB_ **48**, 1.
106. Lehman, I. R., Zimmerman, S. B., Adler, J., Bessman, M. J., Simms, E. S. and Kornberg, A. (1958) _PNAS_ **44**, 1191.
107. Swartz, M. N., Trautner, T. A. and Kornberg, A. (1962) _JBC_ **237**, 1961.
108. Trautner, T. A., Swartz, M. N. and Kornberg, A. (1962) _PNAS_ **48**, 449.
109. Josse, J., Kaiser, A. D. and Kornberg, A. (1961) _JBC_ **236**, 864.

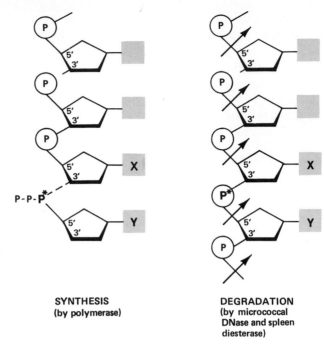

SYNTHESIS
(by polymerase)

DEGRADATION
(by micrococcal
DNase and spleen
diesterase)

FIGURE 4-29
Steps in the nearest-neighbor analysis of DNA. The ^{32}P-label
(indicated as P*) in the α position of dYTP becomes the
nearest neighbor of the terminal X residue and remains
attached to it upon 3′-nucleolytic cleavage.

3′-deoxynucleotides (i.e. Ap̌, Gp̌, Cp̌, Tp̌) in relative amounts that
indicate the frequency with which A was linked to these nucleotides (i.e. Ap̌A, Gp̌A, Cp̌A, Tp̌A). When the analysis is repeated
three times, using successively α-[^{32}P]-labeled dGTP, dCTP, and
dTTP, the remaining twelve nearest-neighbor frequencies can be
determined.

The simplest application of nearest-neighbor frequency analysis
is to the replication of the synthetic copolymers and homopolymers.
With the copolymer poly d(A–T) as template-primer and α-[^{32}P]-
dATP, only Tp̌ was obtained upon hydrolysis, with α-[^{32}P]-dTTP
only Ap̌. This indicates that the sequences were _exclusively_ Tp̌A
and Ap̌T. Since the amounts of A and T in poly d(A–T) are equal, it
follows that the product had the alternating sequence: . . . p̌Ap̌T-
p̌Ap̌Tp̌Ap̌Tp̌Ap̌Tp̌ . . . . When the homopolymer poly dA·poly dT
served as template-primer the results indicated that the sequences
were _exclusively_ Ap̌A and Tp̌T. With sufficient care this method
could detect a dinucleotide sequence as infrequent as one in a
thousand. Within this range no failure in the fidelity of copying
copolymers, homopolymers, and DNAs of defined sequence has
been detected.

Products primed by natural DNAs from diverse sources show

frequencies unique for each DNA[110]. Although these frequencies usually reflect the abundances of the bases in the DNA, there are some interesting exceptions. The frequency of GpC in animal and bacterial DNAs is near random, but the CpG sequence in animal DNAs is extraordinarily low (Table 4-8)[111]. The diagrams at the top

TABLE 4-8
Relative CpG and GpC frequencies[a] in DNA
from animal and bacterial sources

Source of template	CpG	GpC	Ratio: GpC/CpG
Animal cells			
Human spleen	.010	.043	4.3
Rabbit liver	.013	.048	3.7
Chicken red cell	.011	.052	4.7
Crab { main band	.019	.030	1.6
Crab { satellite	.0007	.0015	2.2
Animal viruses			
Shope papilloma	.024	.055	2.3
Polyoma	.018	.052	2.9
Herpes	.109	.108	0.99
Pseudorabies	.146	.132	0.91
Vaccinia	.035	.026	0.74
Bacteria and phage			
M. phlei	.076	.067	0.88
M. luteus	.069	.060	0.87
E. coli	.072	.079	1.10
phage T1	.070	.076	1.08

[a]Subak-Sharpe, H., Bürk, R. R., Crawford, L. V., Morrison, J. M., Hay, J. and Keir, H. M. (1966) CSHS 31, 737.

of the table are included to recall that the CpG sequence in one strand matches a CpG sequence in the complementary strand.

Among the DNA viruses that infect animal cells, the CpG frequency in polyoma and Shope papilloma, two very small viruses with limited genetic content, reflect the pattern of their host. On the other hand, the CpG frequency of three large viruses is distinctly different. These observations have suggested to some investigators a more intimate relationship of the smaller viruses to their host cells in the evolutionary origin of the viruses.

Although the earliest DNA models proposed the two chains oriented in opposite directions, or polarity, it was the analysis of nearest-neighbor frequencies that provided the first strong evidence

110. Swartz, M. N., Trautner, T. A. and Kornberg, A. (1962) JBC 237, 1961.
111. Subak-Sharpe, H., Bürk, R. R., Crawford, L. V., Morrison, J. M., Hay, J. and Keir, H. M. (1966) CSHS 31, 737.

114

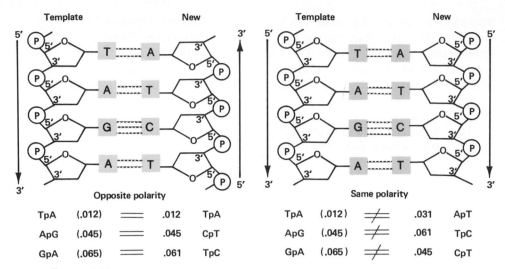

Template		New		Template		New	
Opposite polarity				Same polarity			
TpA	(.012)	==	.012	TpA	TpA	(.012)	≠	.031	ApT
ApG	(.045)	==	.045	CpT	ApG	(.045)	≠	.061	TpC
GpA	(.065)	==	.061	TpC	GpA	(.065)	≠	.045	CpT

FIGURE 4-30
Nearest-neighbor sequences expected from replication of the template with a chain of the opposite polarity (left) or the same polarity (right). (Source: Josse, J., Kaiser, A. D. and Kornberg, A. (1961) *JBC* **236**, 864.)

for this fact[112]. As can be seen from Fig. 4-30, a given template sequence can be copied either in the opposite or the same polarity. The frequencies measured in the enzymatic product showed nearest-neighbor sequences matched by opposite polarity, not by the same polarity. This property of the enzymatically synthesized helix implies that the natural DNA which served as template has the same characteristic. It must be understood that all the information obtained from nearest-neighbor analysis applies directly only to the product in which the labeled triphosphates are incorporated and merely by inference to the natural DNA template. There is an assumption inherent in this method that the template has been replicated extensively and unselectively.

Despite the accuracy with which the base sequences of template-primers were replicated, biological activity in the enzymatically synthesized DNA was not readily reproduced. For example, the transforming activity of *B. subtilis* DNA was not increased by several-fold net synthesis of the DNA[113]; the branched nature of the product has been offered as a reason for the lack of biologic activity. With the demonstration in 1967 that DNA polymerase (twelve years after its discovery) can produce fully infectious copies of circular φX174 DNA[114], several things were made clear. No amino acids, nor unusual nucleotides needed to be added for the in vitro synthesis; the A, T, G, and C nucleotides, in a high grade of purity, sufficed.

112. Josse, J., Kaiser, A. D. and Kornberg, A. (1961) *JBC* **236**, 864.
113. Richardson, C. C., Schildkraut, C. L., Aposhian, H. V., Kornberg, A., Bodmer, W. and Lederberg, J. (1963) **in** *Informational Macromolecules* (H. J. Vogel, V. Bryson and J. O. Lampen, eds.) Academic Press, N.Y. p. 13.
114. Goulian, M. and Kornberg, A. (1967) *PNAS* **58**, 1723; Goulian M., Kornberg, A. and Sinsheimer, R. L. (1967) *PNAS* **58**, 2321.

The 6000-nucleotide stretch of ϕX174 DNA was copied error-free, as inferred from the infectivity of the product. Polymerase action, which by itself could not produce a covalent circle, could be nicely coordinated with ligase action to do so. The success in replicating this rather simple viral chromosome of eight to ten genes opens possibilities for achieving in vitro synthesis of other important DNAs and more complicated chromosomes.

Precisely how the synthetic complementary strand was initiated on the viral template was not resolved by these in vitro studies on ϕX174 replication. The need for a small amount of boiled E. coli extract[115] was later explained by the action of fragments of E. coli DNA in initiating synthesis. Fragments that happened to anneal partially to the viral DNA template had 3'-hydroxyl termini tailored to a correct base-paired fit by the 3' $\rightarrow$ 5' exonuclease of DNA polymerase (Fig. 4-31). Replication by extension of the properly base-

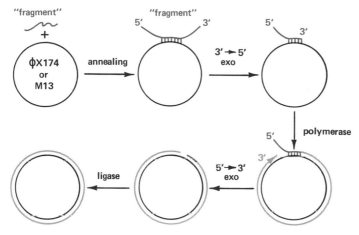

FIGURE 4-31
An oligonucleotide fragment is needed to prime the replication of a single-stranded circle to the replicative form by DNA polymerase I. This sequence of operations illustrates the successive tailoring actions of the 3' $\rightarrow$ 5' exonuclease, polymerase, and 5' $\rightarrow$ 3' exonuclease functions of the enzyme.

paired primer terminus then proceeded around the circle. From the fact that very few nucleotide residues of the priming fragment persisted in the final product (about one residue in five completed circles)[116,117], it is reasonable to attribute to the 5' $\rightarrow$ 3' exonuclease the function of excising the remainder of the fragment.

It is clear now that initiation of DNA synthesis in vivo does not take place through the use of an available population of DNA fragments. Initiation involves the synthesis of an RNA priming fragment

115. Goulian, M. and Kornberg, A. (1967) PNAS **58**, 1723; Goulian M., Kornberg, A. and Sinsheimer, R. L. (1967) PNAS **58**, 2321.
116. Goulian, M. (1968) CSHS **33**, 11.
117. Goulian, M., Goulian, S. H., Codd, E. E. and Blumenfeld, A. Z. (1973) B. **12**, 2893.

that is extended by DNA polymerase and later excised, possibly through the action of the 5′ → 3′ exonuclease (see Chapter 7, Section 5).

18. Physiologic Role[118-120]

One of the most effective tools for assessing the role of an enzyme in vivo is a mutant deficient in it. With this objective, several thousand strains of E. coli from a heavily mutagenized population were examined to find one, the 3478th, whose cell-free extract contained very low levels of DNA polymerase I activity[121]. This residual activity was later identified as a mixture of two distinct enzymes, called DNA polymerases II and III, to be discussed in the next chapter. DNA polymerase I was thought to be entirely absent.

The mutation, designated pol A1[121], is located at about 75 minutes on the E. coli genetic map (Fig. 4-32, Table 4-9). In genetic terms it

118. Okazaki, R., Arisawa, M. and Sugino, A. (1971) *PNAS* **68**, 2954.
119. Lehman, I. R. and Chien, J. R. (1973) *JBC* **248**, 7717.
120. Gross, J. D. (1972) **in** *Current Topics in Microbiology and Immunology*. Springer Verlag, N.Y. **57**, 39.
121. De Lucia, P. and Cairns, J. (1969) *Nat.* **224**, 1164.

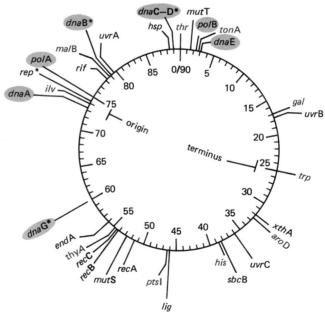

FIGURE 4-32
Genetic map of E. coli with special emphasis on loci involved in DNA replication. The scale, read in minutes, represents the time required for transfer of genes in the chromosome during mating. Values for the starred loci are not as accurately known as the others. See Table 4-9 for further details.

TABLE 4-9
Loci on the *E. coli* genetic map in Figure 4-32

Locus	Minutes	Comments
mut T[a]	1.5	Mutator, AT → CG transversions
dna E[b]	2.5	pol III (pol III*)
pol B[b]	2.5	pol II
uvr B[a]	17.5	Thymine dimer excision
terminus[c]	~25	Terminus of replication
xth A[d]	32	Exonuclease III
uvr C[a]	36	Thymine dimer excision
sbc B[e]	38	Exonuclease I
lig[f]	46	Polynucleotide ligase
rec A[a]	51.5	Recombination
mut S[a]	52.5	Mutator
rec B[a]	54.5	rec BC nuclease
rec C[a]	54.5	rec BC nuclease
end A[g]	56	Endonuclease I
dna G[h]	60-65	DNA replication
dna A[h]	73	DNA replication
rep[a]	74	M13, ϕX174 RF replication
origin[c]	~74	Origin of replication
pol A[a]	75	pol I
rif[a]	78.5	β subunit of RNA polymerase (rifampicin sensitivity)
dna B[i]	~79	DNA replication
uvr A[a]	79.5	Thymine dimer excision
hsp[a]	88.5	DNA restriction and modification endonuclease and methylase
dna C–D[h]	89	DNA replication

Reference loci[a] are: thr, tonA, gal, trp, aroD, his, ctr, thyA, ilv, and mal B.

[a] Taylor, A. L. (1970) *Bact. Rev.* **34**, 155; Taylor, A. L. and Trotter, C. D. (1972) *Bact. Rev.* **36**, 504.
[b] Hirota, Y., Gefter, M. L. and Mindich, L. (1972) *PNAS* **69**, 3238.
[c] Bird, R. E., Louarn, J., Martuscelli, J. and Caro, L. (1972) *JMB* **70**, 549.
[d] Milcarek, C. and Weiss, B. (1973) *J. Bact.* **113**, 1086.
[e] Templin, A., Kushner, S. R. and Clark, A. J. (1972) *Gen.* **72**, 205.
[f] Gottesman, M. M., Hicks, M. L. and Gellert, M. (1973) *JMB* **77**, 531.
[g] Wright, M. (1971) *J. Bact.* **107**, 87.
[h] Wechsler, J. A. and Gross, J. D. (1971) *Mol. Gen. Genet.* **113**, 273.
[i] Carl, P. L. (1970) *Mol. Gen. Genet.* **109**, 107.

is a nonsense mutation[122] suppressed by the *amber* suppressors *su* I[+], *su* II[+], and *su* III[+ 123]; the mutation is recessive to the wild-type *pol* A gene in partial diploids.

The *pol* A mutant grows quite normally but is defective in its capacity to repair the DNA damage of ultraviolet irradiation and of radiomimetic drugs like methylmethane sulfonate. Its sensitivity to these agents is many times greater than normal. Based on a selection of strains defective in DNA repair capacity, a number of additional

122. Gross, J. D. (1972) **in** *Current Topics in Microbiology and Immunology.* Springer Verlag, N.Y. p. 39.
123. Richardson, C. C. Personal communication.

pol A mutants were later isolated, among them a nonsuppressible pol A5[124] and one defective in repair only at elevated temperature, pol A12[125]. The temperature-sensitivity of the DNA polymerase I of pol A12 has provided the best evidence that the pol A locus is the structural gene for DNA polymerase I. Studies of these several pol A mutants furnished insights into the structure of the enzyme which are discussed later in this section.

Although pol A cells grow at a normal rate and support the growth of lysogenic and virulent phages, they do display a number of defects, in addition to their sensitivity to damage by ultraviolet irradiation and methylmethane sulfonate. These are enumerated in Table 4-10. They are generally consistent with a deficiency in filling DNA gaps in a variety of situations.

Because the pol A1 mutant seemed to make DNA normally, despite the apparent lack of DNA polymerase I, the enzyme was widely

---

124. Berg, D. cited in Lehman and Chien, footnote 119.
125. Monk, M. and Kinross, J. (1972) J. Bact. **109**, 971.

TABLE 4-10
Phenotypic defects of pol A mutants[a]

Sensitivity to thymine starvation[b]

Sensitivity to ultraviolet irradiation[c]

Sensitivity to X-ray irradiation[d,e]

Sensitivity to methylmethane sulfonate[f]

Extensive degradation of DNA in repairing ultraviolet damage[f]

Increased frequency of deletions[g]

Increased frequency of recombination[h]

Low viability (5 percent) on nutritional shift up[i]

Col E$_1$ factor not replicated[j]

Col E$_2$ factor unstable[j]

Mini circular plasmid of E. coli 15 not replicated[k]

Poor joining (10 percent) of 10S fragments[l,m]

Single-strand gaps in ϕX174 RF[n]

Poor growth of phage λ, defective in exonuclease (β or γ protein)[o]

Deficiency in phage T4 replication and recombination[p]

---

[a] Gross, J. D. (1972) in Current Topics in Microbiology and Immunology. Springer Verlag, N.Y. p. 39.
[b] Nakayama, H. and Hanawalt, P. C. Personal Communication.
[c] Gross, J. and Gross, M. (1969) Nat. **224**, 1166.
[d] Kato, T. and Kondo, S. (1970) J. Bact. **104**, 871.
[e] Town, C. D., Smith, K. C. and Kaplan, H. S. (1971) Science **172**, 851.
[f] Cooper, P. K. and Hanawalt, P. C. (1972) PNAS **69**, 1156.
[g] Coukell, M. B. and Yanofsky, C. (1970) Nat. **228**, 633.
[h] Konrad, E. B. and Lehman, I. R. Personal communication.
[i] Rosenkranz, H. S., Carr, H. S. and Morgan, C. (1971) BBRC **44**, 546.
[j] Kingsbury, D. T. and Helinski, D. R. (1970) BBRC **41**, 1538; Goebel, W. (1972) NNB **237**, 67.
[k] Goebel, W. and Schrempf, H. (1972) BBRC **49**, 591.
[l] Okazaki, R., Arisawa, M. and Sugino, A. (1971) PNAS **68**, 2954.
[m] Kuempel, P. L. and Veomett, G. E. (1970) BBRC **41**, 973.
[n] Schekman, R. W., Iwaya, M., Bromstrup, K. and Denhardt, D. T. (1971) JMB **57**, 177.
[o] Zissler, J., Signer, E. and Schaefer, F. (1971) in The Bacteriophage Lambda (A. D. Hershey, ed.) CSHL, p. 469.
[p] Mosig, G., Bowden, D. W. and Bock, S. (1972) NNB **240**, 12.

assumed to have no part in replicating the *E. coli* chromosome and the enzyme was at one time called a "red herring" in DNA replication[126]. But assigning physiologic functions on the basis of such data is inherently risky.

Failure to detect an enzyme activity in a cell-free extract means only that. There can be numerous reasons for such failure. The enzyme may be unstable upon extraction, vulnerable to proteolytic enzymes, or tightly bound to a component, in both the soluble and insoluble fractions, that masks its activity. It may require special assay conditions; quantitative estimates of enzyme levels must also be treated with caution. Recently, with appropriate attention to enzyme lability, extracts of all *pol* A mutants were shown to contain measurable levels of DNA polymerase I activity and levels of $5' \rightarrow 3'$ exonuclease approaching that of the wild type[127].

Properties of the polymerases of *amber* and temperature-sensitive mutants are compared in Table 4-11. The DNA polymerase I levels

TABLE 4-11
Polymerase I activities of *pol* A mutants[a]

Activity	*pol* A+ (wild type)	*pol* A1 (amber)	*pol* A12 (temperature sensitive)
1. Amount of polymerase, percent	(100)	1	100
2. Activity ratio: 42°/30°	2.6	2.2	0.1
3. Activity ratio: $(dA)_{800} \cdot (dT_{10})$/nicked DNA	6.0	5.0	0.04
4. Amount of $5' \rightarrow 3'$ exonuclease, percent	(100)	50-100	50-100
5. Activity ratio: 42°/30°	1.6	1.7	1.7
6. Size of $5' \rightarrow 3'$ exonuclease, S_{20w}	5.4	2.8	5.4

[a]Lehman, I. R. and Chien, J. R. (1973) *JBC* **248**, 717.

determined in partially purified preparations obtained from *amber* mutants were about one percent of the levels in wild-type cells; the levels of activity derived from temperature-sensitive cells were near normal, but strikingly labile at 42°. Remarkable features of the *pol* A12 polymerase are its greatly reduced ability to utilize a homopolymer template-primer, indicating another defect in the polymerase activity, and the normal level and temperature response of the $5' \rightarrow 3'$ exonuclease activity.

The $5' \rightarrow 3'$ exonuclease activity was not significantly reduced in the preparations from mutant cells, but the molecular form from *amber* mutants was markedly altered[128] (Table 4-11). It resembled the 2.8 S small fragment obtained by proteolysis of the wild-type enzyme (Section 14 of this chapter). This observation may signify, as suggested in Fig. 4-33, that the presumed *amber* peptide is labile to

126. Editorial (1971) *NNB* **233**, 97.
127. Lehman, I. R. and Chien, J. R. (1973) *JBC* **248**, 7717.
128. Op. cit. in footnote 127.

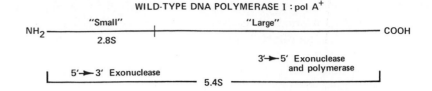

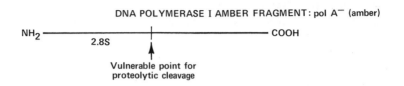

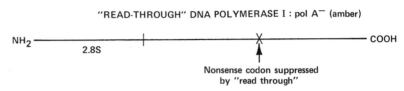

FIGURE 4-33
Suggested organization of the polypeptide chain of DNA polymerase I to account for the properties of the *pol* A amber mutant.

cleavage by endogenous *E. coli* proteases. Expression of polymerase activity by the mutant can be explained by "read through" of the *amber* codon in a small percentage of molecules, as described next.

Suppose one were to accept the recently determined level of polymerase I in *pol* A⁻ extracts as a valid measure of the physiologic level of the enzyme. Should a value as low as one percent be regarded as significant? It may be or it may not. Experience with estimating ligase levels in *E. coli* indicates that mutants with only one percent of the normal level in their extracts show no impairment in growth. However, mutants whose extracts show a level of less than 0.1 percent of normal accumulate DNA in short pieces and die (Chapter 7, Section 6).

An enzyme may perform a vital function in the cell and yet be dispensable, when, for example, another enzyme with similar function substitutes for it. Measurement of <u>overall</u> growth rate is not likely to detect the shifts and altered details in metabolic pathways. On the basis of recent genetic and biochemical data it seems likely that DNA polymerase I does in fact perform an indispensable function in replication, probably that of filling gaps between the short, nascent fragments.

The genetic data in support of the indispensability of DNA polymerase I are of three kinds: (i) failures in attempts to obtain a mutant totally lacking the *pol* A gene[129], (ii) failures in attempts to introduce the *pol* A mutation into cells deficient in recombination or DNA repair (e.g. *rec* A⁻, *rec* B⁻, *rec* C⁻, or *uvr* B⁻)[129,130], and (iii) the result of "read through" suppression[131]. In the *pol* A1 mutant, as in other *amber* mutants (e.g. gal⁻ mutants), some functional protein is made by "read through" of the nonsense codons (Fig. 4-33). An additional mutation, presumably in a ribosomal protein that abolishes the "read through", makes the *pol* A1 mutants inviable. The double mutation can be inferred from the action of streptomycin in the culture medium, which suppresses the ribosomal protein mutation, and thereby restores the "read-through" and viability of the *pol* A1 mutants.

The biochemical evidence for the indispensability of DNA polymerase I is seen in a double mutant, temperature-sensitive for both ligase and polymerase. DNA synthesis stops abruptly and the cells die when shifted to 42°[132]; by contrast, under the same conditions, DNA synthesis continues for several hours in the ligase mutant. In *pol* A mutants, there is an abnormal accumulation of DNA fragments, which increase tenfold (and more) compared to wild-type cells. Knowing that low levels of the enzyme activity persist in *pol* A cells, it seems reasonable that residual levels of polymerase I activity are required for the growth and viability of the cell.

At present, the following proposal may be made for DNA polymerase I function in replication: the enzyme removes the RNA fragment (Chapter 7) that initiates each short sequence of DNA synthesized by DNA polymerase III, and then fills the gap created. In this way, the nick translational activity, seen in the coordination of polymerization with the $5' \rightarrow 3'$ exonuclease function (Fig. 4-21 and Section 13 of this chapter), performs an essential role in replication. A striking recent indication of this role is provided by a temperature-sensitive mutant selected for its hyperactivity in recombination[133]. This mutant maps in the *pol* A region. Extracts have a normal level of polymerase but lack the $5' \rightarrow 3'$ exonuclease activity. At 42°, the mutant accumulates large amounts of 10S DNA fragments and fails to survive.

The physiologic contributions that DNA polymerase I may make to DNA repair, already indicated in the phenotypic behavior of the *pol* A mutant, will be considered again in Chapter 9, Section 2.

129. Gross, J. D. (1972) in *Current Topics in Microbiology and Immunology*. Springer Verlag, N.Y. **57**, 39.
130. Monk, M. and Kinross, J. (1972) *J. Bact*. **109**, 971.
131. Herskowitz, I. Personal communication; Horiuchi, T. and Nagata, T. (1973) *Mol. Gen. Genet*. **123**, 89.
132. Konrad, E. B., Modrich, P. and Lehman, I. R. (1973) *JMB* **77**, 519.
133. Konrad, E. B. and Lehman, I. R. (1974) *PNAS* **71**, 2048.

5

Bacterial and Phage-Induced DNA Polymerases

1. Polymerase II of *E. coli*[1,2]

Discovery of mutants of *E. coli* which had little or no detectable DNA polymerase activity in extracts but nevertheless synthesized DNA at a normal rate (Chapter 4, Section 18) spurred a search for an unknown DNA-synthesizing activity. When conditions were eventually found for preparation of active extracts, and for adequate assays, the existence of two other distinct polymerases was disclosed. They were designated DNA polymerase II[3-5] and DNA polymerase III[6], in the order of their discovery. These are commonly

1. Gefter, M. L., Molineux, I. J., Kornberg, T. and Khorana, H. G. (1972) *JBC* **247**, 3321.
2. Wickner, R. B., Ginsberg, B., Berkower, I. and Hurwitz, J. (1972) *JBC* **247**, 489; Wickner, R., Ginsberg, B. and Hurwitz, J. (1972) *JBC* **247**, 498.
3. Kornberg, T. and Gefter, M. L. (1970) *BBRC* **40**, 1348.
4. Moses, R. E. and Richardson, C. C. (1970) *PNAS* **67**, 674.
5. Knippers, R. (1970) *Nat.* **228**, 1050.
6. Kornberg, T. and Gefter, M. L. (1971) *PNAS* **68**, 761.

referred to as pol II and pol III, and DNA polymerase I as pol I. With a clear idea of their properties, pol II and pol III were demonstrated in extracts of wild-type cells as well. Under assay conditions standardized for pol I, the combined pol II and pol III activity in extracts is usually less than five percent that of pol I. This explains why neutralization of pol I in active extracts of wild-type cells by specific antibodies had failed to disclose the residual activity of pol II and pol III.

Pol II has been purified to homogeneity by procedures similar to those used for pol I. Separation from pol I and pol III was achieved by phosphocellulose chromatography. The enzyme resembles pol I in polymerization properties except in its inability to use a template-primer that is simply nicked or is extensively single-stranded. The optimal template-primer for pol II is a duplex with short gaps. There is no accurate determination of the optimal gap size, and only gross estimates are available; these place it in the range of 10 to 100 nucleotides. Also, unlike pol I, it lacks $5' \longrightarrow 3'$ exonuclease activity.

Other comparisons with pol I, and with pol III, are listed in Table 5-1. The principal differences are in template-primer preference, reaction with sulfhydryl-blocking agents, affinity to phosphocellulose, and resistance to anti-pol I antibody. Pol II separates sharply from pol III on phosphocellulose (Fig. 5-1) and it is also less heat-labile, more resistant to salt, indifferent to ethanol, and effective at lower deoxynucleoside triphosphate concentrations.

The known temperature-sensitive mutants, dna_{ts}, (dna A, dna B, dna C-D, dna E, and dna G) whose products are essential for chromosome replications[7], (see Chapter 7, Section 9) have normal levels of pol II activity. A mutant with no detectable (< 0.5 percent) pol II activity in extracts, pol B⁻, has been obtained by assaying extracts from many surviving colonies of mutagenized pol A1 cells[8]. This pol A⁻ pol B⁻ mutant grows normally at 25° and 42° and supports growth of a variety of phages. The pol B mutation by itself does not make the cells sensitive to ultraviolet irradiation (Fig. 5-2) nor does it affect the frequency of recombination in transductions and mating transfers. Yet, there are clear indications that, in mutants deficient in pol I and III, pol II serves in the repair of ultraviolet lesions[9]. Failure to observe any other deficiencies in the apparent absence of pol II activity does not warrant the conclusion that pol II is dispensable for additional important DNA functions. Investigations of pol I mutants and ligase mutants have demonstrated the importance of low residual levels of activity. Similarly, persistence of still undetected residual pol II activity in the pol B mutant cells may be sufficient in vivo to permit them to function in what appears to be a normal way.

7. Gross, J. (1972) Current Topics in Microbiology and Immunology **57**, 39, Springer-Verlag, N.Y.
8. Campbell, J. L., Soll, L. and Richardson, C. C. (1972) PNAS **69**, 2090.
9. Masker, W., Hanawalt, P. C. and Shizuya, H. (1973) NNB **244**, 242; Tait, R. C., Harris, A. L. and Smith, D. W. (1974) PNAS **71**, 675.

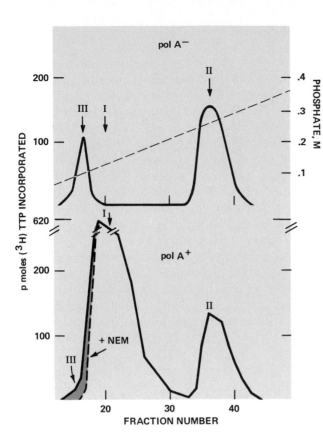

FIGURE 5-1
Separation of DNA polymerases I,
II, and III by phosphocellulose
chromatography. Cell-free extracts
were analyzed. Assays in the presence
of pol III inhibitor N-ethylmaleimide
(NEM) are shown by the heavy
dashed line. The relative absence
of polymerase I in extracts of *pol* A⁻
cells made it possible to detect
polymerases II and III more readily.
(Source: Gefter, M. L., Hirota, Y.,
Kornberg, T., Wechsler, J. A. and
Barnoux, C. (1971) *PNAS* **68**, 3150.)

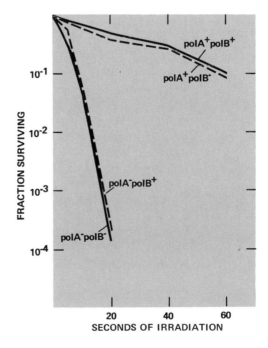

FIGURE 5-2
Sensitivity to ultraviolet irradiation of *pol* A
but not *pol* B mutants. (Source: Campbell,
J. L., Soll, L. and Richardson, C. C. (1972)
PNAS **69**, 2090.)

2. Polymerase III of E. coli[10]

The instability of this enzyme, and its low activity when assayed in solutions of moderate ionic strength, obscured its presence in extracts of even *pol* A$^-$ cells, and delayed its recognition. Not until a fraction with distinctive properties was separated by phosphocellulose chromatography (Fig. 5-1) was the identity of DNA polymerase III established[11]. With precautions to preserve and to assay it under optimal conditions, the pol III activity in extracts of *pol* A$^-$ cells catalyzes DNA synthesis at rates near those of pol I in extracts of *pol* A$^+$ cells. Identification of pol III as the *dna* E gene product has established that it is one of the enzymes essential for DNA replication. It has also been judged essential for one pathway of excision repair of ultraviolet irradiation damage of DNA[12].

Pol III has been purified over 10,000-fold; the best preparations approach homogeneity, as judged by polyacrylamide gel analysis[13,14]. The turnover number of the pure enzyme is estimated to be greater than that of pol I and pol II by factors of 15 and 300, respectively. It is judged that ten molecules of pol III should be sufficient to sustain the in vivo rate of DNA chain growth of about 2000 nucleotides per second.

Pol III resembles pol I in catalytic properties, as does pol II (Table 5-1). The features that distinguish pol III, as well as pol II, from pol I are a preference for a template-primer with short gaps, an inability to utilize single-stranded DNA, inhibition by sulfhydryl-blocking agents, a lack of 5′ $\rightarrow$ 3′ exonuclease activity and insensitivity to antibody to pol I. Pol III differs from both pol II and pol I in its greater instability, sensitivity to salt, stimulation by ethanol, and lower affinity for deoxynucleoside triphosphates.

That pol III plays some essential part in DNA replication became clearer in studies of extracts from *dna*$_{ts}$ *pol* A$^-$ mutants[15]. This survey, done before pol II-defective mutants (*pol* B$^-$) were available, used phosphocellulose chromatography to separate pol III cleanly from pol II. When obtained from *dna* E, but not from other *dna*$_{ts}$ mutants, the enzyme proved markedly labile at 45° (Table 5-2, Fig. 5-3). The table shows among a number of *dna*$_{ts}$ mutants four strains of *dna* E with altered pol III activity: *dna* E3 and *dna* E4 of pronounced thermolability, *dna* E2 with no activity at all, and *dna* E1 of only slight thermolability (its in vivo DNA synthesis is known to slow to a stop only gradually on shift from 30° to 45°).

With these results, the structural gene for pol III could be identified with the *dna* E locus. It is located at 2.5 minutes on the *E. coli* genetic map (Fig. 4-32). Since DNA replication stops in *dna* E mutants when the temperature is raised, and since the pol III isolated from

10. Kornberg, T. and Gefter, M. L. (1972) *JBC* **247**, 5369.
11. Kornberg, T. and Gefter, M. L. (1971) *PNAS* **68**, 761.
12. Youngs, D. A. and Smith, K. C. (1973) *NNB* **244**, 240.
13. Op. cit. in footnote 10.
14. Otto, B., Bonhoeffer, F. and Schaller, H. (1973) *EJB* **34**, 440.
15. Gefter, M. L., Hirota, Y., Kornberg, T., Wechsler, J. A. and Barnoux, C. (1971) *PNAS* **68**, 3150.

TABLE 5-1
Properties of Polymerases I, II, and III of E. coli

	Pol I	Pol II	Pol III
Functions:			
Polymerization: $5' \rightarrow 3'$	+	+	+
Exonuclease: $3' \rightarrow 5'$	+	+	+
Exonuclease: $5' \rightarrow 3'$	+	−	−
Template-primer:			
Intact duplex	−	−	−
Primed single strands[a]	+	−	−
Nicked duplex (poly d(A–T))	+	−	−
Duplex with gaps or protruding single-strand 5' ends of: <100 nucleotides	+	+	+
>100 nucleotides	+	−	−
Polymer synthesis de novo	+	−	−
Activity:			
Effect of salt: percent of optimal 20 mM KCl	60	60	100
50 mM KCl	80	100	50
100 mM KCl	100	70	10
150 mM KCl	80	50	0
Effect of 10 percent ethanol, percent of optimal	40	45	200
K_m for triphosphate	low	low	high
Inhibition by arabinosyl CTP	−	+[b]	−
Inhibition by sulfhydryl (–SH)-blocking agents	−	+	+
Lability at 37°	−	−	+
Inhibition by pol I antiserum	+	−	−
General:			
Size, daltons	109,000	120,000	180,000
Affinity to phosphocellulose: molarity of phosphate required for elution	0.15 M	0.25 M	0.10 M
Molecules/cell, estimated	400	100	10
Turnover number[c], estimated	(1)	0.05	15
Structural genes	pol A	pol B	dna E
Conditional lethal mutant	yes	no	yes

[a] A primed single strand is a long single strand with a short length of complementary strand annealed to it.
[b] Source: Rama Reddy, G. V., Goulian, M. and Hendler, S. S. (1971) NNB **234**, 286.
[c] Nucleotides polymerized at 37°/min/molecule of enzyme, relative to pol I, which is near 1000 (see Chapter 3, Section 4).
+ and − represent the presence and absence, respectively, of the property listed.
▨ = distinction from pol I
▨ = distinction between pol II and pol III

these strains also shows this temperature-lability, we may conclude that pol III, a DNA-synthesizing enzyme, plays an essential part in DNA replication. Considering the very limited template-primer utilization, the sensitivity to physiological salt concentration, and the low affinity for triphosphates of the isolated enzyme in vitro, it seemed that important components were missing from the environment in which the enzyme functions in vivo. Another form of the enzyme has since been discovered which, operating together with additional replication proteins and factors, may prove to be the

TABLE 5-2
Temperature sensitivity of polymerases II
and III from dna_{ts} mutant strains[a]

Strain	pol II	pol III
	Ratio of reaction rates, 45°/30°	
dna A	1.85	1.45
dna B	1.73	1.74
dna C	1.86	1.20
dna D	1.99	1.47
dna E1	1.87	1.0[b]
dna E2	1.70	[c]
dna E3	1.95	< 0.1
dna E4	1.70	< 0.1
dna F	1.55	1.33
dna G	1.75	1.57
dna+	1.92	1.70

[a] Source: Gefter, M. L., Hirota, Y., Kornberg, T., Wechsler,
J. A. and Barnoux, C. (1971) PNAS **68**, 3150.
[b] In vivo DNA synthesis stops slowly on shifting to higher
temperature.
[c] Enzyme activity was not detectable at 30°.

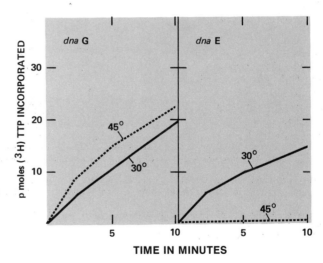

FIGURE 5-3
Influence of temperature on DNA synthesis by preparations
from dna_{ts}G and dna_{ts}E mutants. Only the latter show a
sensitivity to the higher temperature. (Source: Gefter,
M. L., Hirota, Y., Kornberg, T., Wechsler, J. A. and
Barnoux, C. (1971) PNAS **68**, 3150.)

physiological form of pol III involved in growth of the chromosome.
These recent studies of what has been called the "pol III star"
(pol III*) system[16] will be discussed in the next section.

129

SECTION 3:
Polymerase III*
System of E. coli

3. Polymerase III* System of *E. coli*[16]

The more complex form of pol III, which comes closer to representing the physiologically active state, the pol III star (pol III*)
system, was discovered in the investigation of how single-stranded
circular DNAs of the small phages M13 and ϕX174 are converted
to the duplex circular, replicative form. In this initial state of DNA
replication, the virus relies on the enzymatic machinery of the host
cell. Although distinctive host mechanisms exist for initiation of
replication of M13 and of ϕX174 (see Chapter 8, Sections 3, 4) both
employ the pol III* system for chain growth.

Upon fractionation and purification of the enzymes responsible
for the replication of M13 and of ϕX174 DNA, a certain enzyme
fraction with pol III activity was found to be essential. The templates
used were the long, single-stranded viral DNA circles primed with a
short segment of RNA. The active preparation, called pol III* was
distinguished from pol III in the following ways: (i) capacity to
utilize a primed single-stranded DNA as template, (ii) requirement
for an auxiliary protein, called copol III*, in replicating single-
stranded DNA, (iii) larger size than pol III, as judged by gel filtration,
(iv) greater affinity for phosphocellulose, and (v) greater lability.

Pol III* is clearly related to pol III in these ways: (i) product of
dna E gene, as indicated by its thermolability when isolated from
the temperature-sensitive mutant, (ii) similar characteristics of salt
sensitivity, stimulation by ethanol, low affinity for deoxynucleoside
triphosphates, and lack of dependence on copol III* when replicating a duplex template with short gaps, and (iii) conversion of
pol III* to pol III upon heating.

Both pol III* and copol III* have now been isolated in a nearly
homogeneous state. Pol III*, after approximately 40,000-fold purification, appears as a single band of 90,000 daltons in a denaturing
disc gel electrophoresis, and as a single entity about four times that
size in other analytic procedures. Although the data are not con-
clusive, this polymerase molecule is judged to be a tetramer of four
identical 90,000-dalton subunits. Based on estimates of the size of
pol III, pol III* appears to be a dimeric form of pol III. Copol III*, the
protein specifically required by pol III* for utilizing single-stranded
template-primers, also behaves as a pure protein. It is a single polypeptide chain 77,000 daltons in size. Recent studies[16a] show that
pol III* and copol III* are constituted in cell extracts as a holo-

16. Wickner, W., Schekman, R., Geider, K. and Kornberg, A. (1973) *PNAS* **70**, 1764.
16a. Wickner, W. and Kornberg, A., unpublished observations.

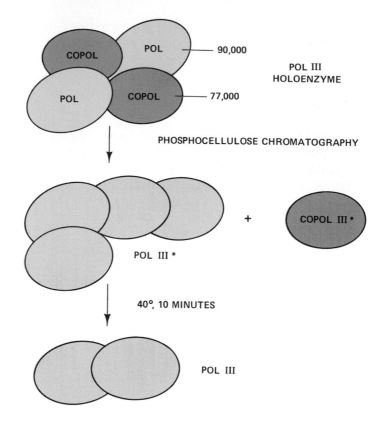

FIGURE 5-4
Forms of DNA polymerase III. The holoenzyme is
an $\alpha_2\beta_2$ tetramer of two subunits each of polymerase III and copolymerase III*. They are resolved by
phosphocellulose chromatography. An oligomeric
form of pol III, called pol III*, remains active in
replicating single-stranded templates provided
that copolymerase III* is present. Pol III* is
"irreversibly" inactivated for this function and
acquires the properties of pol III upon a brief
exposure to an elevated temperature or to
other treatments.

enzyme containing equimolar amounts as an $\alpha_2\beta_2$ tetramer (Fig.
5-4); phosphocellulose chromatography can resolve the holoenzyme.

The mechanism of the pol III* system activity is now under intensive study. An initial stage in its action is formation of a complex
between template-primer, pol III*, and copol III* for which spermidine or E. coli DNA unwinding protein and ATP are essential[17]
(Fig. 5-5). Once the complex is formed and replication is underway,
copol III* appears dispensable. Perhaps copol III* serves pol III*
as the sigma subunit serves RNA polymerase (Chapter 10, Section 3).

17. Wickner, W. and Kornberg, A. (1973) PNAS **70**, 3679.

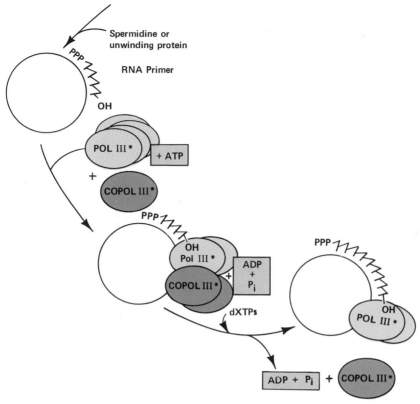

FIGURE 5-5
Stages in a scheme for the start of chain growth on a single-stranded template
primed with a fragment of RNA. A complex formed in the first stage requires
E. coli unwinding protein or spermidine, pol III*, copol III*, and ATP. The
complex isolated by gel filtration, contains ADP and inorganic phosphate (the
products of a specific ATP cleavage) as well as spermidine, pol III*, and copol
III*. In a subsequent stage, the chain grows upon addition of deoxynucleoside
triphosphates; ADP and inorganic phosphate are discharged and chain elongation
is resistant to antibody to copol III*. Thus ATP and copolymerase III* are
required to initiate chain growth but not to sustain it.

4. Polymerases of *Bacillus subtilis*[18,19] and *Micrococcus luteus*[20,21]

Investigation of the DNA polymerase of *B. subtilis* was initiated for
several reasons[22]. For one, attempts to obtain from *E. coli* substantial
amounts of DNA polymerase free of nuclease had been limited by
the relatively low levels of polymerase and high levels of nuclease

18. Ganesan, A. T., Laipis, P. J. and Yehle, C. O. (1973) **in** *"DNA Synthesis In Vitro"*, Steenbock Symposium, Wisconsin (R. D. Wells and R. B. Inman, eds.), University Park Press,
Baltimore, Md., p. 405.
19. Gass, K. B. and Cozzarelli, N. R. (1973) *JBC* **248**, 7688.
20. Litman, R. M. (1968) *JBC* **243**, 6222.
21. Harwood, S. J., Schendel, P. F. and Wells, R. D. (1970) *JBC* **245**, 5614; Hamilton, L., Mahler,
I. and Grossman, L. (1974) *B.* **13**, 1886.
22. Okazaki, T. and Kornberg, A. (1964) *JBC* **239**, 259.

in *E. coli*. For another, there had been no success in demonstrating biologic activity in the DNA product synthesized by *E. coli* polymerase using transforming DNA from *B. subtilis* as template-primer.

In favor of *B. subtilis* was its low nuclease content and high polymerase to nuclease ratio in extracts. It was hoped the *B. subtilis* enzyme might be easier to purify and that the product of replication of its own DNA would indeed be active in transformation. Finally, a developing interest in the biochemistry of sporulation and germination made a comparison of the polymerases of the *B. subtilis* vegetative cell and dormant spore[23] an especially inviting subject.

Although few of these expectations in studying the DNA polymerase of *B. subtilis* were realized, comparisons with the *E. coli* system resulted in some valuable observations.

In purifying *E. coli* polymerase I, separation from nucleic acids is an essential step at the outset and depends in part on a nucleolytic digestion. This nucleic acid removal step failed with *B. subtilis* extracts, presumably because of the low nuclease level. (Partition between polyethylene glycol and polydextran phases was used as a substitute for removing nucleic acids and has since been adopted in other purification procedures.) The *B. subtilis* enzyme[24] proved less stable, more difficult to purify, and no more abundant than *E. coli* polymerase I.

The functional characteristics of the partially purified *B. subtilis* enzyme very much resembled those of *E. coli* pol I (Table 5-3), even

23. Falaschi, A. and Kornberg, A. (1966) *JBC* **241**, 1478.
24. Okazaki, T. and Kornberg, A. (1964) *JBC* **239**, 259.

TABLE 5-3
Properties of polymerase I of *E. coli*, *B. subtilis*, and *M. luteus*

	E. coli pol I	*B. subtilis* pol I	*M. luteus*
Size, daltons	109,000	115,000	110,000
Genetic locus	*pol* A	*pol* A	
Functions:			
Polymerization: 5′ → 3′	+	+	+
Exonuclease: 3′ → 5′	+		+
Exonuclease: 5′ → 3′	+		+
Template-primer:			
Intact duplex	−	−	−
Primed single strands	+	+	+
Nicked duplex	+	+	+
Polymer synthesis, de novo	+	+	+
Activity:			
Product: rapidly renaturable[a]	+	+	+
Ribonucleotide incorporation with Mn^{2+}	+	−	−
Effect of high salt (0.2 M KCl)	−	−	
Inhibition by −SH blocking agents	−	−	−

[a]"Hairpin", many-fold synthesis

the failure to produce biologically active DNA when directed by transforming *B. subtilis* DNA. The only important difference may be in exonucleolytic functions, whose clarification will require further purification of the *B. subtilis* enzyme. An early estimate that the size of the enzyme was only 75,000 daltons may have been erroneous due to cleavage of the polymerase by proteases, which are plentiful in extracts; a recent measurement of 115,000 may be more valid[25].

Recognition of additional DNA polymerases in *E. coli* led to comparable genetic and biochemical studies in *B. subtilis*, and a strikingly similar pattern has emerged (Table 5-4). There are three

TABLE 5-4
Properties of polymerases I, II, and III of *B. subtilis*

		pol I	pol II	pol III
Template-primer:				
Nicked duplex (poly d(A–T))		+	+	−
Duplex with small gaps		+	+	+
Inhibition by single strands		−	+	+
Activity:				
Effect of salt: percent of optimal	15 mM KCl	40	48	100
	100 mM KCl	74	100	40
	250 mM KCl	100	15	4
	350 mM KCl	95	2	<1
Heat lability		−	−	+
Inhibition by –SH blocking agents		−	+	+
Inhibition by arabinosyl CTP		−	+	−
Inhibition by hydroxyphenylazouracil		−	−	+
General:				
Affinity to phosphocellulose, molarity of pH 6.8 phosphate required for elution:		0.20	0.25	0.16
As above, for pH 7.4 phosphate:		0.10	0.15	<0.10
Affinity to DEAE-cellulose, molarity of pH 7.4 phosphate required for elution:		0.25	0.12	0.28

distinctive *B. subtilis* DNA polymerases, named I, II, and III like their counterparts in *E. coli* (compare Tables 5-1 and 5-4).

A mutant sensitive to the mutagen methylmethane sulfonate and deficient in pol I activity was isolated[26]. The gene has been mapped and it can be transferred easily by transformation or transduction. With the mutant, it was easier to detect the presence of two other polymerases which are blocked by sulfhydryl agents, are sensitive to salt inhibition, use poly d(A–T) less well, and are separable on ion exchangers.

Of special interest in determining the role of pol III in replication is the specific inhibitory action of hydroxyphenylhydrazino pyrimidines. The antibiotic 6–(p-hydroxyphenylazo)–uracil (HPUracil)

25. Ganesan, A. T., Yehle, C. O. and Yu, C. C. (1973) *BBRC* **50**, 155.
26. Gass, K. B., Hill, T. C., Goulian, M., Strauss, B. S. and Cozzarelli, N. R. (1971) *J. Bact.* **108**, 364.

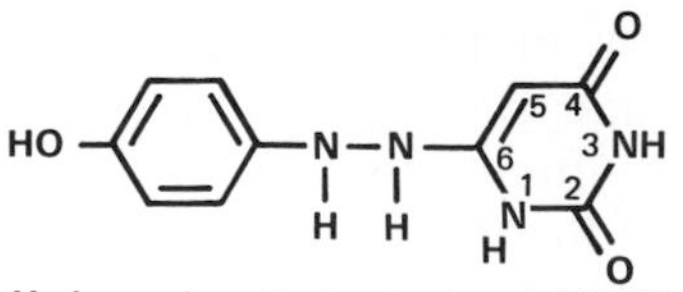

6-(*p*- Hydroxyphenylhydrazino)-uracil (HPUracil)

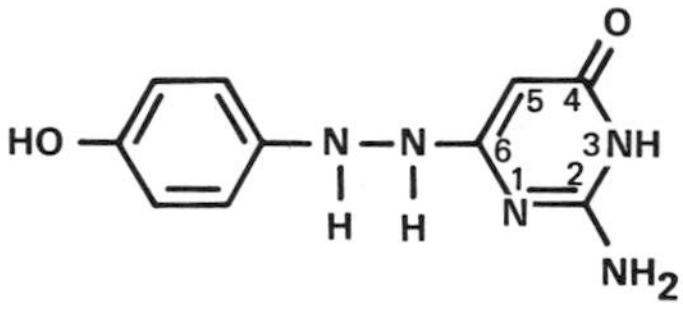

6-(*p* - Hydroxyphenylhydrazino)-isocytosine (HPIsocytosine)

FIGURE 5-6

in its reduced form (Fig. 5-6) inhibits DNA synthesis specifically, and only in certain gram-positive bacteria[27]; *E. coli* is not affected. Inhibition in *B. subtilis* is limited to replicative synthesis. What may be described as repair synthesis and the DNA synthesis associated with replication of certain phages that infect *B. subtilis* are not affected by the drug. Synthesis in permeable cells, considered to be chromosomal replication (Chapter 7, Section 10), is inhibited by HPUracil, providing the reduced form of the drug is used. Of the three separated DNA polymerases of *B. subtilis* only pol III was affected by the drug in very low concentration[28,29]. None of the DNA polymerases of *E. coli*, nor the T4-phage-induced enzyme, was inhibited by reduced HPUracil. Mutants of *B. subtilis* isolated for their resistance to HPUracil were also found to have a pol III activity that was likewise resistant. Additional evidence that links pol III to an essential role in replication of the host chromosome has been the discovery of mutants analogous to the temperature-sensitive *dna* E of *E. coli*; they cease making DNA at elevated temperature and also possess a thermolabile pol III[30].

6-(p-Hydroxyphenylazo)-isocytosine (HPIsocytosine) in reduced form (Fig. 5-6) also inhibits pol III of *B. subtilis*[31]. A suggestion as to how these compounds act is based on the nature of the templates whose replication is inhibited, and the deoxynucleoside triphosphate that specifically antagonizes the effect (Fig. 5-7). The model[32,33] proposes that HPUracil and HPIsocytosine form reversible ternary complexes with the DNA template-primer and the polymerase. The drug binds in the active center of the enzyme in a way that overlaps the deoxynucleoside triphosphate binding site. Binding to the tem-

27. Brown, N. C. (1971) *JMB* **59**, 1.
28. Mackenzie, J. M., Neville, M. M., Wright, G. E. and Brown, N. C. (1973) *PNAS* **70**, 512.
29. Gass, K. B., Low, R. L. and Cozzarelli, N. R. (1973) *PNAS* **70**, 103.
30. Cozzarelli, N. R. and Low, R. L. (1973) *BBRC* **51**, 151.
31. Op. cit. in footnotes 28 and 29.
32. Coulter, C. L. and Cozzarelli, N. R. (1973) *Amer. Cryst. Assoc. Abst.*, Series 2, **1**, 187.
33. Op. cit. in footnote 28.

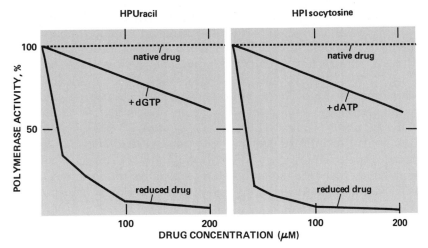

FIGURE 5-7
Effect of arylazopyrimidines on DNA polymerase activity of a partially purified
fraction of *B. subtilis*. HPUracil inhibition is antagonized by dGTP, HPIsocytosine
by dATP. (Source: Mackenzie, J. M., Neville, M. M., Wright, G. E. and Brown,
N. C. (1973) *PNAS* **70**, 512.)

plate is by three H bonds between a C residue in the template and
HPUracil, or a T residue in the template and HPIsocytosine. The
base-pairing scheme in Fig. 5-8 is the only one consistent with the
nuclear magnetic resonance[34] and X-ray data[35].

The mechanism of inhibition by the arylazopyrimidine is not
limited to blocking access of specific deoxynucleoside triphosphates
to the active center of the enzyme. The drug-enzyme-DNA complex
effectively blocks the incorporation of all four deoxynucleoside tri-

---

34. Mackenzie, J. M., Neville, M. M., Wright, G. E. and Brown, N. C. (1973) *PNAS* **70**, 512.
35. Coulter, C. L. and Cozzarelli, N. R. (1973) *Amer. Cryst. Assoc. Abst.*, Series 2, **1**, 187.

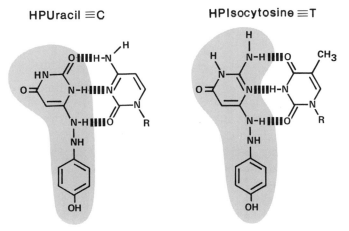

FIGURE 5-8
Models suggested for inhibition by arylazopyrimidines through
base-pairing. (Source: Mackenzie, J. M., Neville, M. M., Wright,
G. E. and Brown, N. C. (1973) *PNAS* **70**, 512; Coulter, C. L. and
Cozzarelli, N. R. (1973) *Am. Cryst. Assoc. Abst.* Series 2, **1**, 187.)

phosphates. Immobilization of the enzyme may depend on an interaction between the phenol ring of the drug and a hydrophobic region unique to the pol III of *B. subtilis*. The latter polymerase, alone among all the *B. subtilis* and *E. coli* DNA polymerases, is inhibited by these drugs.

Preferential inhibition of pol II among the *B. subtilis* polymerases by the arabinosyl analog ara CTP has been observed, as in the case of *E. coli* pol II (Table 5-1). Inasmuch as ara CTP can be covalently fixed to the primer, it behaves as a chain terminator for further growth unless removed by $3' \rightarrow 5'$ exonuclease. Action of this inhibitor may therefore depend on the persistence of the analog on the primer ends of chains, as well as on its capacity to be incorporated.

Micrococcus luteus Polymerase

M. luteus, known earlier as *Micrococcus lysodeikticus* because the cells are easily lysed, was also attractive for enzyme studies of DNA synthesis because of its low nuclease content. As with *B. subtilis*, it was hoped that unresolved questions about *E. coli* DNA polymerase might be better studied with *M. luteus*. Chief among these were how DNA chains are initiated, and what factors are required for synthesis of a biologically active product.

Purification of *M. luteus* polymerase[36,37], however, did not solve either problem. The enzymatic properties are much like those of pol I from *E. coli* and *B. subtilis* (Table 5-3). Special attention has been given to de novo synthesis of polymers by *M. luteus* enzyme, particularly the alternating G–C copolymer, which has never been obtained with the *E. coli* enzyme.

In view of the presence of pol II and pol III in both *E. coli* and *B. subtilis*, it seems likely that these enzymes will be found in *M. luteus* as well.

5. T4-phage-induced Polymerase[38]

The earliest work on the enzymatic synthesis of DNA[39] was designed to explain how a T-even phage, within a few minutes after infecting *E. coli,* induces the replication of its own unique DNA and produces it ten times faster than DNA synthesis in the uninfected host cell. Although the earliest studies on DNA polymerase and its action were carried out with the enzyme from uninfected cells, this work had been motivated and guided by questions posed by phage DNA synthesis (see also Chapter 8).

In order to demonstrate an increased rate of DNA synthesis in extracts of phage-infected cells, it was necessary to use hydroxy-

36. Zimmerman, B. K. (1966) *JBC* **241**, 2035; Litman, R. M. (1968) *JBC* 243, 6222.
37. Harwood, S. J., Schendel, P. F. and Wells, R. D. (1970) *JBC* **245**, 5614; Hamilton, L., Mahler, I. and Grossman, L. (1974) *B*. **13**, 1886.
38. Goulian, M., Lucas, Z. J. and Kornberg, A. (1968) *JBC* **243**, 627.
39. Kornberg, A., Lehman, I. R. and Simms, E. S. (1956) *Fed. Proc.* 15, 950.

methyl deoxycytidine triphosphate as a substrate, because of the rapid destruction of dCTP by phage-induced dCTPase in the crude extracts (Chapter 8, Section 6). The increased rate of DNA synthesis then observed was explained not by an alteration in the level or turnover efficiency of the host enzyme but by a new enzyme[40]. An unusual requirement for single-stranded DNA as template-primer, and a resistance to antibody directed against the host enzyme, indicated that the heightened polymerase activity after infection was new and different. Some years later, the T4-phage-induced enzyme was isolated in pure form[41], identified as the product of a well-analyzed genetic locus (gene 43)[42], and shown to be essential for replication of the phage DNA.

T4-induced polymerase was selected for purification, rather than the T2-induced enzyme, because infection with a certain T4 mutant (in gene 44) prevented the shutoff in synthesis of the polymerase and other phage proteins produced during the early phase of infection. The phage polymerase level and specific activity in crude extracts of this mutant were five to ten times higher than those observed with wild-type T2 or T4 (Fig. 5-9); crude extracts of cells infected for 75 minutes contained the T4-induced enzyme at a specific activity of 78 units per mg of protein, some 25 times the DNA synthesis level of uninfected cells. Intensive genetic investigation[43] of T4 and the availability of a large number of carefully mapped nonsense and temperature-sensitive mutants have also been persuasive in maintaining the focus on T4 for biochemical studies of DNA synthesis and DNA polymerase. However, there are no indications that the DNA polymerases induced by T2 or T6 are significantly different from T4 polymerase.

40. Aposhian, H. V. and Kornberg, A. (1962) *JBC* **237**, 519.
41. Goulian, M., Lucas, Z. J. and Kornberg, A. (1968) *JBC* **243**, 627.
42. deWaard, A., Paul, A. V. and Lehman, I. R. (1965) *PNAS* **54**, 1241.
43. Edgar, R. S. and Epstein, R. H. (1965) *Sci. Amer.* **212**, 70.

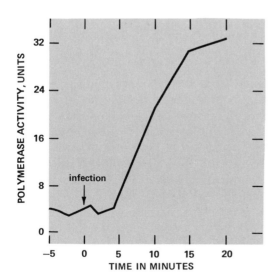

FIGURE 5-9
Polymerase levels in extracts of cells infected with T2 phage. (Source: Aposhian, H. V. and Kornberg, A. (1962) *JBC* **237**, 519.)

DNA polymerase induced by T4 has been purified to homogeneity. The procedure, much like those for other DNA polymerases, leans heavily on phosphocellulose chromatography. The enzyme is a single polypeptide similar in size to pol I of *E. coli* (Table 5-5), but

TABLE 5-5
Comparison of pol I and the phage T4-, T5-, T7-, and SP01-induced polymerases

	Pol I	T4	T5	T7	SP01
Functions:					
Polymerization: $5' \rightarrow 3'$	+	+	+	+	+
Exonuclease: $3' \rightarrow 5'$	+	+		+	
Exonuclease: $5' \rightarrow 3'$	+	−		−	
Template-primer:					
Primed single strands	+	+	+	+	+
Nicked duplex	+	−	−	−	−
Gapped duplex	+	+	+	+	
Polymer synthesis de novo	+	−			
Activity:					
Ribonucleotide incorporation with Mn^{2+}	+	−			
Inhibition by −SH blocking agents	−	+			+
Effect of salt (200mM NaCl compared to 50mM, set at 100)	60	3	400		
Size, daltons	109,000	114,000		80,000	122,000
Single polypeptide chain	yes	yes		yes	yes
Turnover number (relative to pol I)		2			
Genetic locus	*pol* A	gene 43		gene 5	
Phenotype of mutant, *dna*⁻	yes	yes	yes	yes	
Homogeneity of purified enzyme	yes	yes	no	no	no
Number of molecules/cell (est.)	400	600			

different in amino-acid composition (Table 5-6). It has 15 cysteine residues, rather than three, and at least one has a sulfhydryl group essential for activity.

The most striking difference from the host polymerases is the template-primer requirement. T4-polymerase, unlike pol I, cannot use a nicked duplex but requires a primed single-stranded template (Table 5-5), and can utilize such templates whether they have short gaps or long DNA stretches. In its ability to replicate DNA with extensive single-stranded regions, the T4 enzyme differs from pol II and pol III of *E. coli*, which require short gaps. The T4 enzyme displays the same requirements as pol I for a primer terminus, template, and deoxynucleoside triphosphates in the assembly of a chain in the $5' \rightarrow 3'$ direction. Its turnover number under optimal conditions is twice that of pol I, but well short of that of pol III.

TABLE 5-6
Amino acid composition of T4 DNA polymerase[a]

Amino acid	Content moles/112,000 g
Lysine	85
Histidine	16
Arginine	46
Half-cystine	15
Aspartic acid, asparagine	113
Threonine	29
Serine	65
Glutamic acid, glutamine	113
Proline	40
Glycine	61
Alanine	55
Valine	55
Methionine	35
Isoleucine	86
Leucine	59
Tyrosine	47
Phenylalanine	12
Tryptophan	12

[a]Goulian, M., Lucas, Z. J. and Kornberg, A. (1968) *JBC* **243**, 627.

$3' \longrightarrow 5'$ EXONUCLEASES[44]

T4-polymerase does not display $5' \longrightarrow 3'$ exonuclease activity but does have an extraordinarily active $3' \longrightarrow 5'$ exonuclease (Table 5-7) which resembles the proof-reading activity found in pol I (Chapter 4, Section 9). This exonuclease degrades single-stranded DNA to the terminal dinucleotide and has been a useful reagent (as has exonuclease I), for identifying the dinucleotide at the 5' end of a chain.

44. Huang, W. M. and Lehman, I. R. (1972) *JBC* **247**, 3139.

TABLE 5-7
Comparative rates of nuclease activities

	Turnover[a]
T4, $3' \longrightarrow 5'$	4000
Pol I, $3' \longrightarrow 5'$	16
Pol I, $5' \longrightarrow 3'$	27
Pol I, $5' \longrightarrow 3'$ (+ dTTP)	260

[a]Moles hydrolyzed/min/mole enzyme at 37° with $d(T)_{200-400}$ as substrate for the $3' \longrightarrow 5'$ exonuclease and with the $d(T)_{200-400}$ annealed to $d(A)_{4000}$ as substrate for the $5' \longrightarrow 3'$ exonuclease. The presence of dTTP supports concomitant synthesis. The turnover for synthesis by T4 polymerase and pol I under comparable conditions is about 1200 and 600, respectively.

The $3' \rightarrow 5'$ exonuclease has also proved valuable in analysis of the sequence at the 3'-hydroxyl end of a duplex chain[45]. Nuclease action at the end of a frayed duplex, coupled with polymerase action, leads to turnover of the terminus and may be used to identify or label it (Chapter 11, Section 1).

TEMPLATE-PRIMER FUNCTION OF A SINGLE STRAND[46]

A model (Fig. 5-10) for single strand DNA replication, called "hairpin" or "loopback," mentioned in Chapter 3 (Fig. 3-3), is based on

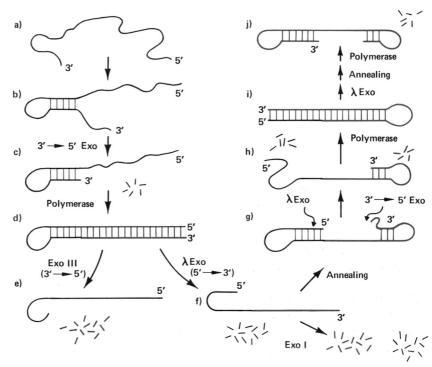

FIGURE 5-10
Scheme for single-stranded DNA functioning as a "hairpin" template-primer for T4 polymerase as indicated by the action of various exonucleases on the polymerase product. (Source: Goulian, M., Lucas Z. J. and Kornberg, A. (1968) *JBC* **243**, 627.)

several aspects of T4 polymerase behavior, namely: (*i*) utilization of a linear, single-stranded template-primer, (*ii*) covalent attachment between the 5' end of the product and the 3' end of the template-primer, (*iii*) facile renaturation of the product, and (*iv*) limitation of synthesis to less than one replication of the template-primer.

According to the model, the new chain starts at the 3'-hydroxyl terminus of the single-stranded template (Fig. 5-10a) which has looped back, the loop facilitated by annealing between regions along the same strand (Fig. 5-10b). An unpaired length of single strand at the 3' end is hydrolyzed by the $3' \rightarrow 5'$ exonuclease back to a base-

45. Englund, P. T. (1971) *JBC* **246**, 3269.
46. Goulian, M., Lucas, Z. J. and Kornberg, A. (1968) *JBC* **243**, 627.

paired primer terminus. The configuration (Fig. 5-10c) now resembles a DNA molecule, which after partial digestion by exonuclease III is known to be an effective template. Chain growth continues until the 5′ terminus of the template is reached, where it stops because the DNA is now fully duplex at the end (Fig. 5-10d).

The effects of specific exonucleases on the structure proposed in the model have been instructive. Exonuclease III in attacking at the 3′ terminus of the duplex product removes only synthesized material (Figs. 5-10e and 5-11a), whereas the phage λ exonuclease (Chapter 9, Section 1) acting on a 5′-helical terminus removes only template nucleotides (Figs. 5-10f and 5-11b). Since both enzymes are inactive

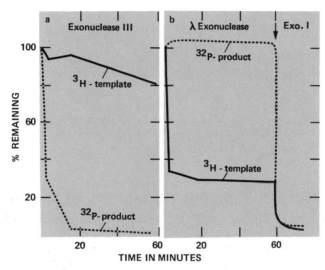

FIGURE 5-11
Susceptibility of the template-primer and the T4-polymerase product (at stage (d) in Figure 5-10) to exonucleases. (a) Exonuclease III. (b) λ Exonuclease. (Source: Goulian, M., Lucas, Z. J. and Kornberg, A. (1968) *JBC* **243**, 627.)

on single-stranded DNA, a certain portion of the template product is in each case spared from attack. (Conversion of the product to single-stranded form, as by the action of λ exonuclease, does make the template-product completely susceptible to exonuclease I, whose action is specific for single-stranded DNA (Fig. 5-11b)).

FURTHER EXONUCLEASE ACTIONS ON THE POLYMERASE PRODUCT
Annealing after λ exonuclease action permits polymerase action to resume (Fig. 5-10g). This sequence explains why addition of λ exonuclease to a replication mixture containing fully replicated DNA and active, but idle, polymerase was followed by a burst of synthesis that paralleled digestion of the template (Fig. 5-12). It is the digestion by λ exonuclease at the 5′ terminus (Fig. 5-10d) that enables the 3′ terminus of the product to loop back and reanneal (Fig. 5-10g).

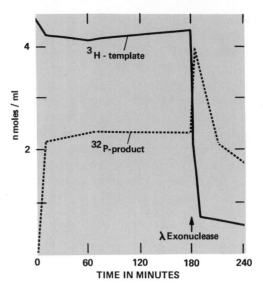

FIGURE 5-12
Action of λ exonuclease on template-primer
and the T4-polymerase product during the
course of replication, as shown in Figure 5-10.
(Source: Goulian, M., Lucas, Z. J. and Kornberg,
A. (1968) *JBC* **243**, 627.)

Trimming of the unpaired 3' end by the $3' \rightarrow 5'$ exonuclease enables
the polymerase to start synthesis again (Fig. 5-10h,i). Subsequent
loss of template can be explained as the result of still another round
of digestion, which becomes possible when the λ exonuclease and
polymerase reach the site of the original loop (Fig. 5-10h,i,j).

Briefly restated, the combined action of λ exonuclease and polymerase on the product of polymerase action entails digestion of
template (Fig. 5-10d), formation of a new loop, and then a wave of
new synthesis (Fig. 5-10g,h). When the loop region has been replicated (Fig. 5-10i), λ exonuclease can resume action, with repetition
of the process, eventually making product material available to this
nuclease (Fig. 5-10j).

PHYSIOLOGIC ROLE

T4-induced polymerase was the first DNA polymerase directly
correlated with chromosome replication, and it remains to this time
the clearest such example[47]. Mutants in gene 43 that fail to induce
and maintain significant levels of DNA polymerase also fail to maintain DNA synthesis even though related enzymes reach normal
levels. Some of these mutants grown at 30° make a polymerase stable
at 42° as well as at 30°, although grown at 42° they make no detectable enzyme. Other gene 43 mutants grown at 30° make a polymerase active at 30° but labile at 42°. These temperature-sensitive

47. de Waard, A., Paul, A. V. and Lehman, I. R. (1965) *PNAS* **54**, 1241.

polymerases show that gene 43 is the structural gene for DNA polymerase. This is buttressed by convincing immunochemical analyses (see Section 7 of this chapter). Evidence that the polymerase coded by gene 43 is involved in DNA synthesis comes from mutator and antimutator effects attributable to gene 43 mutants (Chapter 4, Section 10). Also of interest and deserving further study are certain nonsense mutants that induce the nuclease but not polymerase activity[47a]. It is not known whether the change of structure affects the binding of template or triphosphate, or interferes with the polymerizing step.

An interesting insight into the synthesis and assembly of the polymerase has come from studying the influence of temperature shifts on the behavior of a temperature-sensitive mutant[48]. When E. coli was infected at 43° with the mutant, no DNA synthesis occurred and polymerase activity was not detected (Fig. 5-13). On

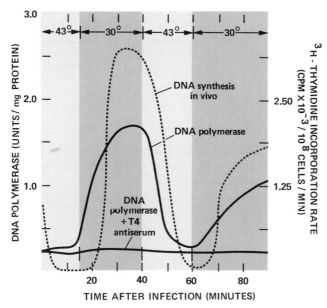

FIGURE 5-13
Activation of latent, temperature-sensitive T4 polymerase by shift-up in temperature. (Source: Swartz, M. N., Nakamura, H. and Lehman, I. R. (1972) Virol. 47, 338.)

lowering the temperature to 30°, polymerase activity appeared immediately and rapidly. The enzyme appeared even when protein and DNA synthesis were blocked. The polymerase that developed at 30° was a phage-induced and not a host-related enzyme, since it was inhibited specifically by T4-polymerase antiserum. The appearance of active enzyme upon shifting to 30° required oxidative

47a. Nossal, N. G. (1969) JBC 244, 218.
48. Swartz, M. N., Nakamura, H. and Lehman, I. R. (1972) Virol. 47, 338.

metabolism: it was blocked by cyanide, azide, or 2,4-dinitrophenol.

A second cycle of shifting up to 43° and down again to 30° produced results similar to the first cycle although the levels of DNA synthesis and DNA polymerase activity were lower, implying that the supply of enzyme precursor had become limiting.

A speculative model to account for these results is shown in Fig. 5-14. An inactive, improperly folded polymerase precursor is syn-

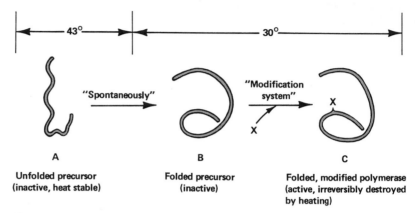

A	B	C
Unfolded precursor (inactive, heat stable)	Folded precursor (inactive)	Folded, modified polymerase (active, irreversibly destroyed by heating)

FIGURE 5-14
Possible stages in the activation of a temperature-sensitive T4 polymerase.
(Source: Swartz, M. N., Nakamura, H. and Lehman, I. R. (1972) *Virol.* **47**, 338.)

thesized at 43°, and upon step-down to 30° is converted to a properly folded, but still inactive, molecule. Active T4 polymerase is produced by an energy-dependent modification of the properly folded, inactive precursor. Whether a similar sequence of events occurs in a wild-type infection is, of course, conjectural; further work with a cell-free system is needed to answer these questions.

6. T5-, T7-, and SP01-induced Polymerases[49, 50]

Initial attempts to find a novel polymerase in *E. coli* upon infection with T5 or T7 phages, as had been found with the T-even phages, were not successful. The smaller size of T5 and T7 DNA, and the absence of unusual bases seemed at that time to explain a total reliance of these phages on the host replication system. However, after adjusting the assay conditions it was possible to show that infection with T5 or T7 phage induces novel polymerase activity.

In the assay for the T5 enzyme[51], a powerful induced DNase was suppressed by high concentrations of DNA and the polymerase was activated by high ionic strength. This made it possible to demonstrate

49. Orr, C. W. M., Herriott, S. T. and Bessman, M. J. (1965) *JBC* **240**, 4652.
50. Grippo, P. and Richardson, C. C. (1971) *JBC* **246**, 6867.
51. Op. cit. in footnote 49.

a tenfold increase in polymerizing activity attributable to a new enzyme. The existence of a T7-induced polymerase was shown by using as host cell the *pol* A⁻ mutant, which has a low level of pol I[52].

The properties of purified T5- and T7-induced polymerases are listed with those of the T4-induced enzyme and pol I in Table 5-5. It is clear at once that the three phage enzymes are quite similar. They are distinguished from pol I in their failure to use nicked DNA. They have limited capacity to fill in single-stranded regions of a template, and depend for priming on "loopbacks", oligonucleotide fragments, or regions denuded by exonuclease III digestion.

Mutants in a defined genetic locus fail to synthesize DNA in vivo and show a parallel failure to develop any distinctive polymerase activity detectable in extracts. These characteristics suggest that polymerase is essential for normal replication of the phage chromosome, but do not sharply define the specific role of the enzyme in any instance. Undoubtedly, the polymerase is part of a more complex replication assembly in which still unidentified enzymes make possible the use of double-stranded template-primer, and contribute in other ways to DNA chain growth (see Chapter 8, Section 5).

Among the other enzymes which are known to be essential for phage replication but whose precise contribution has not been defined are $5' \rightarrow 3'$ exonucleases induced by both T5 and T7 phages[53,54]. The T7 case is especially intriguing because: (i) the nuclease gene maps next to the DNA polymerase I gene and, (ii) the molecular weights of the two gene products (80,000 and 30,000) are so similar to those of the polymerizing and $5' \rightarrow 3'$ exonuclease components of pol I.

SP01, a virulent phage infecting *B. subtilis*, induces a new polymerase and directs the synthesis of DNA with 5-hydroxymethyluracil in place of thymine. Purification of the polymerase[55] shows it to be similar in most physical and functional properties to the polymerases induced by the virulent *E. coli* phages (Table 5-5). The deoxynucleoside triphosphates of uracil, hydroxymethyluracil, bromouracil, and thymine are incorporated by the purified enzyme at identical rates.

7. Polymerase Antibodies

Serological techniques are proving to be as valuable in analyzing the identity, function, and structural relationships of the polymerases as they have been for other enzymes. The principal applications to date have established: (i) the distinction of pol I from pol II and pol III and from phage-induced polymerases, (ii) discrimination of DNA-binding sites within the active center of pol I, (iii) relatedness of

52. Grippo, P. and Richardson, C. C. (1971) *JBC* **246**, 6867.
53. Frenkel, G. D. and Richardson, C. C. (1971) *JBC* **246**, 4839, 4848.
54. Kerr, C. and Sadowski, P. D. (1972) *JBC* **247**, 305.
55. Yehle, C. O. and Ganesan, A. T. (1973) *JBC* **248**, 7456.

pol I among diverse classes of bacteria, (iv) the level of an incomplete T4 polymerase peptide produced in a nonsense mutant and, (v) the direction in which the T4 polymerase gene is transcribed and translated.

Antibodies develop within three to four weeks in rabbits injected with one mg or less of the enzyme. The use of mice for antibody production can reduce the amount of protein required to only 10-20 μg. Nucleases in some sera may interfere with assays of DNA synthesis, but can be removed simply by precipitating the antibody globulin fraction with sodium sulfate (18 percent) and by filtration on DEAE-Sephadex. There are three convenient and sensitive ways in which antibodies have been used: (i) inhibition of enzyme activity, (ii) complement fixation, and (iii) radioimmune competition.

INHIBITION OF ENZYME ACTIVITY

Titration of activity of the enzyme with increasing levels of antibody produces up to 95 percent or greater inhibition of the crude or purified protein. The failure of antibody to pol I to inhibit the polymerase activity induced by phage T2 infection was one of the decisive lines of evidence indicating that a new DNA polymerase was formed, possibly one coded by the phage genome (Fig. 5-15).

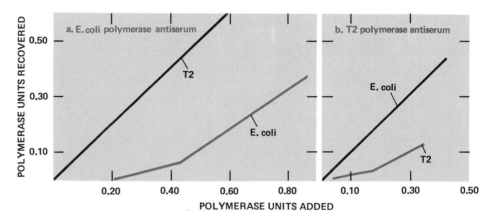

FIGURE 5-15
Selective inhibition of E. coli and T2 polymerase activity by specific antisera. (Source: Aposhian, H. V. and Kornberg, A. (1962) JBC **237**, 519.)

When antibody against the purified T2 polymerase was obtained, its expected failure to inhibit pol I was also demonstrable. Use of pol I antibody also proved crucial in the discovery of pol II and pol III in the *pol* A mutant and in demonstrating the persistence of very low levels of pol I and near normal levels of its component 5' $\longrightarrow$ 3' exonuclease in extracts of the mutant.

Antibody inhibition of enzyme implies that the antibody blocks the catalytic site. However, with enzymes acting on polymeric substrates, the antibody may inhibit activity by preventing access of the polymer to its binding site even though the catalytic site remains free to act on a smaller or more flexible substrate. The data in Table 5-8 show that anti-pol I blocks both pol I and its large proteolytic cleavage fragment (Chapter 4, Section 14). It also blocks the $5' \rightarrow 3'$

TABLE 5-8
Inhibition of various activities of DNA polymerase I by specific antibody

	Polymerization	$5' \rightarrow 3'$ exo	$3' \rightarrow 5'$ exo	
			Double-stranded DNA	Single-strand DNA
Intact enzyme	0.82	1.51	<2.5	>20
Large fragment	1.1		2.3	>20

Values represent microliters of antiserum required to give 50 percent inhibition of 10 pmoles of enzyme.

exonuclease functions of pol I. Of great interest is the failure of the antibody to prevent $3' \rightarrow 5'$ exonuclease action on single-stranded DNA while blocking that on double-stranded DNA. Evidently it is the binding of double-stranded DNA to the enzyme that is impeded by the antibody rather than masking of the active site proper.

COMPLEMENT FIXATION

A group of proteins in serum, collectively called complement, is removed from the serum in antigen-antibody reactions. This "fixation" of complement is a direct measure of the amount of such reaction. The amount of complement in serum can be determined accurately at a micro level in an erythrocyte-anti-erythrocyte reaction in which hemolysis takes place only when complement is present.

Complement fixation techniques permit DNA polymerase in crude extracts as well as in purified form to be measured accurately in the nanogram range (Fig. 5-16). At the "equivalence point" of optimal antigen-antibody interaction, the amount of complement fixed is a measure of the DNA polymerase or any protein antigenically related to it.

Extracts of many bacterial species (Fig. 5-17) contain variable amounts of protein that react with antibody to *E. coli* pol I. Cross-reaction was found in all genera of the class Enterobacteriacea; among them the *Shigella* and *Salmonella* were not distinguishable from *Escherichia*. Extracts of *Proteus vulgaris*, with a markedly different DNA base composition, and those of *B. subtilis*, gram-positive and taxonomically less closely related to *E. coli*, also cross-react. By contrast, complement fixation was not detectable when antibody to pol I was mixed with *E. coli* pol II, pol III, or the polymerase induced by phage T4.

148

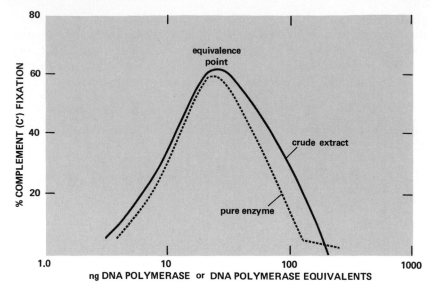

FIGURE 5-16
Antibody titration of DNA polymerase I activity in the pure enzyme or in a crude
extract as measured by complement fixation. (Source: Tafler, S. W., Setlow, P.
and Levine, L. (1973) *J. Bact.* **113**, 18.)

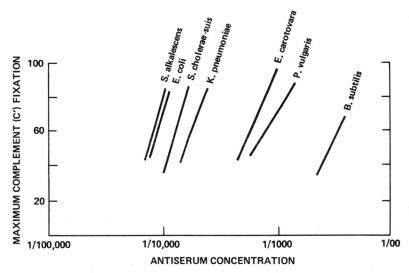

FIGURE 5-17
DNA polymerase I levels in crude extracts of bacteria determined by cross
reaction with antibody against *E. coli* DNA polymerase I. (Source: Tafler,
S. W., Setlow, P. and Levine, L. (1973) *J. Bact.* **113**, 18.)

A pure protein can be labeled (as with ^{125}I) and used to compete for antibody in assays of the unlabeled protein or fragments of the protein still retaining antigenic properties. Loss of the labeled protein in the antigen-antibody precipitate is inversely related to the amount of unlabeled, antigenically similar proteins to which the antibody was first exposed in the assay. This technique has been used even in crude extracts to measure polymerases, as well as proteins and peptides related to them[56]. The assay is effective with 0.01 to 1 μg of pure enzyme; with a crude extract, 0.01 to 1 mg of protein would be required to make up for a 1000-fold enrichment of the enzyme by purification.

Enzymatically inactive or incomplete peptides, called CRM, for cross-reacting material, are generated in nonsense mutants of T4 phage. When mutants in such a series were compared with each other in cross-reactivity with T4-polymerase antiserum, there was a close correlation between the size of CRM fragment determined by gel electrophoresis (down to molecules one-quarter the length of polymerase) with the degree of cross-reactivity. These measurements agreed with the mapping of the mutants by genetic recombination, and this correlation indicated the direction of transcription and translation of gene 43 in the T4 genome.

Antibodies have been prepared against DNA[57] and its components[58]. The considerable base specificity they display may make them highly useful for mapping DNA and delineating the progress of DNA replication.

56. Huang, W. M. and Lehman, I. R. (1972) *JBC* **247**, 7663.
57. Plescia, O. J. and Braun, W., (eds.) (1968) *Nucleic Acids in Immunology*, Springer-Verlag, N.Y.
58. Bonavida, B., Fuchs, S., Sela, M., Roddy, P. W. and Sober, H. A. (1972) *EJB* **31**, 534.

6

Eukaryotic Cell DNA Polymerases

1. Survey of Eukaryotic DNA Polymerases

Until 1970, bacterial cells were thought to have only one DNA polymerase. Accordingly a multiplicity and variety of DNA polymerases in eukaryotic cells was not anticipated, especially because DNA was thought to be restricted to the nucleus and to be replicated only for cell division. Furthermore, work on polymerase in eukaryotic cells was complicated by the relatively low levels of activity and technical difficulties inherent in the assay.

Early attempts to purify and characterize animal cell DNA polymerases were directed at the enzyme in calf thymus gland[1]. This tissue, rich in DNA and obtainable in large quantities, was an obvious choice. However, thymus gland is unusual in possessing an

---

1. Bollum, F. J. (1960) *JBC* **235**, 2399; Yoneda, M. and Bollum, F. J. (1965) *JBC* **240**, 3385.

enzyme, called terminal nucleotidyl transferase (see Section 8 of this chapter), that polymerizes nucleotides *without* template direction, thus interfering with and complicating assays of DNA polymerase.

Now it is clear that eukaryotic DNA polymerases are numerous, in accord with (i) the diversity of their cell locations—nucleus, cytoplasm (cytosol), and mitochondria, (ii) the plural functions performed on DNA—replication, repair, and recombination–and reverse transcription on RNA, (iii) the viral induction of new enzymes, and (iv) the enormous variety of eukaryotic cells and species.

Assays for DNA polymerase activity in cell-free extracts have been taken as a measure of DNA synthesis in vivo because of the close correlation of the assay values with the extent of cell multiplication observed in the tissue. Among tissues that display a wide range in rates of DNA synthesis and chromosome growth, as judged by mitotic activity and by thymidine incorporation rates in vivo, there is good correspondence with in vitro polymerase assay values[2] (Table 6-1). Fetal rat tissues showed considerably higher polymerase

TABLE 6-1
DNA synthesis in vivo, and DNA polymerase activity in extracts of rat tissues

Rat tissue	DNA synthesis[a] in vivo	DNA polymerase activity[a] in extracts
Brain	0.21	0.19
Heart	0.41	0.30
Liver	(1.0)	(1.0)
Kidney	1.2	1.73
Spleen	5.5	4.3
Intestinal mucosa	11.7	9.3

[a]All values are relative to liver. Synthesis in vivo was measured by [^{3}H]-thymidine incorporation.
Source: Baril, E. and Laszlo, J. (1971) *Adv. Enz. Regul.* **9**, 183, Pergamon Press, N.Y.

activities than adult tissues, as did liver tumors (hepatomas) compared with normal liver. Even among those hepatomas categorized as slow-, medium-, or fast-growing, there was correlation of in vivo synthesis rates, DNA polymerase activities, and proliferative states of the tumor[3].

One of the limitations in isolating enzymes from tissues and assessing their physiological functions is the cellular heterogeneity of tissue; the choice of pure cell cultures has an obvious advantage in this respect. Given a particular cell or tissue and a stage in the growth cycle, selection of a fractionation system for isolating the subcellular components must be regarded as inevitably a compromise between obtaining clean organelles or damaged ones. Treat-

2. Baril, E. and Laszlo, J. (1971) *Adv. Enz. Regul.* **9**, 183 Pergamon Press, N.Y.
3. Baril, E. F., Jenkins, M. D., Brown, O. E., Laszlo, J. and Morris, H. P. (1973) *Canc. Res.* **33**, 1187.

ments with detergents, enzymes, or salt washes reduce contamination but the manipulations may cause distortion and leakage of the organelle.

A procedure used to separate subcellular structures of rat liver cells is outlined in Fig. 6-1[4]. The cells were disrupted in a loose-fitting, piston-style homogenizer. Nuclei were separated first by

4. Baril, E. F., Brown, O. E., Jenkins, M. D. and Laszlo, J. (1971) B. **10**, 1981.

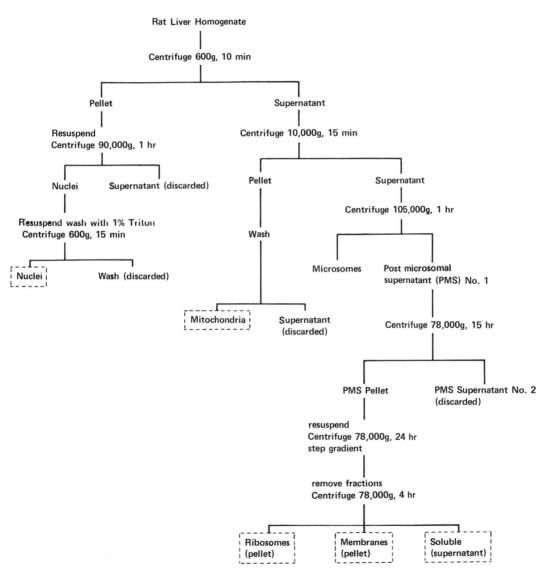

FIGURE 6-1
Procedure for separation of subcellular fractions of rat liver for DNA polymerase activities. (Source: Baril, E. F., Brown, O. E., Jenkins, M. D. and Laszlo, J. (1971) B. **10**, 1981.)

low-speed centrifugation, and then the mitochondria, microsomes (sacs of the membranous endoplasmic reticulum), ribosomes, and smooth cytoplasmic (endoplasmic) membranes (lacking ribosomes) were fractionated at successively higher speeds.

Assays of DNA polymerase activity in these subcellular structures indicated that the nuclei contain only a minor part of the activity and the mitochondria even less (Fig. 6-2). The bulk of the activity appeared in a cytoplasmic fraction associated with smooth membranes. There were also differences observed in utilization of native or denatured DNA as template-primer; however distinctions among DNA polymerases in crude extracts on this basis may be entirely misleading. Nucleases in crude enzyme fractions may convert inactive to active template and vice versa; endogenous DNA and RNA, polyanions, metal ions, and other interfering components in crude extracts have significant influences on polymerase behavior. In short, little or no reliance should be given characterizations of polymerases in crude extracts that depend largely on the relative utilization of different template-primers.

Some general statements may be made about the enzymes purified from various liver fractions and comparable enzymes from hepatoma, thymus gland, and cultured fibroblasts (cells that generate fibrous tissue):

(i) All the DNA polymerases appear to have the same basic requirements and mechanism for chain extension as the *E. coli* enzymes, but lack the $3' \rightarrow 5'$ exonuclease activity.

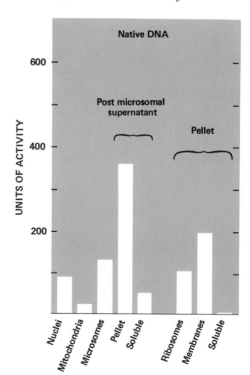

FIGURE 6-2
DNA polymerase activities among subcellular fractions of rat liver, measured with native DNA template-primer.

155

SECTION 2:
Cytoplasmic
Polymerase:
the Large
Polymerase

(ii) The preponderant DNA polymerase activity is found in _large, heterogeneous molecules,_ with sedimentation coefficients of 6 to 8 S (molecular weight near 200,000). It is found chiefly in the _cytoplasmic fraction_ although it appears also in extracts of nuclei, especially during chromosome growth.

(iii) The characteristic _nuclear enzyme_ is isolated as a _very small,_ basic molecule with a sedimentation coefficient in a range near 3.4S (molecular weight, 40,000). It may also be found in cytoplasm.

(iv) A direct structural relationship between large and small polymerases has been suggested[5] and some antigenic similarity has been claimed[6].

(v) The DNA polymerase in mitochondria is distinct in functional and structural properties from either of the foregoing.

(vi) RNA-directed DNA polymerases (reverse transcriptases) of both viral and host-cell origin are distinct from DNA-directed enzymes.

(vii) The terminal nucleotidyl transferase of calf thymus is restricted to that gland and is unique in its lack of dependence on template direction for base selection.

2. Cytoplasmic Polymerase: the Large Polymerase[7-9]

The principal DNA polymerase activity in rat liver[10] and in hepatomas[10] has been found in a cytoplasmic fraction associated with smooth membranes. The enzymes purified extensively both from human cell cultures[11,12] and animal tissues[13] share certain basic features that warrant grouping them now in a single category. Important differences will undoubtedly be found among them after closer study.

The size of the cytoplasmic enzymes, as judged by sedimentation and Sephadex filtration, is 200,000 to 230,000 daltons[14], with sedimentation coefficients in the range 6S to 8S (Table 6-2). With chromatography and other standard procedures, 500- to 1000-fold purification of the enzyme from cultured human cells (KB variety) has been achieved[15]; yet the enzyme may still be quite impure.

In functional properties, the large cytoplasmic enzymes (Table 6-3) were found to resemble DNA polymerases II and III of _E. coli_, for example, in sensitivity to sulfhydryl-blocking agents and salt, and in their marked preference for a gapped template. Although these

---

5. Hecht, N. B. (1973) _NNB_ **245**, 199; Lazarus, L. H. and Kitron, N. C. (1973) _JMB_ **81**, 529.
6. Chang, L. M. S. and Bollum, F. J. (1972) _Science_ **175**, 1116.
7. Baril, E. F., Brown, O. E., Jenkins, M. D. and Laszlo, J. (1971) _B._ **10**, 1981.
8. Wang, T. S.-F., Sedwick, W. D. and Korn, D. (1974) _JBC_ **249**, 841.
9. Weissbach, A., Schlabach, A., Fridlender, B. and Bolden, A. (1971) _NNB_ **231**, 167.
10. Op. cit. in footnote 7.
11. Op. cit. in footnote 8.
12. Op. cit. in footnote 9.
13. Hecht, N. B. and Davidson, D. (1973) _BBRC_ **51**, 299.
14. Holmes, A. M. and Johnston, I. R. (1973) _FEBS Lett._ **29**, 1.
15. Op. cit. in footnote 8.

TABLE 6-2
Physical properties of cytoplasmic and nuclear DNA polymerases

	Cytoplasmic	Nuclear
Size	Large	Small
Molecular weight	200,000–230,000	40,000
S value	6–8	3.4
Homogeneity	No	Yes
Increase with mitotic activity	Yes	No
Isoelectric pH	5.6	9.2
Adsorption to DEAE-cellulose	Yes	No
Phosphate concentration required for elution from phosphocellulose	0.2 M	0.5 M
Stability: activity loss at 45°[a]	100 percent after 5 minutes	50 percent after 15 minutes

[a]Values vary depending on source, purity, and conditions.

enzymes utilized a ribo polymer, poly(rA), as primer, they did not use it as a template.

The large size, acidic isoelectric point, and lability of the cytoplasmic polymerase distinguish it clearly from the small, basic nuclear polymerase (see Table 6-2 and Section 3 of this chapter). The two also differ in important physiological features. Yet paradoxically, the two types cannot be distinguished unambiguously by

TABLE 6-3
Functional properties of cytoplasmic and nuclear DNA polymerases

	Cytoplasmic (large)	Nuclear (small)
"K_m" of DNA, μg/ml	50	7
K_m of Mg^{2+}, mM	10	10
K_m of triphosphate, μM	50	160
Inhibition by:		
0.2 M KCl	90 percent	None
p-hydroxymercuribenzoate, 30 μM,	90 percent	None
ethidium bromide, 20 μM	60 percent	None
Activity with one triphosphate	Small	Moderate
Template-primer:		
gapped DNA	(100)	(100)
native DNA (nicked)	1	2
denatured DNA	3	3
3'-phosphoryl termini	0	0
ribo template·deoxy primer	No	No
deoxy template·ribo primer	Yes	Yes
activated calf thymus (gapped) (pH 7.2, Mg^{2+})	Yes	No
$(dC)_{500}·(dG)_{5-12}$ (pH 8.6, Mn^{2+})[a]	No	Yes
Exonuclease 3' → 5'	No	No
Exonuclease 5' → 3'	No	No
Exchange of PP_i and dNTP	No	No

[a]Note these special assay conditions

157

SECTION 2:
Cytoplasmic
Polymerase:
the Large
Polymerase

their subcellular location. A DNA polymerase of the cytoplasmic type has been extracted from the nucleus and a nuclear-type enzyme has been identified in the cytoplasmic ribosome fraction. The significance of the occurrence of DNA polymerases in cytoplasm is still uncertain.

Perhaps the most important indication of the physiological role of the large DNA polymerases is the sharply increased level found in rat liver with heightened mitotic activity. The activity of the large enzyme increased in the cytoplasmic membrane fraction and in the nuclear extract, in parallel, during liver regeneration after hepatectomy (Fig. 6-3), and also in hepatomas. By contrast, levels of the small polymerase during liver regeneration (Fig. 6-3) and in liver tumors were unchanged. The large polymerase was thus

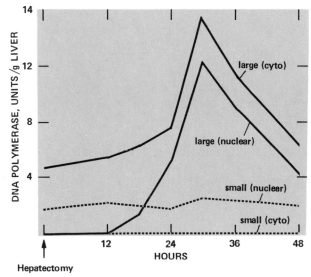

FIGURE 6-3
Levels of large and small polymerases obtained from cyto-plasmic and nuclear liver fractions during liver regeneration after hepatectomy in the rat. (Source: Baril, E. F., Jenkins, M. D., Brown, O. E., Laszlo, J. and Morris, H. P. (1973) *Canc. Res.* **33**, 1187.)

clearly associated with replicative activity and its presence in cyto-plasm in nonproliferating tissue may denote the site of its synthesis and storage. According to this interpretation, the enzyme is trans-ferred, in some undefined way, to the nucleus when the need for replicative activity is signalled.

Other enzymes involved in DNA synthesis, in addition to the large polymerase, were found to be physically associated with the large polymerase in the postmicrosomal smooth membrane fraction[16]. Thymidine kinase, ribonucleotide reductase, and thymidylate syn-

16. Baril, E., Baril, B., Luftig, R. B. and Elford, H. (1975) *EJB* (in press).

thetase were found in this fraction in regenerating and fetal livers and in hepatomas but not in normal tissue (Table 6-4). These four enzymes appeared to be part of an assembly that withstood salt treatments sufficient to dissociate electrostatically bonded complexes. This behavior suggests that strong hydrophobic interactions

TABLE 6-4
Enzymes in postmicrosomal pellet[a] of proliferating tissues[b]

Tissue	DNA polymerase	Thymidine kinase	Ribonucleotide reductase
		Units/mg	
Normal rat liver	1.02	0.0	0.0
Regenerating rat liver: (hours after hepatectomy)			
5	1.08	0.0	0.0
18	1.53	0.0	0.0
24	2.78	10.1	12.1
40	3.44	16.4	31.9
Fetal rat liver (18-day)	4.30	74.0	178
Hepatoma (Novikoff tumor)	6.50	60.0	115

[a]The postmicrosomal pellet was obtained by centrifugation of the postmicrosomal supernatant at $78,000 \times g$ for 15–20 hours. *Thymidylate synthetase* assayed in other experiments showed patterns comparable to those found for the kinase and reductase.
[b]Source: Baril, E., Baril, B., Luftig, R. B. and Elford, H. (1974) EJB (in press).

hold the complex and membrane together. Upon treatment with butanol and detergents, the membrane was dissociated but the enzymes were still recovered fully active in particles 85-120 Å in diameter[17].

The location of these enzymes in postmicrosomal membrane fragments, and only at times when DNA synthesis was active in the liver cell, suggests that these enzymes have an essential part to play in DNA replication. The multienzyme ribosome complex for protein synthesis may be a model for such complexes for the biosynthetic events in replication and in transcription.

3. Nuclear Polymerase: the Small Polymerase[18-21]

Nuclear DNA polymerases obtained from human cell cultures[22,23], rat liver[24,25], and tumor cells[26] have proved to be distinctly dif-

17. Baril, E., Baril, B., Luftig, R. B. and Elford, H. (1975) EJB (in press).
18. Baril, E. F., Brown, O. E., Jenkins, M. D. and Laszlo, J. (1971) B. **10**, 1981.
19. Wang, T. S.-F., Sedwick, W. D. and Korn, D. (1974) JBC **249**, 841.
20. Weissbach, A., Schlabach, A., Fridlender, B. and Bolden, A. (1971) NNB **231**, 167.
21. Chang, L. M. S. and Bollum, F. J. (1972) B. **11**, 1264; (1973) JBC **248**, 3398; Chang, L. M. S. (1973) JBC **248**, 3789.
22. Op. cit. in footnote 19.
23. Op. cit. in footnote 20.
24. Op. cit. in footnote 18.
25. Berger, H. Jr., Huang, R. C. C. and Irvin, J. L. (1971) JBC **246**, 7275.
26. Tsuruo, T., Satoh, H. and Ukita, T. (1972) BBRC **48**, 769.

ferent in physical properties from the predominant DNA polymerase extracted from cytoplasmic cell fractions (Tables 6-2, 6-3). Purified nuclear polymerases from human[27,28] and animal sources[29,30] were similar in their small size, sedimenting with a coefficient near 3.4 S (40,000 daltons), their basic isoelectric point, and their stability to agents that inactivate the cytoplasmic enzyme. For convenience they are called the small polymerases rather than nuclear enzymes, because they are not located exclusively in the nucleus; a measurable portion of the small enzyme, indistinguishable from the nuclear polymerase, was found in the cytoplasmic ribosome fraction.

Although interaction of a small basic protein with ribosomes might be an artifact of disrupting the cell, great care had been taken in the fractionation procedures to keep the nuclei intact. The significance of the presence of the small enzyme in cytoplasm is therefore not clear, unless, as suggested for the large polymerase, the enzyme is made and stored temporarily in the cytoplasm.

Enzymatic properties of the small polymerase resembled those of the large polymerase. However, the physiological functions of the small enzyme appeared to be distinct. The levels of the small enzyme in rat tissues, unlike those of the large enzyme, remained unchanged even during intense proliferation[31,32]. Despite this lack of correlation with replicative activity, the small polymerase may still perform a replicative function that does not appear to be essential because it is present in excess. It may also fill some other role in DNA metabolism, as in repair, recombination, or differentiation.

The small size of nuclear DNA polymerases is remarkable; all other polymerases isolated thus far are large, over 10^5 daltons. The large fragment of *E. coli* polymerase I, which carries out polymerization and proofreading exonuclease action, has a mass of 75,000 daltons; an enzyme of comparable size has also been isolated from *B. subtilis*. Inasmuch as the small nuclear polymerase carries out template-directed polymerization with a mass half that size, it will be interesting to learn from more careful structural studies of this class of enzymes how the complex polymerization reaction is performed.

4. Mitochondrial Polymerases[33,34]

Mitochondria contain a small, circular, duplex DNA not found elsewhere in the cell. It codes for the 16S and 23S RNA of the mitochondrial ribosome[35] and for a limited number of transfer RNAs

27. Wang, T. S.-F., Sedwick, W. D. and Korn, D. (1974) *JBC* **249**, 841.
28. Weissbach, A., Schlabach, A., Fridlender, B. and Bolden, A. (1971) *NNB* **231**, 167.
29. Baril, E. F., Brown, O. E., Jenkins, M. D. and Laszlo, J. (1971) *B.* **10**, 1981.
30. Chang, L. M. S. and Bollum, F. J. (1972) *B.* **11**, 1264; (1973) *JBC* **248**, 3398; Chang, L. M. S. (1973) *JBC* **248**, 3789.
31. Baril, E. and Laszlo, J. (1971) *Adv. Enz. Regul.* **9**, 183, Pergamon Press, N.Y.
32. Chang, L. M. S., Brown, McK. and Bollum, F. J. (1973) *JMB* **74**, 1.
33. Meyer, R.R. and Simpson, M. V. (1970) *JBC* **245**, 3426.
34. Tibbetts, C. J. B. and Vinograd, J. (1973) *JBC* **248**, 3367, 3380; Fry, M. and Weissbach, A. (1973) *B.* **12**, 3602.
35. Ojala, D. and Attardi, G. (1972) *JMB* **65**, 273; Lederman, M. and Attardi, G. (1973) *JMB* **78**, 275.

which may prove to be the most important for synthesis of hydrophobic mitochondrial proteins. Because of its limited size, the DNA can code for only a few of the many proteins found in mitochondria[36].

DNA synthesis in mitochondria of rat liver, yeast, and the slime mold, *Physarum polycephalum,* occurs in a different time pattern from the cycle of DNA synthesis in the nucleus. Intact mitochondria in vitro incorporate DNA precursors by a mechanism resembling that of the DNA polymerases of bacteria and the small and large eukaryotic polymerases.

Rat liver[37], cultures of HeLa cells[38,39] (cells that stem from a human cervical cancer), and yeast[40] have been sources for mitochondrial polymerase. The rat liver enzyme has been purified 3500-fold but its low yield and high lability have delayed full purification and characterization. Rat liver mitochondrial polymerase is solubilized from the mitochondrial membranes by strong salt (1 M NaCl) or the detergent deoxycholate (0.5 percent). Its subsequent behavior on ion-exchange adsorbents (DEAE-cellulose and hydroxylapatite) separates and distinguishes it from nuclear and cytoplasmic polymerases, as does its effective utilization of denatured DNA as template-primer. Another feature unique to the rat liver mitochondrial enzyme is its strong stimulation by salt; KCl and NaCl, at 0.15 M, increase the rate tenfold, an effect resembling that seen with the T5-phage-induced polymerase (Table 5-5). A slight stimulation has been observed with the polymerase from calf liver or HeLa cell mitochondria, but none with the enzyme from yeast mitochondria nor with any of the bacterial or mammalian polymerases.

Two other features of liver mitochondrial polymerase noted in early investigations deserve comment. A strong preference for mitochondrial DNA as template-primer observed in the crude preparations has not been seen with the partially purified mitochondrial enzyme; and a special sensitivity of the mitochondrial enzyme of rat liver to inhibition by the mutagenic dyes ethidium bromide and acriflavin has now been observed with the large polymerase of human KB cells[41] as well.

Purification of the mitochondrial polymerase from HeLa cell cultures demonstrates it to be one of _four_ distinctive DNA polymerases isolated from this single cell line[42]. It is distinct from the large cytoplasmic and the small nuclear DNA polymerases and the RNA-directed enzyme described in Section 7 of this chapter. Compared to the cytoplasmic DNA polymerase (Table 6-5), the mitochondrial enzyme is smaller and has several other distinguishing features. The most striking of these is a relatively poor utilization

36. Ojala, D. and Attardi, G. (1972) *JMB* **65**. 273; Lederman, M. and Attardi, G. (1973) *JMB* **78**, 275.
37. Meyer, R. R. and Simpson, M. V. (1970) *JBC* **245**, 3426.
38. Tibbetts, C. J. B. and Vinograd, J. (1973) *JBC* **248**, 3367, 3380.
39. Fry, M. and Weissbach, A. (1973) *B*. **12**, 3602.
40. Wintersberger, U. and Wintersberger, E. (1970) *EJB* **13**, 20.
41. Korn, D, personal communication.
42. Op. cit. in footnote 39.

TABLE 6-5
Mitochondrial and cytoplasmic DNA polymerases of HeLa cells

	Mitochondrial	Large cytoplasmic
Molecular weight	106,000	250,000
Influence of salt	Slight stimulation	No effect, or inhibition
Inhibition by — SH-blocking agents	No	Yes
Template-primer:		
Salmon sperm DNA, activated by nicking	(100)	(100)
Salmon sperm DNA, native	7	0.2
Salmon sperm DNA, denatured	0	1.5
Mitochondrial DNA	12	0.0
$(dG)_{12}$·poly dC^a	6	600.
Endonuclease activity	Yes	No
Heat inactivation, 45° for 20 min.	98 percent	20 percent
ATP inhibition, 2.5 mM	13 percent	96 percent

^{a}The assay conditions here differ from those in Table 6-3.

of $(dG)_{12}$·poly dC, one of the most favored template-primers for the other DNA polymerases isolated from HeLa cells. The mitochondrial enzyme preparation was able to copy duplex DNA, including circular rat liver mitochondrial DNA by virtue of an endonuclease activity present in minute but significant amounts. Unlike the mitochondrial DNA polymerase obtained from rat liver, the HeLa cell enzyme did not show striking stimulation by salt nor preference for a denatured template-primer.

A 40-fold increase in the level of mitochondrial DNA polymerase was observed when the protozoan *Tetrahymena pyriformis* was subjected to irradiation with ultraviolet light or X-rays, to thymine starvation, or to ethidium bromide[43]. Damage to DNA by these agents can be corrected by excision repair (Chapter 9, Section 2). Induction of the polymerase following exposure to DNA-damaging agents requires RNA and protein synthesis. Inhibition of polymerase induction by cycloheximide, an inhibitor of nonmitochondrial protein synthesis, and not by chloramphenicol, an inhibitor of mitochondrial protein synthesis, indicates nuclear direction of translation of the enzyme. The principal polymerase activity of normal cells does not change in amount with induction of the mitochondrial enzyme. Upon induction, the mitochondrial enzyme becomes the predominant polymerase in cell extracts rather than the minor one. The induced mitochondrial enzyme has been purified and characterized and found to differ sharply from the nonmitochondrial enzyme in physical and functional properties (Table 6-6). Of special interest in this eukaryotic system is how DNA damage leads specifically to the induction of the gene that codes specifically for a mitochondrial DNA polymerase.

43. Westergaard, O., Marcker, K. A. and Keiding, J. (1970) *Nat.* **227**, 708; Westergaard, O. (1970) *BBA* **213**, 36.

TABLE 6-6
Tetrahymena pyriformis polymerases

Properties	Mitochondrial polymerase	Principal polymerase
Cellular level after DNA damage	40-fold increase	No change
Optimal Mg^{2+}, μM	55	5
Template-primer	Denatured DNA	Native DNA
NaCl, 0.2 M	Stimulates	Inhibits
Duration of synthesis	Sustained	Brief
Size	Small	Large

5. Virus-induced Polymerases[44,45]

DNA viruses that infect animal cells vary widely, as do bacterial viruses, in size and consequently in genetic content. The very small tumor viruses, such as polyoma and SV40, contain only about 5000 base pairs and must rely almost entirely on host replication systems for their multiplication. The very large viruses, such as herpes and vaccinia, contain about 200,000 base pairs. As with the T-even phages of comparable size, they very likely contain many genes that code for DNA replication proteins and are specific for the multiplication of viral DNA. Detailed information will be needed about the host DNA polymerases that are exploited, and the viral DNA polymerases that are induced, in order to understand viral reproduction and transformation of cells to tumor growth. Moreover, this information may provide incisive probes into the mechanism of replication of the cell's own chromosome. In this connection, the discovery that herpes virus multiplies in the nucleus and that vaccinia (pox) virus is assembled in the cytoplasm may also offer important clues to the subcellular distribution of host replication processes.

Polyoma and SV40 multiply in the nucleus, and from one to several copies of their DNA are assimilated into the host chromosome. Details of how they are replicated (Chapter 8, Section 9) and how they recombine with host DNA are not accurately known and the specific steps have yet to be defined. There is no evidence, as in certain phage infections, for induction of a new DNA polymerase.

Herpes simplex, whose natural host cells are in the mucous membranes of the human eye, mouth, and genitalia, causes either an abortive, latent infection or a productive, virulent one. Replication of this virus in the nucleus is not restricted, as with some viruses, to the DNA-synthesis phase (S phase) of the cell cycle[46] (Fig. 6-4), implying that the virus can initiate replication of its DNA under its own control. Onset of viral DNA replication is preceded by inhibition

44. Keir, H. M. (1968) *Symposia of Society of General Microbiology*, **18**, 67.
45. Weissbach, A., Hong S.-C. L., Aucker, J. and Muller, R. (1973) *JBC* **248**, 6270.
46. Cohen, G. H., Vaughan, R. K. and Lawrence, W. C. (1971) *J. Virol.* **7**, 783.

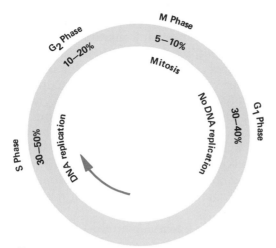

FIGURE 6-4
Cell growth cycle showing the DNA replication S
phase.

of cellular DNA synthesis and by massive increases in the levels of
several enzymes, including DNA polymerase, DNase, thymidine
kinase, and deoxycytidine kinase[47]. Each of these highly increased
enzyme activities displays properties distinct from those of the
enzymes of the uninfected cells, and immunological evidence
suggests that the information for their synthesis is indeed coded in
the viral genome[48].

The herpes-induced DNase can be separated from the predominant cellular DNase by DEAE-cellulose chromatography[49]. It acts
exonucleolytically on both native and denatured DNA releasing
5'-mononucleotides mainly from the 3' terminus, in contrast to the
endonucleolytic action of the cellular DNase.

The virus-induced DNA polymerase is in the size range of the
large cellular DNA polymerase but can be distinguished from the
large and small polymerases in these ways: (i) activation by high salt
concentrations (0.2 M KCl), (ii) higher affinity for DEAE-cellulose
and lower affinity for hydroxylapatite, (iii) susceptibility to herpes
antiserum, and (iv) a distinctive associated exonuclease. The homopolymer pair poly $dC \cdot dG_{12}$ is by far the best template-primer[50].
Viral exonuclease remains associated with viral polymerase through
all purification steps, although a peak of polymerase-free exonuclease is obtained from hydroxylapatite chromatography[51]. This
association may be due to the presence of polymerase and one of the
exonucleases within the same polypeptide chain, or to an association
of subunits such that the polymerase subunit is active only when the
exonuclease subunit is bound to it.

47. Perera, P. A. J. and Morrison, J. M. (1970) *BJ* **117**, 21P; Hay, J., Moss, H. and Halliburton, I. W.
(1971) *BJ* **124**, 64.
48. Keir, H. M. (1968) *Symp. Soc. Gen. Microbiol.*, **18**, 67.
49. Morrison, J. M. and Keir, H. M. (1968) *J. Gen. Virol.* **3**, 337.
50. Weissbach, A., Hong S.-C. L., Aucker, J. and Muller, R. (1973) *JBC* **248**, 6270.
51. Paton, R. D. and Morrison, J. M. (1969) *BJ* **144**, 39P; Morrison, J. M. (personal communication).

Cellular DNA polymerases were not detected in the chromatograms of infected cells, not even when a mixture of control and infected cells was used. This implies that specific inhibition of cellular polymerases can take place, which may explain the inhibition of cellular DNA synthesis after virus infection. The interaction of cellular polymerases, viral polymerases, and exonucleases, and other herpes-induced, DNA-related enzymes may form a system that serves as a useful model for eukaryotic DNA replication. There is evidence of many copies of a herpes-type virus (Epstein-Barr) in the cellular DNA of Burkitt's lymphoma, a human cancer[52]. Incorporation of the DNA of the viral genome into the host chromosome may prove to be a highly significant factor in directing mitotic activity to a proliferative state[53].

Although much less is known about the DNA polymerase induced by vaccinia virus than the herpes enzyme, it also appears to have an associated exonuclease activity[54]. This nuclease, or a similar activity found in the virus particles, has been implicated in the prompt inhibition of DNA synthesis in the infected cell[55].

6. RNA-directed Polymerases — Viral Reverse Transcriptases[56-60]

RNA tumor viruses have been of long-standing biologic interest for various reasons. They produce cancer in animals; they multiply in cells without interrupting cell division; and they have genetic information both for virus production and cell transformation. This inherited information is called the _provirus_ and is now thought to be contained in the chromosome as a DNA copy of the RNA genome (Fig. 6-5). This DNA copy is derived by transcription of viral RNA into DNA by a unique kind of DNA polymerase contained in the virus[61].

These RNA viruses produce muscle cancer in chickens (Rous sarcoma) and monkeys, leukemia in birds (avian myeloblastosis, AMV), mice (Rauscher leukemia), and cats, and mammary tumors in mice and monkeys. The RNA genome is a large, segmented complex of about 10 million daltons (70S). It is composed of four medium-size units of about 2.5 million daltons (35S) and smaller RNA segments hydrogen-bonded in a structure that is partly duplex but, at the time of this writing, is still largely undefined. The 35S RNA

52. Wolf, H., zur Hausen, H. and Becker, V. (1973) NNB **244**, 245.
53. Tooze, J. (ed.) (1973) The Molecular Biology of Tumour Viruses, CSHL, p. 470.
54. Citarella, R. V., Muller, R., Schlabach, A. and Wiessbach, A. (1972) J. Virol. **10**, 721.
55. Pogo, B. G. T. and Dales, S. (1973) PNAS **70**, 1726.
56. Temin, H. M. and Baltimore, D. (1972) Adv. Virus Res. **17**, 129.
57. Sarin, P. S. and Gallo, R. C. (1973) **in** Intl. Rev. Sci. Series in Biochem. (K. Burton, ed.) Vol. 6 Ch. 8, Butterworth and Med. Tech. Publish., Oxford.
58. Bishop, J. M. (1973) **in** DNA Synthesis in vitro, (R. D. Wells and R. B. Inman eds.) Steenbock Symposium. Univ. Park Press, Baltimore, Md., p. 341.
59. Green, M. (1973) Prog. N. A. Res. and Mol. Biol. **13**.
60. Op. cit. in footnote 53, pp. 502–699.
61. Op. cit. in footnote 56.

165

SECTION 6:
RNA-directed
Polymerases—
Viral Reverse
Transcriptases

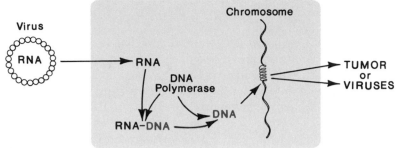

FIGURE 6-5
Scheme illustrating the DNA copy of an RNA tumor virus genome integrated
into the host chromosome as a "provirus".

segments are considered to be intact subunits unique to the virus,
whereas twelve to twenty 4S molecules and three or four 5S molecules resemble a sampling of cellular tRNA and the 5S RNA. Tracts
of poly A (about 200 residues) terminate the 3′ ends of the 35S
subunits.

In addition to RNA and the RNA-directed DNA polymerases,
many other enzymes have been found in the virus preparation; these
in most instances seem to be cellular contaminants. A ribonuclease
H, distinctive in its capacity to degrade the RNA of an RNA-DNA
duplex, appears to be part of the polymerase and will be mentioned
later in this section. Replication of the RNA passes through a
RNA-DNA hybrid stage before duplex DNA is produced. Inasmuch
as the organization of the viral RNA is so poorly understood, most
aspects of the replication events must also remain in doubt. However, several details worked out in recent studies are of considerable
interest.

Replication in vitro is primed by certain of the 4S components (of
cellular tRNA origin) found in the virus. Among the replication
products these 4S units, terminated by an adenylate residue, have
been found covalently linked to deoxyadenylate in each of several
avian viruses examined. The sequence of the nascent DNA was
unique for at least the first eight residues[62]:

$$-\underbrace{pApApAp}_{\text{primer RNA}}\underbrace{TpGpApApGpC}_{\text{nascent DNA}}-$$

The early products appear as single strands, 5–8S in size in alkaline
velocity gradients, and a large fraction renatures readily as hairpin
structures. Yet virtually the entire 70S genome is replicated[63].
Removal of RNA from the DNA-RNA hybrid, presumably by the
endogenous ribonuclease H, provides the DNA template required
for synthesis of duplex DNA in the later stages. In these duplex

62. Taylor, J. M., Faras, A. J., Varmus, H. E., Goodman, H. M., Levinson, W. E. and Bishop, J. M.
    (1973) B. 12, 460; Taylor, J. M., Garfin, D. E., Levinson, W. E., Bishop, J. M. and Goodman,
    H. M. (1974) B. 13, 3159.
63. Faras, A. J., Taylor, J. M., McDonnell, J. P., Levinson, W. E. and Bishop, J. M. (1972) B. 11,
    2334; and succeeding paper.

products a considerable amount of the DNA represents reiteration of certain viral sequences.

The viral, RNA-directed DNA polymerase has been purified from a variety of RNA tumor viruses even though the exceedingly small amounts of source material available has made the isolation of a homogeneous enzyme difficult. The enzymes from mouse, cat, and monkey viruses have been reported to be a single polypeptide of 70,000 daltons. The enzymes isolated from chicken viruses (Rous sarcoma)[64,65], and AMV[66,67,68,69], appear to be principally near 160,000 daltons in size and composed of subunits 65,000 and 105,000. The smaller subunit possesses the DNA polymerase and RNase H activities, whereas the larger one has no known function. Immunological cross-reaction[70] was found among the mouse, cat, and monkey enzymes on the one hand, and the enzymes from chicken on the other. Neither group cross-reacted with enzymes from mouse and simian mammary viruses.

The viral polymerases resemble all the known DNA polymerases in the $5' \rightarrow 3'$ direction of synthesis, and in requiring a template, a 3'-hydroxyl primer terminus, the four deoxyribonucleoside triphosphates, and Mg^{2+}. The data in Table 6-7[71] show the need for template and primer and the effectiveness of homopolymers, especially the ribopolymers. There is a sharp preference for poly rA·oligo dT over poly dA·oligo dT. Surprisingly, the ribopolymer poly rU was ineffective whereas the deoxypolymer poly dC was utilized. The results in Table 6-8[72] demonstrate the capacity of the enzyme to utilize natural

TABLE 6-7
Utilization of homopolymer template-primers[a]
by RNA-directed viral polymerase

Primer	Templates and relative rates			
	polyribo		polydeoxy	
dT_{10}	rA	100	dA	1
dA_{14}	rU	1	dT	1
dG_{12}	rC	50	dC	25
dC_{14}	rI	25	dI	1

[a]No activity was observed with any primer alone or with any template alone.

TABLE 6-8
Utilization of natural template-primers by RNA-directed polymerase

	$(dT)_{12-18}$	
	Absent	Present
Poly rA	<1	(100)
70S RNA	6	25
70S RNA, denatured	<1	17
HeLa rRNA	<1	<1
Globin mRNA	<1	108
Ovalbumin mRNA	<1	60

64. Faras, A. J., Taylor, J. M., McDonnell, J. P., Levinson, W. E. and Bishop, J. M. (1972) B. 11, 2334; and succeeding paper.
65. Garapin, A. C., Varmus, H. E., Faras, A. J., Levinson, W. E. and Bishop, J. M. (1973) Virol. 52, 264.
66. Hurwitz, J. and Leis, J. (1972) J. Virol. 9, 116; and succeeding paper.
67. Flugel, R. M. and Wells, R. D. (1972) Virol. 48, 394.
68. Canaani, E. and Duesberg, P. (1972) J. Virol. 10, 23.
69. Grandgenett, D. P., Gerard, G. F. and Green, M. (1973) PNAS 70, 230.
70. Scolnick, E. M., Parks, W. P., Todara, G. J. and Aaronson, S. A. (1972) NNB 235, 35. Nowinski, R. C., Watson, K. F., Yaniv, A. and Spiegelman, S. (1972) J. Virol. 10, 959. Parks, W. P., Scolnick, E. M., Ross, J., Todaro, G. J. and Aaronson, S. A. (1972) J. Virol. 9, 110.
71. Baltimore, D. and Smoler, D. (1971) PNAS 68, 1507.
72. Bishop, J. M., personal communication.

167

SECTION 6:
RNA-directed
Polymerases—
Viral Reverse
Transcriptases

RNAs as templates provided an oligomer of dT is present to serve as primer. Even the excellent 70S RNA template-primer was several times more effective, and various mRNAs proved to be excellent templates when $(dT)_{12-18}$ was present. The known presence of a polyadenylate tail at the 3' end of mRNA and its occurrence at the 3' end of the 4S units in the 70S RNA complex very likely explain this primer function of a dT oligomer.

The viral enzyme bears some resemblance to the DNA polymerases II and III of *E. coli* in a failure to use nicked DNA duplexes and in a preference for a DNA template-primer with short single-strand gaps. The enzyme fills such gaps completely as judged by the capacity of ligase to seal the chains.

The viral polymerase is most clearly distinguished from the large and small cellular DNA-dependent polymerases by its capacity to utilize ribopolymers, particularly poly $rC \cdot dG_{12}$, under standard conditions[73]. It is unique among all DNA polymerases in two respects: (i) the capacity to utilize the 70S viral RNA complex, thereby producing a DNA product that hybridizes specifically and completely with this RNA, and (ii) possession of an inherent ribonuclease H activity[74,75]. This RNase H is an exonuclease which specifically degrades the RNA of a RNA-DNA hybrid. It does so from either the 5' → 3' or the 3' → 5' direction. The oligonucleotide products are two to eight residues long, each terminated in a 3'-hydroxyl and 5'-phosphate. The enzyme is considered an exonuclease, unlike an RNase H activity of animal cells, because it fails to act on RNA linkages in a covalently closed duplex circle.

The polymerase activity can be measured in virus particles rendered permeable to substrates by damaging the lipoprotein envelope with nonionic detergents or diethyl ether. It is remarkable how similar the polymerase activities of the virus particle and the activity of the purified enzyme are, even in the replication of the complicated 70S viral RNA. Nevertheless it should be evident that differences in replication among the related RNA viruses, and the progress of infection in various cells, may depend upon still unrecognized variations in enzyme patterns among the viruses and the cells they infect.

The capacity of viral DNA polymerase to replicate 70S viral RNA and the general template preference it displays among the synthetic homopolymers for a _ribo_ over a _deoxy_ chain distinguish this enzyme from other DNA polymerases. The viral enzyme is therefore called an RNA-directed DNA polymerase or even a reverse transcriptase. Yet, the enzyme in the virus particle does copy DNA as well as RNA templates and produces a duplex DNA product, as does the highly purified enzyme. It should also be recalled that the classic DNA-directed DNA polymerases can utilize ribohomopolymer templates, sometimes even preferentially over deoxypolymers (see Chapter 4,

73. Flugel, R. M. and Wells, R. D. (1972) *Virol.* **48**, 394.
74. Baltimore, D. and Smoler, D. (1972) *JBC* **247**, 7282.
75. Leis, J. P., Berkower, I. and Hurwitz, J. (1973) *PNAS* **70**, 466.

Section 16). Preferences among secondary structures of DNA templates vary among different forms of the AMV polymerase. Certain preparations are more active than others in utilizing nicked DNA and in repairing lengthy gaps in DNA[76]. Conceivably, when the behavior of RNA-directed viral polymerases under a variety of conditions is better understood, the distinctions between them and the DNA-directed enzymes will appear to be even less sharp than they seem now.

Since animal cells infected with tumor viruses contain the RNA-directed polymerase in association with a 70S RNA, a similar activity has been sought in human leukemic cells where the viral etiology of leukemia is still uncertain. Evidence has been obtained for a DNA polymerase with the distinctive features of the viral enzyme[77]: (i) location in a particulate fraction associated with a 70S RNA complex, (ii) capacity to replicate the RNA of the complex, (iii) synthesis of a DNA product that sediments with the complex (but sediments in alkali at 5S) and, (iv) a product that hybridizes specifically with RNA of the Rauscher leukemia virus. This enzymatic activity has been reported in leukemic cells of patients with acute or chronic leukemia and in brain tumors[78]. A reverse transcriptase isolated from a patient with acute leukemia had the immunological as well as biochemical features of the enzymes found in primate and mouse leukemia viruses[79].

Do animal cells without any history of viral infection possess an RNA-directed DNA polymerase? Reasons for expecting that such an enzyme might exist and the evidence for it are presented in the next section.

7. RNA-directed Polymerases– Cellular Reverse Transcriptases[80–82]

Following the discovery of the RNA-directed DNA polymerase in the tumor-causing RNA viruses, a similar enzyme was sought in human neoplastic cells not known to harbor these viruses. As described in the preceding section, such an enzyme was found in leukemic cells. These revelations of active systems for RNA-directed DNA synthesis suggested that conversion of RNA messages to DNA

76. Leis, J. and Hurwitz, J. (1972) PNAS **69**, 2331.
77. Kufe, D., Hehlmann, R. and Spiegelman, S. (1973) PNAS **70**, 5.
78. Baxt, W., Hehlmann, R. and Spiegelman, S. (1972) NNB **240**, 72; Cuatico, W., Cho, J.-R. and Spiegelman, S. (1973) PNAS **70**, 2789.
79. Todaro, G. J. and Gallo, R. C. (1973) Nat. **244**, 206.
80. Sarin, P. S. and Gallo, R. C. (1973) in Intl. Rev. Sci. Series in Biochem. (K. Burton, ed.) Vol. 6 Ch. 8, Butterworth and Med. Tech. Publish., Oxford.
81. Scolnick, E. M., Aaronson, S. A., Todaro, G. J. and Parks, W. P. (1971) Nat. **229**, 318.
82. Fridlender, B., Fry, M., Bolden, A. and Weissbach, A. (1972) PNAS **69**, 452; Bolden, A., Fry, M., Muller, R., Citarella, R. and Weissbach, A. (1972) Arch. B.B. **153**, 26.

occurs in normal cellular processes, as for example in differentiation, gene amplification, or in immunogenic responses. However, the lack of definitive basic information about the various kinds of _DNA-directed_ polymerases in animal cells made it very difficult, until recently, to establish the existence of a distinctive RNA-directed enzyme. Once progress had been made in the purification and characterization of the several kinds of DNA-directed DNA polymerases found in a single cell or tissue, it became clear that normal animal cells also have DNA polymerases that may be categorized as RNA-directed.

Such enzymes have been fractionated from extracts of animal tissues[83] and cultured HeLa cells[84]. The RNA-directed polymerase has template preferences, physical properties, and other features sharply different from DNA polymerases isolated from the nuclei, cytoplasm, and mitochondria of these cells[85,86]. The enzyme does resemble the DNA-directed polymerases in all of its basic requirements, including a template-primer preference for natural DNA over RNA. The major distinction rests in an excellent utilization of ribohomopolymer templates such as polyadenylate.

The cellular RNA-directed polymerase appears to differ from the viral enzyme in its preference for Mn^{2+} and in a failure to use natural RNAs as templates. One further distinction may emerge in the nature of the associated RNase H activity. That of the virus behaves as an exonuclease and appears to be physically inseparable from the polymerase, whereas the cellular RNase H behaves like an endonuclease and its interaction with the polymerase remains to be determined.

A DNA polymerase that appears to be strangely different from both the known RNA-directed and DNA-directed enzymes has been found in cytoplasmic fractions of "fresh" cells obtained directly from normal chick embryos[87,88], and from human blood lymphocytes[89] stimulated to synthesize DNA by certain plant proteins (phytohemagglutinins). Synthesis in crude extracts is prevented by RNase, thus implicating endogenous RNA either as a template or primer. However, upon being purified, the enzyme has the properties of a DNA-directed polymerase, preferring DNA to DNA-RNA hybrid synthetic polymers, and failing to transcribe either the 70S viral RNA complex or naturally occurring single-stranded RNA.

169

SECTION 7:
RNA-directed
Polymerases—
Cellular Reverse
Transcriptases

83. Scolnick, E. M., Aaronson, S. A., Todaro, G. J. and Parks, W. P. (1971) _Nat._ **229**, 318.
84. Fridlender, B., Fry, M., Bolden, A. and Weissbach, A. (1972) _PNAS_ **69**, 452; Bolden, A., Fry, M., Muller, R., Citarella, R. and Weissbach, A. (1972) _Arch. B.B._ **153**, 26.
85. Sarin, P. S. and Gallo, R. C. (1973) **in** _Intl. Rev. Sci. Series in Biochem._ (K. Burton, ed.) Vol. 6 Ch. 8, Butterworth and Med. Tech. Publish., Oxford.
86. Op. cit. in footnote 84.
87. Stavrianopoulos, J. G., Karkas, J. D. and Chargaff, E. (1972) _PNAS_ **69**, 2609.
88. Kang, C.-Y. and Temin, H. M. (1972) _PNAS_ **69**, 1550; (1973) _NNB_ **242**, 206.
89. Bobrow, S. N., Smith, R. G., Reitz, M. S. and Gallo, R. C. (1972) _PNAS_ **69**, 3228.

8. Terminal Nucleotidyl Transferase[90, 91]

This enzyme, detected so far only in extracts of the thymus gland[92] and in plants[93], behaves like a DNA polymerase in synthesizing a DNA chain by $5' \rightarrow 3'$ polymerization of 5'-deoxynucleoside triphosphates. Unlike a DNA polymerase the enzyme neither requires nor responds to a template and may even dispense with a primer. The usefulness of the enzyme to the cell is still a mystery, although the role of the thymus cells in antibody synthesis suggests some novel genetic chemistry in which the enzyme could conceivably participate. However, it is for its usefulness to the biochemist in the synthesis of defined homopolymers that the enzyme deserves our attention. In this regard, it ranks with polynucleotide phosphorylase, whose random assembly of ribonucleotides has been so important in the synthesis of homo- and mixed RNA polymers.

The enzyme has been found in both nuclear and cytoplasmic fractions. The solubilized enzyme from cytoplasm differs from the nuclear in its inability to incorporate ribonucleotides into DNA and in a preference for dATP among the triphosphates. The cytoplasmic enzyme has been purified and studied in great detail. It is a small basic protein of molecular weight 33,000, and is dissociable into two subunits of 8,000 and 26,000.

The most remarkable feature in the procedure for purification of the cytoplasmic enzyme was the relative absence of enzyme activity when crude thymus fractions were assayed. There was a twofold increase in activity after extracts were concentrated with ammonium sulfate and a 25-fold greater total activity (referred to the crude extract) among fractions collected from a sucrose gradient. Failure to assay the transferase in the crude extract could not be due simply to inhibition by nucleases since the DNA polymerase activity in the extract could be assayed readily. Possibly, the enzyme in the crude extract was part of a larger complex in which its activity was so modified as to be virtually inactive when measured by the standard incorporation assay.

In phosphate buffer, the enzyme was shown to require a primer, at least as large as a trinucleotide[94]. Such a trinucleotide required not only a free 3'-hydroxyl group to serve as a primer terminus for extension, but a free 5'-phosphate as well: d(pApApA) and d(pppTpTpT) were primers, but d(ApApA) was not. In cacodylate buffer, polymerization took place without a primer after some delay. The rate and extent of incorporation of nucleotides was greatly influenced by buffer, metal ion, and enzyme concentrations. All deoxynucleoside 5'-triphosphates, including a variety of unnatural

90. Bollum, F. J. (1962) *JBC* **237**, 1945.
91. Chang, L. M. S. and Bollum, F. J. (1971) *JBC* **246**, 909.
92. Chang, L. M. S. (1971) *BBRC* **44**, 124.
93. Sahai Srivastava, B. I. (1972) *BBRC* **48**, 270.
94. Hayes, F. N., Mitchell, V. E., Ratliff, R. L., Schwartz, A. W. and Williams, D. L. (1966) *B.* **5**, 3625.

bases, were polymerized, with varying V_{max} and K_m values. With a primer present, the polymer lengths were 50 to 500 residues.

The enzyme has been used in many ways. As one example, block copolymers have been made in which a homopolymer chain of one nucleotide is followed by one or more stretches of another kind of nucleotide. In this way, the polymer $d(pT)_6(pG)_{10}(pA)_{313}$ was obtained[95]. In other studies, cellulose to which a short oligomer is attached in ester linkage through its 5'-phosphate was extended by terminal transferase and then used successfully for assay of ligase and studies of its mechanism[96] (Fig. 6-6) The same principle has been applied to DNA polymerase studies[97] and for the assay of a nicking enzyme (endonuclease). The use of terminal transferase in the artificial assembly of genes into a chromosome will be discussed later (Chapter 11, Section 5).

It has also been possible by careful adjustment of conditions to add one or a very few nucleotide residues, in random distribution, to the end of a synthetic polymer or natural DNA. At low ratios of

95. Ratliff, R. L. and Hayes, F. N. (1967) BBA 134, 203.
96. Cozzarelli, N. R., Melechen, N. E., Jovin, T. M. and Kornberg, A. (1967) BBRC 28, 578.
97. Jovin, T. M. and Kornberg, A. (1968) JBC 243, 250.

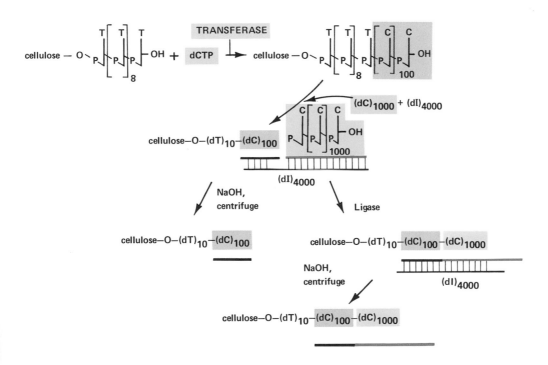

FIGURE 6-6
Scheme illustrating how oligomers linked to cellulose by terminal transferase can be used for detection and assay of DNA ligase. Covalent linkage of the $(dC)_{1000}$ polymer to the oligomer associated with cellulose was readily monitored in the cellulose pellet obtained by sedimenting the ligase incubation mixture in alkali.

deoxynucleoside triphosphate to primer, the distribution of chain lengths is determined by the ratio. In this way, a duplex homopolymer with one of the 3' primer termini containing a mismatched base (i.e. $d(A)_{4000} \cdot d(T)_{200}d(C)_1$) has been made and used to demonstrate the _proofreading_ function of the $3' \rightarrow 5'$ exonuclease of _E. coli_ DNA polymerase I (Chapter 4, Section 9).

An important requirement of terminal transferase is that the primer be single-stranded. Inert duplex DNA was converted by phage λ exonuclease (Table 9-1) to an active primer, by a $5' \rightarrow 3'$ exonucleolytic action that created a single-strand 3' terminus. This maneuver was an essential step in the genetic engineering of DNA duplexes into novel chromosomal arrangements (Chapter 11, Section 5). Also illustrative of the need for a single-stranded primer was the failure of terminal transferase to synthesize poly dG chains, which tend to form duplex structures. This difficulty was readily overcome by using N-acetyl dGTP, which could be polymerized to chains 300 residues in length; the acetyl groups could later be removed from the guanine residues with alkali[98].

---

98. Lefler, C. F. and Bollum, F. J. (1969) _JBC_ **244**, 594.

1. Basic Rules of Replication[1-6]

What are now generally agreed to be the basic principles of DNA replication will surely be altered, even in important details. Yet some of the patterns now perceived are at least a useful basis for future work; these will be listed here and expanded in subsequent sections.

I REPLICATION IS A SEMICONSERVATIVE PROCESS
Each strand of the parental DNA duplex is conserved and copied by base-pairing with matching nucleotides and the result is two

1. *Replication of DNA in Micro-organisms* (1968) CSHS **33**.
2. Richardson, C. C. (1969) ARB **38**, 795.
3. Goulian, M. (1971) ARB **40**, 855.
4. Klein, A. and Bonhoeffer, F. (1972) ARB **41**, 301.
5. Wells, R. D. and Inman, R. B. (eds) (1973) *DNA Synthesis In Vitro*, Steenbock Symposium Univ. Park Press, Baltimore.
6. Gross, J. D. (1971) **in** "Current Topics in Microbiology and Immunology," **57**, 39.

duplexes identical to the parental one. Base-pairing and chain extension create a growing point, or fork, at which the parental duplex is unwound (Fig. 7-1). To relieve the torque of unwinding large rod-like

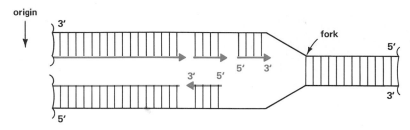

FIGURE 7-1
Diagram of a unidirectional replicating fork.

or circular elements of the chromosome, a swivel in the duplex is needed. The mechanism of this swivel can only be suggested at present. For instance, an endonuclease might introduce a break in one strand of the duplex, allowing free rotation about a phosphodiester bond in the other strand to relieve the tension; a joining enzyme (ligase) might later reseal the bond. Alternatively a single enzyme might accomplish both the break and reunion, conserving the energy of the phosphodiester linkage (see Section 9 of this chapter).

II REPLICATION HAS DIRECTION

Replication proceeds from a given point on the chromosome either in one or in both directions. Bidirectional movement involving two growing points or forks appears to be the most common (Fig. 7-2).

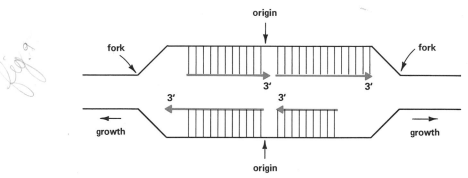

FIGURE 7-2
Diagram of a bidirectional replicating fork.

The movement of two forks away from each other creates a loop which appears as a "bubble" or "eye" in electron micrographs. It does not follow, however, that the speed or extent of movement of the two forks is the same in both directions.

III REPLICATION STARTS AT A UNIQUE POINT ON BACTERIAL AND VIRAL CHROMOSOMES

By starting another copy at a unique point, called the origin, before the preceding one is completed, the chromosome of a rapidly growing bacterium may have several forks and increase its rate of replication. By contrast, replication in eukaryotic chromosomes begins at many origins located at different positions along the chromosomal DNA. As a result of the large number of simultaneously active forks, the overall rate of replication in eukaryotic chromosomes may be faster than in bacteria, even though the rate of movement of the replicating fork may be an order of magnitude slower.

IV REPLICATION OF BOTH STRANDS PROCEEDS BY THE ADDITION OF NUCLEOTIDE MONOMERS IN THE $5' \longrightarrow 3'$ DIRECTION

The addition of nucleotide monomers in the $5' \longrightarrow 3'$ direction proceeds in accordance with the requirements of DNA polymerases (Chapter 3, Section 4; Fig. 7-1). Consequently, replication at the monomer level of nucleotides cannot occur simultaneously along both antiparallel strands in the region of the fork; a transient single-stranded region is created on one side of the duplex while the other strand is being replicated. Although replication appears to be simultaneous on both strands when measured autoradiographically or by genetic techniques, these methods resolve the process only at the micron level, whereas the polymerization of nucleotide monomers is measured at the angstrom level. Electron microscopy with a resolving power of about 100 nucleotides has clearly demonstrated the existence of single-stranded regions on one side of the growing fork.

V REPLICATION OCCURS IN SHORT, DISCONTINUOUS PULSES

The short fragments, about 100 nucleotides long in animal cells and 1000 to 2000 in prokaryotes, are later joined to the main body of the growing chains.

VI REPLICATION AT THE LEVEL OF SHORT FRAGMENTS IS INITIATED BY THE PRODUCTION OF A SHORT SEGMENT OF RNA TO SERVE AS A PRIMER FOR DNA POLYMERASE

This RNA region is later excised, and the gap thus created is filled with DNA. The nascent pieces, or replication fragments of DNA, are then joined to the growing chromosome.

An important distinction needs to be made between the *origin* of replication seen as a fork or eye at one or several points on the chromosome and the actual initiation of DNA chains. Very little is known, and can therefore be discussed, about the factors that create the origin and set replication in motion. Nor can more be said about the details of the termination of replication, as for example, when two replicating forks moving in opposite directions on a circular chromosome approach each other at a point that will signal a division into the progeny chromosomes.

2. Origin and Direction of the Replicating Fork[7-9]

The replicating chromosome has been observed in a variety of ways and in several organisms. From these observations emerges a basically similar pattern with interesting variations.

One of the first insights into the structure of the replicating chromosome came from the use of autoradiography[10]. In this technique, emissions from (^{3}H)-thymidine-labeled DNA produce a grainy image of the labeled structure on a photosensitive emulsion. When the entire circular chromosome of E. coli was labeled for an interval during growth, and then isolated and displayed for autoradiographic observation, the following remarkable picture emerged.

Most simply represented as in Fig. 7-3, the replicating E. coli chromosome appeared as a complex circular structure with three segments. Based on the grain density of the segments, a scheme of semiconservative replication starting at a fixed origin was proposed. Portions of the chromosome in which both strands were synthesized (and therefore labeled with ^{3}H in the successive rounds) have double the grain density of those portions in which one of the strands served

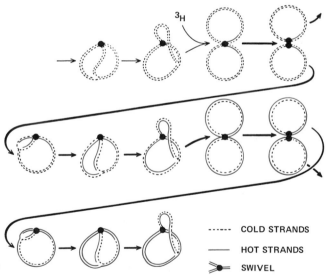

FIGURE 7-3
Stages in the semiconservative replication of the circular chromosome of E. coli starting at a single origin as visualized by relative grain densities. Theta structures are intermediates. (Courtesy of Dr. J. Cairns.)

7. Schnös, M. and Inman, R. B. (1970) JMB **51**, 61; (1971) **55**, 31.
8. Gyurasitis, E. B. and Wake, R. G. (1973) JMB **73**, 55; Wake, R. G. (1973) JMB **77**, 569.
9. Wolfson, J., Dressler, D. and Magazin, M. (1972) PNAS **69**, 499; also (1972) PNAS **69**, 998.
10. Cairns, J. (1963) JMB **6**, 208.

as a template in the first round. The intermediates in this picture of the replicating chromosome, in view of its resemblance to the Greek letter θ upon planar projection, has been called a "theta" structure.

At the time the structure was first observed, the presence of two forks suggested a model[11] (Fig. 7-4) in which replication starts on the circular duplex at some fixed point or origin, and then advances as a replication fork _in one direction_. The distance between the grow-

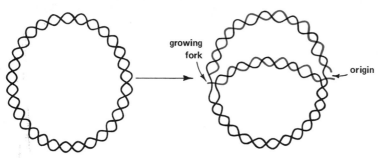

growing fork

origin

FIGURE 7-4
Model of a circular chromosome undergoing semiconservative replication in one direction as first proposed by Cairns.

ing fork and the fork at the origin measures the extent of replication. When replication is complete, the two duplex circles are detached in some manner.

Subsequent studies have disclosed that replication is generally bidirectional and have required revision of this earlier unidirectional model. These data were obtained by: (i) autoradiography[12], (ii) electron microscopy[13, 14], and (iii) genetic analysis [15].

AUTORADIOGRAPHY[12]
B. subtilis spores were germinated in a medium containing thymine of low specific radioactivity. During this period of about 90 minutes in which vegetative cells develop, DNA synthesis does not occur. Toward the end of this period, when DNA replication starts, the germinated spores were switched for 5 to 20 minutes into a medium containing thymine of higher specific radioactivity. The unreplicated portion of the DNA was unlabeled and therefore not visible autoradiographically. The different levels of specific radioactive labeling made it possible to distinguish the nascent regions of replication loops by their higher grain density. Labeling in one region of the loop indicates unidirectional growth, in two opposite regions, bidirectional growth (Fig. 7-5). The majority of loops were observed to have the labeling expected of bidirectional replication movement (Fig. 7-6). Based on the size to which the loops expanded with time of incubation, the growth rates were judged to be approximately equal in each direction until at least 50 percent of the chromosome had been replicated.

11. Op. cit. in footnote 10.
12. Gyurasitis, E. B. and Wake. R. G. (1973) _JMB_ **73**, 55; Wake, R. G. (1973) _JMB_ **77**, 569.
13. Schnös, M. and Inman, R. B. (1970) _JMB_ **51**, 61; (1971) **55**, 31.
14. Wolfson, J., Dressler, D. and Magazin, M. (1972) _PNAS_ **69**, 499; also (1972) _PNAS_ **69**, 998.
15. Bird, R. E., Louarn, J., Martuscelli, J. and Caro, L. (1972) _JMB_ **70**, 549.

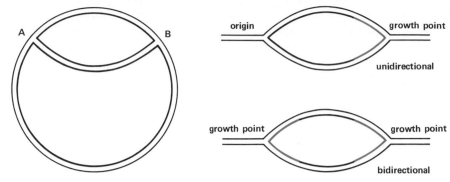

FIGURE 7-5
Nascent regions (of heavy grain density) anticipated in autoradiographic studies of the replicating region of a chromosome undergoing unidirectional or bidirectional growth. Grain density at forks A and B is decisive.

Earlier studies[16] of the order of replication of genetic markers had suggested unidirectional growth of the B. subtilis chromosome from a defined locus, ade 16, for a distance of at least 70 percent of its length. It is difficult to reconcile these genetic and autoradiographic findings. However, genetic evidence for bidirectional replication near the initiation site has since been reported[17]. Convincing autoradiographic evidence of bidirectional replication in E. coli has also been obtained[18], adding to the generality of the observations with germinating B. subtilis spores.

ELECTRON MICROSCOPY[19]

Improved methods for isolating replicating bacteriophage chromosomes and for their fine physical mapping made it possible to establish the precise origin of replication and the direction and manner of fork movement from that point.

Cells growing in a medium containing heavy isotopes, ^{2}H and ^{15}N, were infected with phage λ in which the duplex DNA was labeled with ^{32}P. The ^{32}P label served to distinguish the viral DNA from the host DNA, and was a marker for isolating the replicating forms of the viral DNA. λ molecules circularize upon entering the cell. In heavy-isotope medium they developed hybrid (light : heavy) densities upon replication. These molecules, when isolated at various stages of replication, distributed themselves between fully light and hybrid densities in a CsCl equilibrium gradient, which separates on the basis of density.

The molecules were then collected and examined in the electron microscope. The replicating molecules were all seen to be circular

16. Sueoka, N. and Quinn, W. G. (1968) CSHS 33, 695.
17. Sueoka, N., Matsushita, T., Ohi, S., O'Sullivan, A. and White, K. in Wells, R. D. and R. B. Inman (eds) (1973) DNA Synthesis In Vitro, Steenbock Symposium Univ. Park Press, Baltimore p. 385.
18. Prescott, D. M. and Kuempel, P. L. (1972) PNAS 69, 2842.
19. Schnös, M. and Inman, R. B. (1970) JMB 51, 61; (1971) 55, 31; Wolfson, J., Dressler, D. and Magazin, M. (1972) PNAS 69, 499; also (1972) PNAS 69, 998.

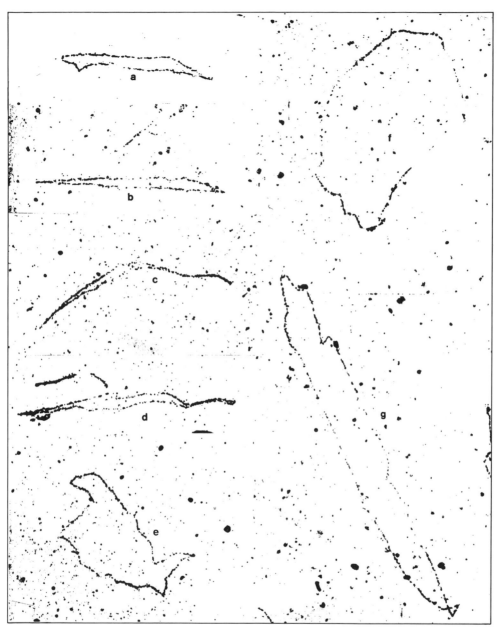

FIGURE 7-6
Autoradiographs of chromosomal loops in *B. subtilis* (germinating spores) indicating bidirectional growth. (a) Loop with an even grain density. (b-g) Loops with heavy grain density at both ends or forks. Scale is 100μm. (Gyurasitis, E. B. and Wake, R. G. (1973) *JMB* **73**, 55. Courtesy of Professor R. G. Wake.)

and most of them contained forks. They clearly resembled the auto-radiographic image of the replicating *E. coli* chromosome. Although the distance between the forks varied, two of the three segments in the structure were always equal in length. The sum of the average length of the equal segments plus that of the remaining one added up to 17 μ, the length of a mature λ chromosome.

With this remarkable image of the replicating λ molecule in mind, two important questions could be posed. Is the origin located at a unique place? Is progress from the origin in only one direction or in both?

The technique of *denaturation mapping*[20] was used to answer these questions by providing markers on the chromosome. If replication is unidirectional, and one fork always remains at the origin, the distance between it and a marker should remain constant throughout the cycle of replication. However, if replication is bidirectional the distance between the marker and both forks of replication should vary with the extent of replication (Fig. 7-7).

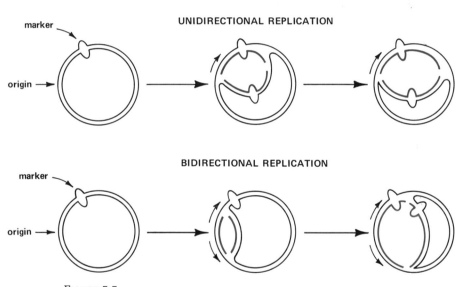

FIGURE 7-7
Denaturation mapping locates markers which can be used to determine the direction of chromosome growth (see text).

Upon treatment of the mature linear chromosome of phage λ at pH 11.05 for 10 minutes at 25°, the A–T-rich regions of the DNA melt or denature whereas the G–C-rich regions remain in the duplex form. Renaturation of the A–T regions was prevented by the presence of formaldehyde, which prevents base-pairing of the amino group of adenine. The melted regions of λ DNA were seen in the electron microscope as single-strand bubbles distributed along the entire 17 μ length of the chromosome (Fig. 7-8). By determining the loca-

20. Inman, R. B. (1966) *JMB* **18**, 464.

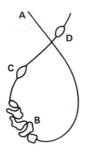

FIGURE 7-8
Diagram of a partially denatured DNA molecule as observed in the electron microscope. The simplified example in the upper diagram is a molecule with single sites of denaturation at *C* and *D*, a multiple site at *B*, and one crossover. (Courtesy of Professor R. B. Inman.)

tion and size of these bubbles in many hundreds of molecules, these significant features of the λ duplex were distinguished. (i) The molecule has a distinctive pattern that can be correlated with the left and right ends of the genetic map, and (ii) there are major zones of melting with fine and distinctive profiles (Fig. 7-9).

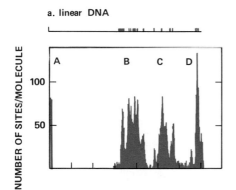

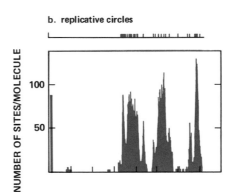

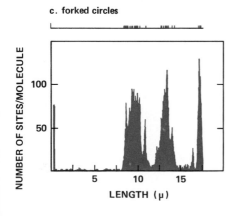

FIGURE 7-9
Denaturation maps of phage λ DNA. Distribution and size of the denatured regions are the same for (a) mature linears, (b) simple circles, and (c) replicating forked circles. (Courtesy of Professor R. B. Inman.)

The pattern or regions of A–T melting obtained in the denaturation mapping technique provide the markers with which to analyze the replicating chromosome.

Replicating circular λ molecules showed the same denaturation map as vegetative linear ones (Fig. 7-9). The theta structures isolated by buoyant density separation showed, on each of the two equal segments, denaturation patterns identical and predictable from their position on the vegetative chromosome. From mapping molecules at many states of replication up to 90 percent of completion, it became clear that replication (i) starts at a particular point, 18.3 percent from the designated right end of the molecule, and (ii) proceeds at an equal rate in both directions.

The denaturation mapping technique applied to T7 DNA[21], which does not form a circle on replication, revealed both a unique starting point, 17 percent from the left end, and bidirectional replication from this point (Fig. 7-10). By contrast, in phage P2, which resembles λ in having a circular replicative form, replication proceeds unidirectionally from a fixed point, 11.7 percent from the right end[22].

GENETIC ANALYSIS

Does the very much larger chromosome of E. coli also have a single origin for replication, and is replication bidirectional as in B. subtilis and in λ? Genetic and biochemical analyses of E. coli[23], as well as recent audioradiographic studies[24], have now revised the original proposal of unidirectional progress of a single replicating fork, and favor instead a bidirectional movement of both forks around the circle.

The origin of replication has been fixed near the ilv locus, at 74 minutes on the standard genetic map of E. coli (Fig. 4-32). Replication progresses in opposite directions at the same rate, and the two growing forks meet at a point 180° from the origin near the trp locus at 25 minutes on the map.

Conditionally lethal E. coli mutants, with lesions in gene products required for chromosome initiation, can be relieved of this defect if phage P2 DNA is incorporated or integrated somewhere in the host chromosome[25]. This phenomenon is called integrative suppression. Inasmuch as this phage DNA can insert at several different sites in the coli chromosome, the phage locus may provide a starting signal wherever it finds itself in the host chromosome. When inserted, the P2 origin leads to a bidirectional replication of the E. coli chromosome, even though replication of the free viral chromosome is unidirectional.

EUKARYOTIC CHROMOSOMES

Bidirectional replication also prevails in the more complex chromo-

21. Wolfson, J., Dressler, D. and Magazin, M. (1972) PNAS **69**, 499; also (1972) PNAS **69**, 998.
22. Schnös, M. and Inman, R. B. (1971) JMB **55**, 31.
23. Bird, R. E., Louarn, J., Martuscelli, J. and Caro, L. (1972) JMB **70**, 549.
24. Prescott, D. M. and Kuempel, P. L. (1972) PNAS **69**, 2842.
25. Lindahl, G., Hirota, Y. and Jacob, F. (1971) PNAS, **68,** 2407.

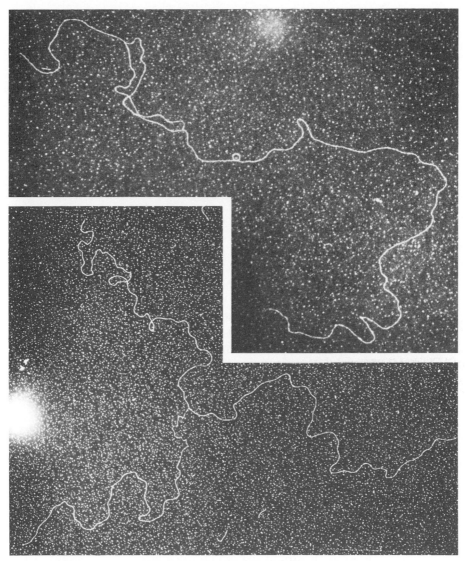

FIGURE 7-10
Electron micrographs of replicating phage T7 molecules. (Top) An expanding
"eye" advances toward both ends of the molecule. (Bottom) A "Y" form in which
the left end of the molecule has been replicated and the other fork advances
toward the right end; the growing fork has a single-stranded template region about
3,000 residues long. Contour length of the T7 chromosome is $12\,\mu$m. (Courtesy
of Professor D. Dressler.)

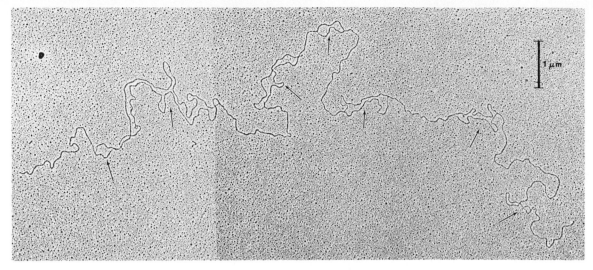

FIGURE 7-11
Electron micrograph of replicating *Drosophila* DNA showing multiple "eyes"
(bubbles). (Courtesy of Henry Kriegstein.)

somes of yeast[26], *Drosophila*[27], and mammals[28]. Multiple starts form
loops which are seen as bubbles or eyes along the chromosome in
electron micrographs (Fig. 7-11). Based on the number of starts in a
length of chromosome, the size of these replicating units in eukaryotic chromosomes is generally nearer that of a virus than a bacterium.
The unit is near 100,000 base pairs in the forms mentioned, but may
reach *E. coli* size in certain tissues of the newt (*Trituris*).

3. Fine Structure and Movement
   of the Fork[29-32]

If replication takes place in a $5' \longrightarrow 3'$ direction on both strands of a
duplex by the same polymerase mechanism, this mechanism cannot
occur simultaneously and synchronously on both strands. In the
region of the fork, an asymmetric single-stranded section must exist,
if only transiently, in one strand of the duplex in the region where
the opposite strand is being replicated.

Using a technique with the electron microscope that distinguishes
clearly between single-stranded and duplex DNA, single-strand
template regions have been seen at the replicating fork of phage λ[33],

26. Callan, H. G. (1973) *CSHS* **38**, 195.
27. Kriegstein, H. J. and Hogness, D. S. (1974) *PNAS* **71**, 135.
28. Huberman, J. A. and Riggs, A. D. (1968) *JMB* **32**, 327; Huberman, J. A. and Tsai, A. (1973)
    JMB **75**, 5.
29. Op. cit. in footnote 27.
30. Inman, R. B. and Schnös, M. (1971) *JMB* **56**, 319.
31. Wolfson, J. and Dressler, D. (1972) *PNAS* **69**, 2682.
32. Blumenthal, A. B., Kriegstein, H. J. and Hogness, D. S. (1973) *CSHS* **38**, 205.
33. Op. cit. in footnote 30.

phage T7[34], and *Drosophila*[35]. These regions were almost invariably limited to one side of the fork (Fig. 7-12). In λ, the regions ranged in length up to 0.4μ, suggesting that approximately 1000 base pairs is the size of this single-stranded unit; the size in T7 is considerably larger. What signals the initiation of DNA replication along these single-stranded stretches has not yet been made clear.

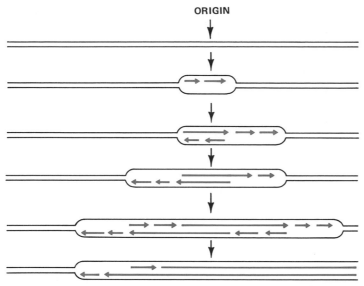

FIGURE 7-12
Model for movement of a growing point (fork), in two directions, indicating the location of single-stranded regions. The limiting size of such regions, up to 0.4 μm long in phage λ, suggest a "replication unit". With progress of replication, the single-stranded regions are replicated and the nascent fragments are joined.

There may be complexities of structure at the growing fork that have not been predicted. For example, replicating molecules of phages P2 and T4, and *Drosophila*[36] were observed in the electron microscope with an extra single strand (a "whisker") protruding at the fork. This single strand was very likely the nascent strand displaced by the reannealing strands of the parental helix ("branch migration"), thereby setting back the fork by that distance. Whether such a structure is an obligatory stage in replication with special significance seems doubtful. There is likely some in vivo mechanism that prevents an extent of branch migration that would displace nascent strands and interfere with replication.

An interesting variant of a replicating fork, called a rolling circle[37], has been proposed and will be discussed in detail in Chapter 8,

34. Wolfson, J. and Dressler, D. (1972) *PNAS* **69**, 2682.
35. Kriegstein, H. J. and Hogness, D. S. (1974) *PNAS* **71**, 135.
36. Op. cit. in footnote 35.
37. Gilbert, W. and Dressler, D. (1968) *CSHS* **33**, 473; Dressler, D. (1970) *PNAS* **67**, 1934; Knippers, R., Whalley, J. M. and Sinsheimer, R. L. (1969) *PNAS* **64**, 275.

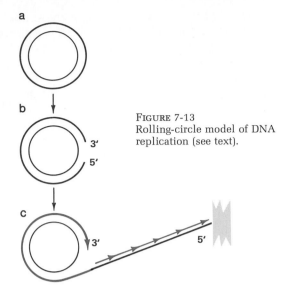

FIGURE 7-13
Rolling-circle model of DNA
replication (see text).

Section 4. A nick in one strand of a circular duplex is the starting point for covalent extension of the 3′-hydroxyl terminus (Fig. 7-13). While replication proceeds around the circle, the 5′ end of the strand is displaced as a growing length of single strand and anchored in some manner. The latter may be replicated by initiation of a new strand that also grows in 5′ → 3′ direction. The best evidence for this proposal was the production of extra-long φX174 viral strands from the replicative intermediate (Chapter 8, Section 4). Structures resembling a rolling circle were seen in the electron microscope and viral strands of greater than unit length were isolated after denaturing the replicative intermediate. Rolling-circle intermediates have also been observed in the amplification of ribosomal RNA genes during development of *Xenopus* (toad) oocytes[38].

Still another intriguing scheme of replication[39, 40] involves a structure called a D-loop, first found in mitochondrial DNA of animal cells[41]. Replication of these circular duplexes, about five microns long, is conveniently studied in mouse L-cell lines that lack the cytoplasmic thymidine kinase but maintain the mitochondrial one. It was possible, by labeling the mitochondrial DNA at a specific radioactivity fifty times that of the nuclear DNA[42], to confirm and extend electron microscopic observations of the D-loop as an early stage in replication[43] (Fig. 7-14).

The model for mitochondrial DNA replication in Figure 7-15 assumes that the twist in the supercoiled duplex circle is transiently relieved, so that a new strand can be initiated at a fixed point[44],

38. Hourcade, D., Dressler, D. and Wolfson, J. (1973) *PNAS* **70**, 2926.
39. Tibbetts, C. J. B. and Vinograd, J. (1973) *JBC* **248**, 3367, 3380.
40. Robberson, D. L. and Clayton, D. A. (1972) *PNAS* **69**, 3810.
41. Borst, P. (1972) *ARB* **41**, 333.
42. Berk, A. J. and Clayton, D. A. (1973) *JBC* **248**, 2722.
43. Berk, A. J. and Clayton, D. A. (1974) *JMB* **86**, 801.
44. Grossman, L. I., Watson, R. and Vinograd, J. (1973) *PNAS* **70**, 3339.

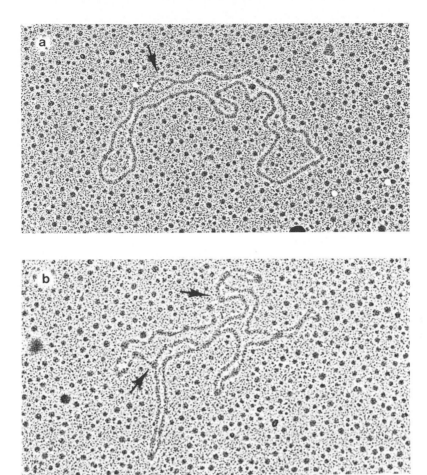

FIGURE 7-14
Electron micrograph of D-loop of replicating mitochondrial DNA. Arrows
denote a small D-loop in (a) and the boundaries of the displaced single
strand in an expanding D-loop in (b). Contour length of the DNA is 5 μm.
(Courtesy of Professor D. Clayton.)

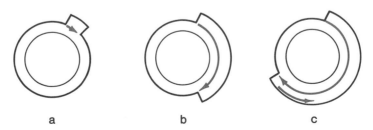

FIGURE 7-15
Model for replication of mitochondrial DNA. The D-loop started in
(a) is expanded in (b); only later is replication of the displaced
parental ("heavy") strand initiated, as in (c). (Grossman, L. I.,
Watson, R. and Vinograd, J. (1973) PNAS **70**, 3339.)

using as template one strand of the duplex. The complementary strand, idle as a template, appears as the ballooned-out D-loop in the electron microscope (Fig. 7-14a, 7-15a). As replication proceeds, the D-loop expands (Fig. 7-14b, 7-15b). Isolated structures containing D-loops have distinctive denaturation-renaturation behavior. If the D form is denatured, the newly synthesized strand is released, and the original circular duplex is restored upon renaturation. Only at a later stage is replication initiated on the displaced single-stranded loop (Fig. 7-15c). Progress of replication thus involves a staggered copying of the strands, each in a unidirectional manner. The relatively slow rate of mitochondrial DNA synthesis in animal cells, consuming about 100 minutes for completion, may explain why at any given time one finds D-loops in various stages of replication.

Mitochondrial DNA is known to be sensitive to alkaline cleavage. Based on kinetic measurements of chain scissions induced by alkali and by RNase H (an enzyme that degrades the RNA of a DNA-RNA hybrid), this sensitivity could be attributed to ribonucleotides in the DNA[45]. Approximately ten residues were judged to be present, interspersed among the 15,000 base pairs, possibly the vestigial remnants of priming events in replication (see Section 5).

4. Synthesis in Short Steps[46]

When replicating phage T4 DNA was isolated, the most recently synthesized, or nascent, DNA was found in short pieces called replication fragments (Okazaki fragments). They sedimented at about 10S, representing chain lengths of 1000–2000 residues. Nascent DNA in a large variety of viral, bacterial, and animal forms has since been characterized as consisting of short fragments of this size range in prokaryotic systems and of considerably smaller size in eukaryotic cells.

To find the T4 replication pieces before they are joined to the main body of growing chains, the following conditions must be observed:

(i) low temperature of growth to reduce replication rate, e.g. 8° for T4 phage growth in *E. coli*,

(ii) DNA precursor of high specific radioactivity that enters DNA rapidly and directly, for example, ^{3}H-thymine or thymidine, and

(iii) brief exposure of the cells to the labeled precursor (a pulse) for 5 to 60 seconds at the low temperature, and

(iv) efficient quenching of the pulse.

Under these conditions, *virtually all* the precursor label can be captured as small pieces but, after a prolonged pulse or after a subsequent exposure to a large concentration of unlabeled pre-

45. Grossman, L. I., Watson, R. and Vinograd, J. (1973) *PNAS* **70**, 3339.
46. Sugino, A. and Okazaki, R. (1972) *JMB* **64**, 61; Okazaki, R., et al., **in** Wells, R. D. and Inman, R. B. (eds) (1973) *DNA Synthesis In Vitro*, Steenbock Symposium Univ. Park Press, Baltimore, p. 83.

cursors (a chase), the radioactive precursor is found exclusively in high-molecular-weight DNA.

In mutant cells deficient in ligase or in DNA polymerase I, or in both, the replication fragments accumulate and may comprise the bulk of the newly synthesized DNA.

The indication that <u>both</u> strands of the duplex are replicated discontinuously comes from the recovery of virtually all the nascent DNA in short fragments. This evidence was confirmed in the following way. Isolated replication fragments were annealed to the template strands which had been separated on the basis of their density. The short pieces were found in most instances annealed to both strands to an equal extent in replicating E. coli, and phages λ, P2, and P4. The result clearly fits with discontinuous replication on both strands of the duplex DNA. This interpretation may not appear definitive in bidirectional replication, i.e. one fork might replicate strand A continuously and strand B discontinuously, while the fork growing in the opposite direction replicated strand A discontinuously and strand B continuously. However, if all the pulse-labeled material (nascent DNA) can be recovered as short fragments, there is no room for a substantial fraction of continuous synthesis. Were asymmetry in the amount of DNA fragments annealed to the two template strands observed, it could be attributable to a slower rate of joining of the nascent fragments on strand A compared to strand B or to relatively continuous replication on strand B due to fewer initiations[47].

With these indications that discontinuous replication occurs on both strands of the template, it is important to determine the direction of growth on each template strand. This has been done as follows (Fig. 7-16): a short pulse of ^{3}H-thymidine was administered to cells in which the replicating DNA was already being labeled with ^{14}C-thymidine. The 10S replication fragments were isolated and their 5′ ends were tagged with ^{32}P, by first removing the 5′-phosphate terminus with phosphatase and then restoring it with polynucleotide kinase using γ-[^{32}P]-ATP. The triply labeled 10S fragment (^{14}C, ^{3}H, ^{32}P) was then degraded by nuclease. Using one nuclease that degrades DNA from the 3′ end and another from the 5′ end, it was possible to establish whether there was preferential release of ^{32}P, or ^{3}H, or both, relative to ^{14}C, and then to infer at which end DNA chain growth had taken place, i.e. the 5′ end or the 3′ end.

Results showed that the nucleotides added last were at the 3′ end. Growth was 5′ → 3′, and this was equally true with fragments annealing to both strands of the duplex template. Also, by determining the rate at which the short pieces became ^{3}H-labeled in their entire length, the in vivo rate of nucleotide addition could be calculated (Table 7-1). The rates calculated for phage T4 chain growth proved to be about twice the known rate of movement of the replicating fork, suggesting that there is a pause between the com-

---

47. Olivera, B. M. and Bonhoeffer, F. (1972) NNB **240**, 233; Herrmann, R., Huf, J. and Bonhoeffer, F. (1972) NNB **240**, 235.

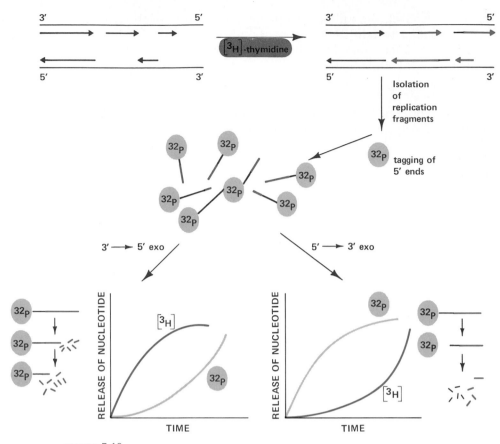

FIGURE 7-16
Outlines of procedures that demonstrate the direction of growth of the small
replication fragments. ¹⁴C labels the fragment, ³H its most nascent portion,
and ³²P the 5'-terminus. (See text for details.)

pletion of one fragment and the start of the next. Thus, DNA replication may be discontinuous in time as well as in space.

TABLE 7-1
Rates of T4 chromosome growth and replication-fragment
elongation

Temperature	Fork movement, nucleotides/second	Fragment elongation, nucleotides/second
8°	10	20
14°	40	100
25°	250	

Synthesis of DNA in short steps requires a phosphodiester joining
action to link the replication fragments to one another and to the
growing chromosome. Furthermore, excision of the RNA primer of
each fragment (see next section) leaves gaps to be filled before
joining. When these gap-filling and joining operations, required in a

discontinuous mode of synthesis, are deficient in mutant cells, there is, as expected, a large accumulation of the 10S fragments. In *E. coli* mutants that lack detectable ligase activity in extracts and are inviable at a restrictive temperature, DNA synthesis proceeds but leads to accumulation of 10S replication fragments. In *pol* A⁻ cells, deficient in DNA polymerase I, short pieces and gaps between the pieces are ten times more numerous than in wild-type cells[48].

The transient appearance of replication fragments in normal replication and their accumulation in ligase- or polymerase-I-deficient cells has also been given another interpretation[49]. It has been argued that DNA synthesis is continuous and that the fragments are a secondary result of nuclease attack on the newly synthesized DNA, which may be especially susceptible. However, the production of 10S pieces by nuclease action cannot explain the preferential and progressive labeling toward the 3' end of the chain, nor (as will be seen in the next section) can it explain the presence of RNA at the 5' end of each chain.

5. Initiation of a DNA Chain[50-53]

DNA polymerases isolated from bacterial and animal sources are known to extend primer DNA chains, and in so doing show remarkable fidelity in matching each base of the template regardless of the source of primer or template. However, in the absence of primer, no known DNA polymerase is able to start a new chain (Chapter 3).

Several alternatives were considered. One was that new DNA chains are never started but that all DNA synthesis takes place by covalent extension of pre-existing DNA chains. Thus, an oligonucleotide fragment of a DNA chain or an end produced by endonucleolytic scission of a chain might serve as primer. Searches for a suitable pool of DNA primer fragments and for specific nicking enzymes were not successful. Alternatively, another DNA polymerase might be found with the capacity to recognize a starting signal on a DNA chain and to orient an initiating nucleotide. No enzyme has as yet been found to do this job.

Still another alternative involves RNA polymerase, an enzyme that can recognize a specific DNA sequence, such as the promoter regions for transcription (Chapter 10, Section 3), and can start a new RNA chain. In doing so, a hybrid duplex is formed: the DNA template strand with the RNA transcript. Furthermore, DNA polymerases have been shown to be able to extend a polynucleotide chain containing a ribonucleotide primer terminus (Chapter 4, Section 16). In

48. Okazaki, R., Arisawa, M. and Sugino, A. (1971) *PNAS* **68**, 2954.
49. Werner, R. (1971) *Nat.* **230**, 570; (1971) *NNB* **233**, 99.
50. Brutlag, D., Schekman, R. and Kornberg, A. (1971) *PNAS* **68**, 2826.
51. Wickner, W., Brutlag, D., Schekman, R. and Kornberg, A. (1972) *PNAS* **69**, 965.
52. Schekman, R., Wickner, W., Westergaard, O., Brutlag, D., Geider, K., Bertsch, L. L. and Kornberg, A. (1972) *PNAS* **69**, 2691.
53. Sugino, A., Hirose, S. and Okazaki, R. (1972) *PNAS* **69**, 1863; Sugino, A. and Okazaki, R. (1973) *PNAS* **70**, 88.

this way, a brief transcriptional action by RNA polymerase might provide an RNA primer for DNA replication. This priming piece of RNA could later be recognized and excised by nuclease action. Thus, a de novo initiation event catalyzed by RNA polymerase would be an essential step for the start of DNA synthesis (Fig. 7-17).

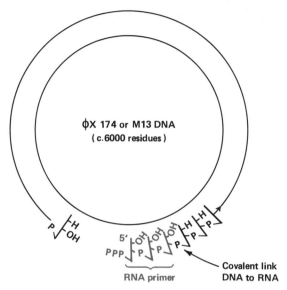

FIGURE 7-17
A scheme for RNA-primed initiation of synthesis of a strand complementary to the viral DNA of φX174 or M13. A short chain of RNA is extended covalently by DNA. A short gap remains between the 3′ DNA terminus and RNA at the 5′ end.

The first step in bacteriophage M13 DNA synthesis, the conversion of the single-stranded, circular DNA to a duplex, replicative form (RF) (see Chapter 8, Section 3), provided an excellent system to test this possibility. This step does not require the expression of viral genes but uses host cell enzymes. Were one of these enzymes RNA polymerase, then rifampicin, an antibiotic drug that specifically inhibits initiation of transcription by RNA polymerase, should prevent the first step in M13 DNA synthesis. This, in fact, has been observed[54]. Rifampicin completely inhibited conversion of the phage single-stranded DNA to the duplex form. If RNA polymerase were not involved in the reaction, rifampicin should have no effect on the M13 DNA replication.

Rifampicin also promptly and profoundly inhibited multiplication of the duplex forms if administered five minutes after infection, a time when such multiplication normally occurs. The effect of rifampicin on DNA synthesis was not a secondary result of inhibition

54. Brutlag, D., Schekman, R. and Kornberg, A. (1971) PNAS **68**, 2826.

of synthesis of messenger RNA, and consequently of protein synthesis, because chloramphenicol, a direct inhibitor of protein synthesis, did not prevent the initial conversion of the phage single strands to duplexes, but inhibited multiplication of the duplexes only after some delay.

The inference that the rifampicin inhibition was localized to RNA polymerase was also supported by failure of the drug to inhibit M13 DNA synthesis in a host mutant with RNA polymerase resistant to rifampicin. Additional proof that the RNA polymerase contributed to DNA synthesis by synthesizing an RNA priming fragment came from studies of the conversion of the M13 single strand (SS) to the duplex RF in vitro. The product contains a small gap in the synthesized complementary strand and is called RF II to distinguish it from a fully circular RF I duplex. The evidence for RNA priming obtained from these studies can be summarized as follows.

(i) Inhibition by rifampicin, streptolydigin, and actinomycin, all inhibitors of RNA polymerase; lack of inhibition by rifampicin in extracts of mutants with rifampicin-resistant RNA polymerase.

(ii) Requirement for all four ribonucleoside triphosphates,

(iii) Conversion of single strands to RF in two stages: an initial stage (A) of RNA synthesis produces a primed single strand which could be isolated and converted to RF in a second stage (B) in the absence of ribonucleoside triphosphates (except for ATP), and in the presence of rifampicin,

$$\text{SS} \xrightarrow[\text{RNA Synth.}]{A} \text{``Primed'' SS} \xrightarrow[\text{DNA Synth.}]{B} \text{RF II.}$$

(iv) Demonstration of a phosphodiester linkage of deoxynucleotide to a ribonucleotide in the isolated RF, in stoichiometric equivalence to the number of RF molecules formed. ^{32}P was transferred from α-(^{32}P)-deoxynucleoside triphosphates to a ribonucleotide (2'- and 3'-AMP) upon alkaline cleavage of the product (Fig. 7-18).

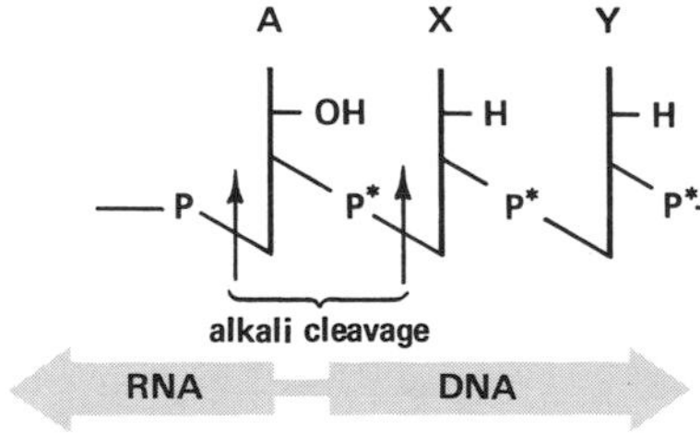

FIGURE 7-18
Procedure for demonstrating covalent linkage between the RNA primer and the DNA chain. Transfer of ^{32}P from an α-labeled deoxynucleoside triphosphate to a ribonucleotide is observed after cleavage by alkali.

(v) Persistence of an RNA segment at the 5' end of the complementary strand, as inferred by requirements for closure of the RF II product to an RF I. To produce an _alkali-stable_ RF I, with an implied absence of ribonucleotides, E. coli DNA polymerase I was required for gap-filling and for the excising action of its 5' $\longrightarrow$ 3' exonuclease;

either T4 or *E. coli* ligase (Section 6) sealed the circle. To produce an *alkali-labile* RF I, T4 DNA polymerase, which fills gaps but lacks the excising function of a $5' \rightarrow 3'$ exonuclease, was required; T4 ligase, which can utilize both DNA and RNA chains, could serve, but *E. coli* ligase, which joins only DNA chains could not[55] (Fig. 7-19).

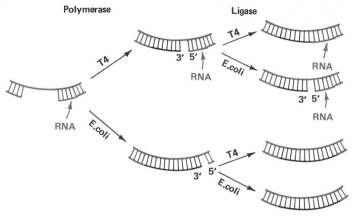

FIGURE 7-19
Scheme illustrating the use of T4 and *E. coli* DNA polymerases and ligases in filling the gap and closing the circle in the enzymatically synthesized RF II. An alkali-stable RF I (lacking RNA) was produced with *E. coli* polymerase and either ligase; an alkali-labile RF I (containing RNA) was produced with T4 polymerase and T4 ligase (see text).

Extension of the M13 findings to other systems has led to the isolation of DNA replication fragments and replicative intermediates formed in vivo that still retain the RNA priming segment. These will be discussed further. Providentially, it was inhibition of phage M13 DNA synthesis by rifampicin that was chosen for testing the RNA priming hypothesis. Had the choice been ϕX174, an apparently similar phage, results would have been discouragingly negative. It turns out that the conversion of the ϕX174 single strand to a duplex form does not depend on the classical RNA polymerase and is therefore not inhibited by rifampicin. Nevertheless, ϕX174 DNA synthesis is primed by an RNA fragment, in this case synthesized by a previously unrecognized RNA-synthesizing system[56] (Chapter 8, Section 4).

Involvement in ϕX174 DNA synthesis of a novel RNA polymerase, distinguished from the classic RNA polymerase by resistance to rifampicin inhibition, has a major significance. It may also be responsible for the RNA priming of the 10S replication DNA fragments produced during replication of the *E. coli* chromosome, this

55. Westergaard, O., Brutlag, D. and Kornberg, A. (1973) *JBC* **248**, 1361.
56. Schekman, R., Wickner, W., Westergaard, O., Brutlag, D., Geider, K., Bertsch, L. L. and Kornberg, A. (1972) *PNAS* **69**, 2691.

priming also known to be unaffected by rifampicin. The requirement for *E. coli* replication proteins (*dna*A, *dna*B, *dna*C-D, *dna*E, and *dna*G gene products; see Section 9 of this chapter) in ϕX174 replication as well as the resistance to rifampicin inhibition suggest that ϕX174 replication may be exploiting the major replication system of the *E. coli* chromosome.

RNA involvement in the initiation of a DNA chain appears now to be a mechanism of widespread use in bacterial and animal cells (Table 7-2). Evidence for priming is clear in some instances and remains to be established in others. In *E. coli*, RNA polymerase is responsible for synthesis of the RNA transcript in M13 and *col* E1 replication. Rifampicin inhibition in some instances, as in F plasmid replication in *E. coli*, suggests but provides no direct evidence of priming. Rifampicin-insensitive synthetic systems operate in other

TABLE 7-2
Initiation of DNA synthesis by RNA

Evidence for priming

1. M13: single strand conversion to replicative form[a]
2. ϕX174: single strand conversion to replicative form[b]
3. Colicinogenic factor, *col* E1[c]
4. Nascent *E. coli* replication fragments[d]
5. Polyoma DNA replication in animal nuclei[e]
6. Nascent replication fragments in tumor cells and lymphocytes[f]
7. Replication in *Physarum polycephalum*, a slime mold[g]

No direct evidence for priming

1. M13: replicative form multiplication[a]
2. Origin of *E. coli* chromosome replication[h]
3. Origin of λ replication[i]
4. Replication of F plasmid in *E. coli*[j]
5. Replication of miniplasmid in *E. coli* 15[k]

[a]Brutlag, D., Schekman, R. and Kornberg, A. (1971) *PNAS* **68**, 2826.
[b]Schekman, R., Wickner, W., Westergaard, O., Brutlag, D., Geider, K., Bertsch, L. L. and Kornberg, A. (1972) *PNAS* **69**, 2691.
[c]Blair, D. G., Sherratt, D. J., Clewell, D. B. and Helinski, D. R. (1972) *PNAS* **69**, 2518; Williams, P. H., Boyer, H. W. and Helinski, D. R. (1973) *PNAS* **70**, 3744; Sakakibara, Y. and Tomizawa, J. (1974) *PNAS* **71**, 802.
[d]Sugino, A., Hirose, S. and Okazaki, R. (1973) *PNAS* **69**, 1863; Sugino, A. and Okazaki, R. (1973) *PNAS* **70**, 88.
[e]Magnusson, G., Pigiet, V., Winnacker, E-L., Abrams, R. and Reichard, P. (1973) *PNAS* **70**, 412.
[f]Sato, S., Ariake, S., Saito, M. and Sugimura, T. (1972) *BBRC* **49**, 827; Fox, R. M., Mendelsohn, J., Barbosa, E. and Goulian, M. (1973) *NNB* **245**, 234.
[g]Huberman, J. A., Horwitz, H., Minkoff, R. and Wagar, M. A. *CSHS* **38**, 233.
[h]Lark, K. G. (1972) *JMB* **64**, 47.
[i]Dove, W. F., Inokuchi, H. and Stevens, W. F. (1971) in *The Bacteriophage Lambda* (A. D. Hershey, ed.) *CSHL* p. 747; Hayes, S. and Szybalski, W. (1973) *Fed. Proc.* **32**, 529 abs.; Dahlberg, J. E. and Blattner, F. R. (1973) *Fed. Proc.* **32**, 664 abs.
[j]Bazzicalupo, P. and Tocchini-Valentini, G. P. (1972) *PNAS* **69**, 298.
[k]Messing, J., Staudenbauer, W. L. and Hofschneider, P. H. (1972) *NNB* **238**, 202.

instances. Isolation of the RNA primer covalently linked to DNA has been successful in three in vivo systems.

(i) *E. coli* replication fragments isolated from growing cells (with rigid precautions to exclude ribonuclease, alkali, and other factors that might degrade RNA) retained a segment of RNA at the 5′ end of the DNA chain[57]. The RNA was 50 to 100 residues long, representing up to about 10 percent of the lengths of the DNA fragments which ranged from 400 to 2000 residues. Most remarkably, there is considerable specificity of the ribonucleotide and the deoxynucleotide in the region of the covalent junction (Fig. 7-20). The base speci-

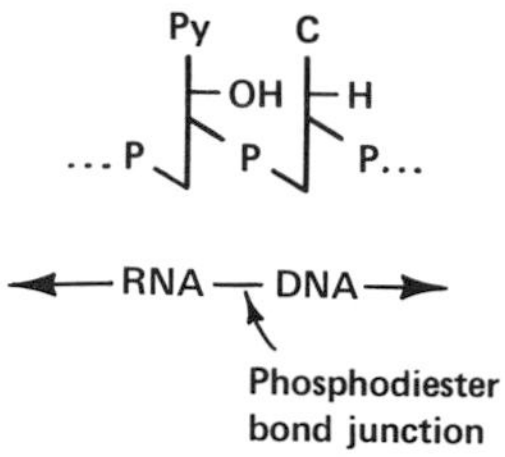

FIGURE 7-20
Sequence at the covalent junction of RNA to DNA in the small replication fragment of E. coli. (Hirose, S., Okazaki, R. and Tamanoi, F. (1973) *JMB* **77**, 501.)

ficity at the nucleotide linkage of RNA to DNA argues for a unique and recurring sequence in the template that signals a switch from RNA transcription to DNA replication.

(ii) *Col* E1 plasmids, which accumulate in large numbers in *E. coli* inhibited by chloramphenicol, contained a short segment of RNA in one of the two strands of the twisted circle[58]. It may be inferred that these poisoned cells were deficient in the system normally present to excise the RNA segment that primes DNA replication (see Chapter 8, Section 7).

(iii) Tumor cells and lymphocytes were the source of nascent DNA replication fragments containing RNA in covalent linkage[59]. Also, in nuclei that sustain replication of polyoma DNA[60], the very short replication fragments were found to have ribonucleotides in phosphodiester union with the DNA. The slime mold *Physarum* also showed similar evidence of RNA priming of discontinuously synthesized DNA[61].

RNA polymerase and transcriptional syntheses are involved in the origin of the *E. coli*[62] and phage λ chromosomes[63], but it is not known yet whether the RNA produced serves a priming function,

57. Hirose, S., Okazaki, R. and Tamanoi, F. (1973) *JMB* **77**, 501; Williams, P. H., Boyer, H. W. and Helinski, D. R. (1973) *PNAS* **70**, 3744.
58. Blair, D. G., Sherratt, D. J., Clewell, D. B. and Helinski, D. R. (1972) *PNAS* **69**, 2518; Williams, P. H., Boyer, H. W. and Helinski, D. R. (1973) *PNAS* **70**, 3744.
59. Sato, S., Ariake, S., Saito, M. and Sugimura, T. (1972) *BBRC* **49**, 827; Fox, R. M., Mendelsohn, J., Barbosa, E. and Goulian, M. (1973) *NNB* **245**, 234.
60. Magnusson, G., Pigiet, V., Winnacker, E.-L., Abrams, R. and Reichard, P. (1973) *PNAS* **70**, 412; Hunter, T. and Francke, B. (1974) *JMB* **83**, 123.
61. Huberman, J. A., Horwitz, H., Minkoff, R. and Wagar, M. A. (1973) *CSHS* **38**, 233.
62. Lark, K. G. (1972) *JMB* **64**, 47.
63. Dove, W. F., Inokuchi, H. and Stevens, W. F. (1971) **in** *The Bacteriophage Lambda* (A. D. Hershey, ed.) *CSHL* p. 747; Hayes, S. and Szybalski, W. (1973) *Fed. Proc.* **32**, 529 abs.; Dahlberg, J. E. and Blattner, F. R. (1973) *Fed. Proc.* **32**, 664 abs.

or a structural role in organizing the chromosome (see Chapter 1, Section 6) or possibly both.

With the recognition of RNA priming of DNA chain initiation, at least three important questions need to be answered about the detailed mechanism (Fig. 7-21). (i) What signals the RNA start? Are

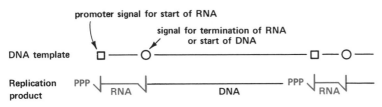

FIGURE 7-21
A scheme suggesting the presence of signals for initiation and termination of the RNA primer, and for initiation of the small DNA replication fragment.

there promoter-like regions at the replication origin that generate the forks? Are there distinctive signals at regular intervals on the template for initiation of the discontinuous synthesis of the small fragments? Or, is there simply a competition between RNA synthesis starts and DNA extensions? (ii) What determines the size of the RNA segment and its specific termination, as suggested by the presence of a distinctive nucleotide at the RNA-DNA linkage in the M13 RF and the *E. coli* replication fragments? (iii) What system is used for excising the RNA segment? One possibility is the $5' \rightarrow 3'$ exonucleolytic function of DNA polymerase I, which has the advantage that close coordination of this activity with the polymerizing function also takes care of filling the gap created by the excision (Chapter 4, Section 18)[64,65]. Another possibility is that the job is done by a nuclease such as RNase H[66,67], first discovered in animal cells and more recently in *E. coli*. The enzyme recognizes RNA-DNA hybrid regions and specifically degrades the RNA component.

6. Linking the Short Pieces: Ligase[68-70]

Discontinuous replication requires an enzyme that can join the replication fragments to the main body of the chromosome. This

64. Keller, W. and Crouch, R. (1972) *PNAS* **69**, 3360.
65. Roychoudhury, R. and Kössel, H. (1973) *BBRC* **50**, 259; Roychoudhury, R. (1973) *JBC* **248**, 8465.
66. Op. cit. in footnote 64.
67. Berkower, I., Leis, J. P. and Hurwitz, J. (1973) *JBC* **248**, 5914; Miller, H. I., Riggs, A. D. and Gill, G. N. (1973) *JBC* **248**, 2621; Henry, C. M., Ferdinand, F.-J. and Knippers, R. (1973) *BBRC* **50**, 603.
68. Gumport, R. I. and Lehman, I. R. (1971) *PNAS* **68**, 2559; Modrich, P., Anraku, Y. and Lehman, I. R. (1973) *JBC* **248**, 7495.
69. Harvey, C. L., Gabriel, T. F., Wilt, E. M. and Richardson, C. C. (1971) *JBC* **246**, 4523.
70. Konrad, E. B., Modrich, P. and Lehman, I. R. (1973) *JMB* **77**, 519; Gottesman, M. M., Hicks, M. L. and Gellert, M. (1973) *JMB* **77**, 531; Lehman, I. R. (1974) *Science* **186**, 790.

enzyme is apparently DNA ligase; its function is essential for normal chromosome replication and viability of the cell.

Even before the need for ligase in discontinuous replication became clear, such an enzyme was postulated for the breakage and reunion events in DNA recombination, and the swivel action of a replicating circular chromosome. The need for ligase was also seen for repair of DNA (Chapter 10, Section 2), where a sealing operation should follow the excision and patching of lesions.

Several different assays were successful in the nearly simultaneous discoveries of the enzyme (Fig. 7-22). The first one dealt with phage λ DNA, a linear duplex with cohesive or "sticky" ends which join and form a doubly nicked circle[71] (see Chapter 8, Section 7); the assay measured sealing of the nicked circle to a completely covalent one. Another assay utilized a pair of synthetic polymers in which many short chains of poly dT labeled with ^{32}P at the 5' end were aligned along a long chain of poly dA[72]. These labeled ends are susceptible to removal by phosphomonoesterase. Joining of the short chains was measured by the loss of ends susceptible to phosphomonoesterase. Another way to observe the joining reaction depended on using a short DNA chain immobilized on cellulose; when additional short pieces were linked to it they became sedimentable with the cellulose[73]. Still another assay, in current use, determines conversion of an open poly d(A–T)[74] or poly d(I–C) chain to a closed circular form no longer susceptible to exonuclease action.

Ligases from E. coli and B. subtilis[75] use DPN as cofactor, whereas the enzymes induced by T4[76] and T7[77] phage infections, and those found in animal cells depend on ATP[78]. E. coli ligase has been isolated in pure form; it is a single polypeptide with molecular weight of 75,000. Insights into the reaction mechanism have been made possible by having the pure enzyme in hand.

The mechanism involves three discrete reactions (i) formation of an enzyme intermediate by transfer of the adenyl group of the coenzyme to the ϵ-NH_2 of a lysine residue in the enzyme; the other product is NMN when DPN is the cofactor, and PP_i when the cofactor is ATP; (ii) adenylyl activation of the 5'-phosphoryl terminus of DNA by transfer of the adenyl group from the enzyme; and (iii) phosphodiester bond formation by attack of the 3'-hydroxyl terminus of the DNA on the activated 5'-phosphoryl group with release of AMP (Fig. 7-23).

The ligase reaction is demonstrably reversible at each step, as shown by: (i) exchange of radioactive NMN (E. coli ligase) or PP_i

71. Gellert, M. (1967) PNAS 57, 148; Gefter, M. L., Becker, A. and Hurwitz, J. (1967) PNAS 58, 240.
72. Olivera, B. M. and Lehman, I. R. (1967) PNAS 57, 1426, 1700; Weiss, B. and Richardson, C. C. (1967) PNAS 57, 1021.
73. Cozzarelli, N. R., Melechen, N. E., Jovin, T. M. and Kornberg, A. (1967) BBRC 28, 578.
74. Modrich, P. and Lehman, I. R. (1970) JBC 245, 3626.
75. Laipis, P. J., Olivera, B. M. and Ganesan, A. T. (1969) PNAS 62, 289.
76. Weiss, B. and Richardson, C. C. (1967) PNAS 57, 1021.
77. Masamune, Y., Frenkel, G. D. and Richardson, C. C. (1971) JBC 246, 6874.
78. Lindahl, T. and Edelman, G. M. (1968) PNAS 61, 680; Tsukada, K. and Ichimura, M. (1971) BBRC 42, 1156; Sambrook, J. and Shatkin, A. J. (1969) J. Virol. 4, 719.

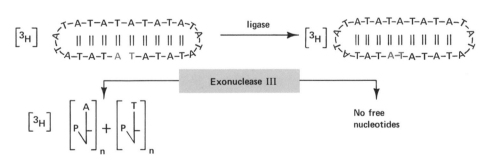

FIGURE 7-22
Procedures used in the assay of ligase. (a) Sealing of the two nicks in a cohered
phage λ circle causes the DNA to sediment rapidly in an alkaline velocity gradient.
(b) Sealing of oligomers labeled with ³²P at the 5′ end renders the ³²P insusceptible
to phosphomonoesterase. (c) Sealing of a labeled oligomer to DNA immobilized
on cellulose converts the labeled oligomer to a readily sedimentable form, despite
treatment with alkali. (d) Sealing of nicked poly d (A-T) to a closed circular
form renders it insusceptible to exonuclease action.

(phage-induced ligase) into DPN or ATP, respectively, in the absence of DNA; (ii) formation of DPN from DNA-adenylate, NMN, and enzyme in the presence of NH_4^+; and (iii) nicking or relaxation of a covalently closed, superhelically twisted DNA circle with AMP[79]; the nicked species is the adenylylated DNA; the relaxed form is the nicked circle rejoined after relief of the superhelical twist (Fig. 7-23).

Ligase action resembles that of DNA polymerase, but is distinctive in its details. For example, a 5'-triphosphate-terminated chain, in which pyrophosphate rather than adenylate is the activating group, does not serve as an intermediate; it is not attacked by the 3'-hydroxyl terminus of the adjacent chain.

The phage-induced and host-cell ligases differ in their cofactor requirement, in activation of the *E. coli* enzyme by NH_4^+ and in substrate specificities (Table 7-3). The phage ligase[80], but not the

TABLE 7-3
Substrates for ligases

	T4 ligase	*E. coli* ligase
Poly dA · oligo dT	yes (100)[a]	yes
Poly rA · oligo dT	yes (2)[a]	no
Poly dT · oligo rA	yes (2)[a]	no
Poly dA · oligo rU	no	no
Number of base pairs required at strand break	≥6	>10
End-to-end joining of duplex	yes	no

[a]Percentage values relative to that with poly dA · oligo dT set at 100.

E. coli enzyme, can link DNA chains aligned on an RNA chain, RNA chains aligned on a DNA chain, as well as an RNA to a DNA chain when both are aligned on DNA. The phage enzyme also tolerates somewhat shorter helical lengths on either side of a break and can even align and link the base-paired ends of two duplex chains[81]. These remarkable properties have been exploited in laboratory syntheses and manipulations of DNA (Chapter 11).

The functions performed by ligase in the cell, as judged by the behavior of mutants, are consistent with those observed in in vitro studies. The fivefold increase in ligase level induced by T4 phage infection is abolished by a mutation in the T4 ligase gene (gene 30)[82]. DNA synthesis starts at a normal rate but is nearly arrested within a few minutes. The DNA piles up in short pieces and very few phages are produced[83]. Attempts by T4 molecules to recombine result in DNA with gaps and branches. Evidently the host cell ligase does not

79. Modrich, P., Lehman, I. R. and Wang, J. C. (1972) *JBC* **247**, 6370.
80. Fareed, G. C., Wilt, E. M. and Richardson, C. C. (1971) *JBC* **246**, 925; Kleppe, K., van de Sande, J. H. and Khorana, H. G. (1970) *PNAS* **67**, 68.
81. Sgaramella, V. (1972) *PNAS* **69**, 3389.
82. Fareed, G. C. and Richardson, C. C. (1967) *PNAS* **58**, 665.
83. Sugimoto, K., Okazaki, T. and Okazaki, R. (1968) *PNAS* **60**, 1356; Newman, J. and Hanawalt, P. (1968) *JMB* **35**, 639.

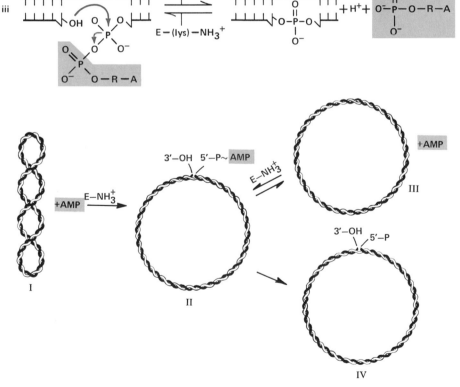

FIGURE 7-23
Sequence of three equations representing steps in the DNA ligase reaction (i–iii,
above) and an illustration (below) of its reversal; a superhelical circle (I) is relaxed
or nicked in the presence of AMP to produce an adenylylated enzyme intermediate
(II) which may be rejoined after relief of the superhelical twist (III) or hydrolyzed (IV).

suffice. However, some *E. coli* mutants[84] will support the normal reproduction of T4 gene 30⁻; the level of host cell ligase in these mutants is about five times that of wild-type *E. coli*.

E. coli mutants with a temperature-sensitive ligase (lig_{ts} 7) accumulate vast amounts of replication fragments at 42°, the restrictive temperature, and rapidly lose viability[85]. Extracts of this strain contain very low levels of a thermolabile ligase[86]. Interestingly, a "leaky" (less complete) mutant (*lig* 4)[84], extracts of which contain only 1–3 percent of the normal level of enzyme, still grows at a nearly normal rate at 42°, although it is more sensitive to damage by ultraviolet light. But such mutants are utterly unable to support DNA synthesis or phage production by the T4 ligase mutants.

E. coli contains about 300 molecules of ligase per cell. A crude calculation of the turnover number of the enzyme (25 joining events per minute at 30°) and the number of sealing events in replication (200 per minute per cell) suggests that only 2–3 percent of the molecules are required to sustain replication. Such a value is consistent with the near-normal growth rate of a mutant (*lig* 4) possessing a correspondingly low level of ligase and the deficiency of the mutant in repair of ultraviolet damage and other events that require additional ligase action.

The value of nearly 300 ligase molecules per cell is close to the estimated number of pol I molecules. Although this similarity may be fortuitous, there are reasons to believe that the two enzymes collaborate in the filling and sealing of gaps between segments of DNA. Conceivably the two enzymes may even form a complex assembly in the cell to carry out the integrated function.

The following discussion illustrates some complexities and subtleties in the intracellular function of ligases that remain to be explained, for example, the influence of T4 phage rII gene product.

Mutants in rII reproduce in *E. coli* B and produce large plaques, but they fail to multiply in *E. coli* K12, lysogenic for λ[87]. No other mutants make this distinction. To digress briefly, the T4 phage rII locus may be recalled for its role in the classic studies of ultrafine genetic mapping[87]. It is ironic that the protein coded by this gene remained elusive for so long after the functions of virtually all the other T4 genes had been assigned. Proof of the colinearity of gene and polypeptide chain[88], which might have come from the rII gene product, had to wait ten years for the fine-structure analysis of *E. coli* tryptophan synthetase and its genes, and the major coat protein in phage T4.

Now, important clues link the rII gene product to the in vivo role of ligase[89]. A mutation in the rII gene suppresses a mutation in the phage ligase gene (Table 7-4)[90]. Whereas the gene 30 mutant fails

84. Gellert, M. and Bullock, M. L. (1970) *PNAS* **67**, 1580.
85. Pauling, C. and Hamm, L. (1968) *PNAS* **60**, 1495.
86. Modrich, P. and Lehman, I. R. (1971) *PNAS* **68**, 1002.
87. Benzer, S. (1955) *PNAS* **41**, 344; (1961) *PNAS* **47**, 403.
88. Yanofsky, C. (1965–1966) *Harvey Lectures* **61**, 145; Stretton, A. D. and Brenner, S. (1965) *JMB* **12**, 456.
89. Berger, H. and Kozinski, A. W. (1969) *PNAS* **64**, 897.
90. Ebisuzaki, K. and Campbell, L. *Virol.* **38**, 701 (1969); Karam, J. D. (1969) *BBRC* **37**, 416.

TABLE 7-4
Suppression of ligase mutation by mutation in
the rII gene[a]

	E. coli B	E. coli B ligase⁻
T4 rII⁻	+	+
T4 ligase⁻	−	−
T4 rII⁻ ligase⁻	+	−

[a]+ and − signs denote phage growth, and failure of phage growth
respectively, in the wild-type and ligase⁻ mutant of E. coli.

to reproduce in *E. coli* B, introduction of a second mutation in the
rII gene (rII⁻, gene 30⁻) permits the phage to reproduce normally;
this suppression does not work if the host cells are themselves
deficient in ligase. These and additional observations have led to
the proposal that rII gene product is, or controls, a nuclease involved
in the morphogenesis of the phage. According to one proposal the
rII nuclease, by making many adventitious nicks in DNA, involves
and diverts a significant fraction of both T4 and host ligase molecules. By elimination of the damaging nuclease effects in the rII⁻
mutant, the available ligase can be spared for essential functions.
This explanation implies that rII gene product has a morphogenetic
role and that host ligase can partially supplement phage-induced
ligase.

Thus, nuclease actions may expose a cell's dependence on ligase[91],
while reserve or auxiliary supplies of ligase may obscure it. Ligase
is, according to all the studies done with phage and bacterial
mutants, an essential component in the processes of DNA replication, repair, and recombination.

7. Unwinding Proteins[92]

Although proteins to facilitate unwinding of the double helix in
advance of the replicating fork had been postulated and sought for
many years, the first suitable possibility appeared with the isolation
of the product coded by gene 32 of phage T4 and known to be essential for replication of T4 DNA[93].

The gene 32 protein[94], prototype of this class of proteins, is called
an unwinding protein because of its demonstrable capacity to convert a duplex DNA to single strands at a temperature 40° below the
midpoint of the melting temperature transition (T_m) for the DNA.
This property results from tight binding to single-stranded DNA
with little, or weak, binding to double-stranded DNA or RNA.

91. Sadowski, P. D. (1974) *J. Virol.* **13**, 226.
92. Alberts, B. and Frey, L. (1970) *Nat.* **227**, 1313; Alberts, B. (1971) **in** *Nucleic Acid-Protein
    Interactions* (D. W. Ribbons, J. F. Woessner, and J. Schultz, eds.) North-Holland, Amsterdam,
    p. 128.
93. Epstein, R. H., Bolle, A., Steinberg, C. M., Kellenberger, E., Boyde La Tour, E. and Chevalley,
    S. (1963) *CSHS* **28**, 375; Kozinski, A. W. and Felgenhauer, Z. Z. (1967) *J. Virol.* **1**, 1193.
94. Alberts, B. and Frey, L. (1970) *Nat.* **227**, 1313.

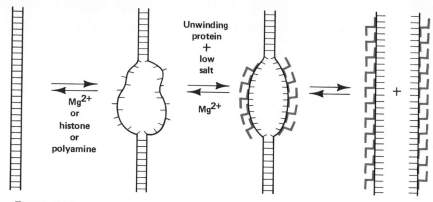

FIGURE 7-24
Unwinding protein and other agents influence the equilibrium between the duplex and single-stranded forms of DNA.

The binding to single strands occurs in regions of duplex DNA that undergo transient melting or "breathing" even at temperatures 40° below the melting midpoint. The energy of binding drives the melting to completion (Fig. 7-24). Because A–T-rich regions melt first, a denaturation map of phage λ (see Section 3 of this chapter), similar to that produced with alkali or heating, can be obtained by using an unwinding protein[95]. The binding involves the DNA backbone and shows no sequence specificity beyond that for low-melting regions.

Gene 32 protein, with a molecular weight of only 35,000, can bind a stretch of ten nucleotides because of its elongated shape; protein molecules are bound in tight alignment on the DNA chain. At high concentrations, the protein aggregates to form large multimeric complexes.

The binding is highly cooperative (Fig. 7-25) in that protein molecules line up successively in close juxtaposition along a DNA strand. Each protein molecule, because of favorable protein-protein interactions, binds adjacent to another protein molecule rather than at an isolated stretch of DNA. When viral single-stranded circles were mixed with an amount of gene 32 protein sufficient to bind one third of the DNA, one-third of the circles were completely covered with protein and appeared extended in the electron microscope[96]; the rest were bare and appeared collapsed.

Preferential binding of the single-stranded backbone by an unwinding protein, besides melting a duplex, can also have the reverse effect of facilitating base-pair alignment between single strands, a reaction of great importance in recombination (Chapter 9, Section 3) and also in replication. Presented with single-stranded DNA, the unwinding protein extends the chain into a conformation which increases the rate of reannealing of homologous strands under ionic conditions that permit the duplex to remain stable—even in the

95. Sigal, N., Delius, H., Kornberg, T., Gefter, M. L. and Alberts, B. (1972) PNAS **69**, 3537.
96. Delius, H., Mantell, N. J. and Alberts, B. (1972) JMB **67**, 341.

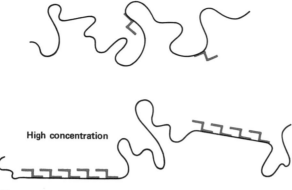

Low concentration

High concentration

FIGURE 7-25
Diagram illustrating cooperative binding of the unwinding
protein to DNA. (Courtesy of Professor B. Alberts.)

presence of the unwinding protein (Fig. 7-24). In this way, the rate
of DNA renaturation may be increased as much as 1000-fold and the
rate of polymerase replication ten-fold.

GENE 32 PROTEIN

The technique of DNA-cellulose chromatography, a type of affinity
chromatography using denatured DNA fixed to cellulose[97], made it
possible to separate from an extract of T4-infected cells some twenty
proteins with affinity for single-stranded DNA. The proteins could be
graded in strength of binding by elution with polyanions or salt
solutions. Among these proteins the most tenacious was the gene 32
product. It resists elution by dextran sulfate and requires 0.6 M to
2 M NaCl for detaching it from the DNA. This protein was lacking
from extracts of cells infected with gene 32 amber mutants. In these
infections there is neither replication nor recombination of the
phage DNA.

Gene 32 protein is apparently required throughout replication. In
contrast to the phage-induced enzymes, it is needed not in catalytic
quantities, but in amounts stoichiometric with the DNA made[98].
Thus it serves a structural function, even though not incorporated
into the mature phage. By binding to individual template strands at
the replication fork, gene 32 protein may effect a local unwinding,
protect the single strands, and extend the DNA into a conformation
optimal for exposure of the bases (Fig. 7-26). This possibility is
supported by observations that purified T4 polymerase is stimulated
by gene 32 protein in replicating the single-stranded template, that
the stimulation is specific for T4 polymerase (Table 7-5), and that
this protein and polymerase interact directly[99].

97. Alberts, B. and Herrick, G. (1971) *Meth. Enzym.* **22**, 198.
98. Sinha, N. K. and Snustad, D. P. (1971) *JMB* **62**, 267.
99. Huberman, J. A., Kornberg, A. and Alberts, B. M. (1971) *JMB* **62**, 39.

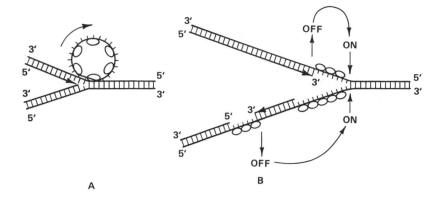

FIGURE 7-26
Schemes for an unwinding protein action at a replication fork. Unwinding
protein molecules act cooperatively in melting the template and orienting the
single strand for replication in the growing fork, and are then recycled.
(Courtesy of Professor B. Alberts.)

At levels of ionic strength and Mg^{2+} presumed to be present in
the cell, the T_m for T4 DNA would be about 85°. The action of gene 32
protein alone would therefore be inadequate to keep the duplex
pried apart at 30° or 37° in advance of the moving fork. There may
be contributions by still unknown proteins which, together with
gene 32 protein molecules, participate in a ribosome-like replication
assembly. The self-association of gene 32 protein into dimers[100],
even in 2 M KCl, has suggested a model in which DNA reannealing
is prevented and replication is facilitated.

It has been estimated that 60 replication forks act in a T4-infected
cell to maintain the very rapid rate of T4 DNA replication. Since
infected cells contain about 10,000 gene 32 protein molecules, each
replication assembly should contain about 170 molecules of gene 32
protein. Such an assembly unit could serve in each replication fork
as a moving scaffold and be used continuously until the replication

TABLE 7-5
Specificity of stimulation
of DNA polymerases by unwinding proteins

DNA polymerase	Unwinding proteins		
	T4 (gene 32)	T7	E. coli
T4	+	−	−
T7		+	−
E. coli I	−	−	−
II	−	−	+
III	−	−	−
III*	−		+

+ denotes stimulation; − denotes no stimulation.

100. Carroll, R. B., Neet, K. E. and Goldthwait, D. A. (1972) PNAS **69**, 2741.

of each DNA molecule is completed. The amount of gene 32 protein made would thus limit the rate and quantity of DNA production by determining the number of forks operating at a given time.

GENE 5 PROTEIN[101]

The protein coded by gene 5 of M13 phage is essential for production of viral single strands from the duplex replicative forms. It has a molecular weight of 10,000 and binds four nucleotides. Gene 5 protein coats the entire DNA. Thus 1500 molecules are required for each 6000-nucleotide strand. The steady-state population of about 100 viral single strands produced during infection thus involves about 150,000 gene 5 protein molecules. In mutants with a temperature-sensitive defect in the protein, viral strands do not accumulate but are instead diverted to the duplex replicative form.

Gene 5 protein may serve in a replicating fork and promote synthesis of, and use of, viral strands for assembly into phage. It may do so in one or more of the following ways: (i) protection of the viral strand against nucleases, (ii) prevention of the use of the viral strand as a template for conversion to a duplex by DNA polymerase or in unproductive binding by RNA polymerase, or (iii) conferral of a particular conformation that leads to encapsidation of the DNA by the coat protein stored in the inner membrane of the cell.

E. coli AND T7 PROTEINS[102]

Knowledge of the gene 32 and gene 5 proteins and their functions in DNA replication led to successful searches for unwinding proteins in extracts of uninfected E. coli and of cells infected with T7. The proteins were isolated by DNA-cellulose chromatography and characterized by their stimulation of a homologous DNA polymerase. The protein purified from E. coli stimulated replication of single-stranded DNA by DNA polymerase II and the pol III* system[103]; the protein induced by T7 infection stimulated only the T7 DNA polymerase (Table 7-5). However, due to lack of mutants defective in genes coding for either the host or T7 unwinding protein, it is difficult to be certain of their physiologic functions. The small number of unwinding protein molecules found in uninfected E. coli is in contrast to the many gene-32 protein molecules found in T4-infected cells. This can be reconciled with the few replication forks per E. coli chromosome as compared to the estimated 60 in a T4-infected cell. The properties of the gene 32, gene 5, and E. coli unwinding proteins are compared in Table 7-6.

EUKARYOTE UNWINDING PROTEIN

A protein isolated from the nuclei of higher plants (Lilium) and mammalian spermatocytes has properties resembling those of gene

101. Oey, J. L. and Knippers, R. (1972) JMB **68**, 125; Alberts, B., Frey, L. and Delius, H. (1972) JMB **68**, 139; Pratt, D., Laws, P. and Griffith, J. (1974) JMB **82**, 425.
102. Sigal, N., Delius, H., Kornberg, T., Gefter, M. L. and Alberts, B. (1972) PNAS **69**, 3537; Reuben, R. C. and Gefter, M. L. (1973) PNAS **70**, 1846.
103. Geider, K. and Kornberg, A. (1974) JBC **249**, 3999.

TABLE 7-6
Comparison of properties of unwinding proteins

	T4 Gene 32	M13 Gene 5	E. coli[a]
Molecular weight, native	35,000	10,000	74,000
Native form	monomer	monomer	tetramer
Isoelectric pH	5.5	–	6.0
Binding:			
DNA > RNA	yes	yes	yes
single strand > double strand	yes	yes	yes
Nucleotides bound/monomer	10	4	8
Weight ratio in binding:			
protein/DNA	12	8	9
Nucleotide packing when bound to protein	4.6 Å	–	1.8 Å
Cooperative binding	yes	yes	yes
Copies/cell	10,000	100,000	800
Copies/replicating fork	170	–	270
Biologic function	Replication recombination	RF → single strand	Replication

"Source of data on weight, form, and isoelectric pH: Weiner, J. H., Bertsch, L. I. and Kornberg, A. (1975) *JBC* **250,** 1972.

32 protein[104]. It has a molecular weight of 32,000, binds single-stranded DNA and promotes renaturation. Its location in the nuclear membrane, associated with lipids, has made purification difficult. Having been found only during the stage of meiosis, its relationship to recombinational events of the chromosome has been suggested.

An unwinding protein has also been isolated from somatic animal cells, but its properties remain to be determined[105].

8. Histones[106]

Histone proteins, in contrast to unwinding proteins, stabilize DNA in the duplex state (Fig. 7-24). Because of their basic nature, histones and polyamines (e.g. spermidine and spermine) can neutralize the phosphate groups of DNA and contribute to its organization in the doubly helical form. Although the functional roles of histones and polyamines are still not known, there is no doubt that histones, at least, are essential for the structure of chromosomes in nuclei of eukaryotic cells, which invariably contain them. Histones also contribute to the structure of DNA packaged in germ cells and to the organization of replicative forms of viruses that infect animal cells.

Earlier conjectures that histones are responsible for the fine regulation of replication and transcription have been weakened by

104. Hotta, Y. and Stern, H. (1971) *Develop. Biol.* **26,** 87; (1971) *NNB* **234,** 83.
105. Salas, J. and Green, H. (1971) *NNB* **229,** 165; Herrick, G. and Alberts, B., personal communication.
106. DeLange, R. J. and Smith, E. L. (1972) *Accounts of Chemical Research,* **5,** 368; (1971) *ARB* **40,** 279; Huberman, J. A. (1973) *ARB* **42,** 355.

the existence of only a very few kinds of histone molecules, as defined by their primary amino acid sequences. Five major histone fractions are found in calf thymus, for example, classified according to their relative arginine and lysine contents (Table 7-7).

TABLE 7-7
Characterization of calf thymus histones[a]

Class	Fraction	Minimum number of subfractions	Ratio of lysine to arginine	Total residues	Molecular weight
Very lysine-rich	I (f1)	3–4	20	~215	~21,000
Lysine-rich	IIb1 (f2a2)	2	1.25	129	~14,500
	IIb2 (f2b)	0–2	2.5	125	13,774
Arginine-rich	III (f3)	3	0.72	135	15,324
	IV (f2a1)	4	0.79	102	11,282

[a]Source: DeLange, R. J. and Smith, E. L. (1972) *Accts. of Chem. Res.* **5**, 368; (1971) *ARB* **40**, 279.

Histones, especially the arginine-rich histones, are markedly similar in different tissues and in widely different species. A most remarkable instance of the conservation of amino acid sequence in widely diverging lines of evolution is seen in a comparison of histone IV from many sources. This 102-residue histone, isolated from calf thymus and pea seedling, differs in only two residues, a substitution of valine for isoleucine and of lysine for arginine; it is identical in the rat, pig, and cow. This unusual conservation of sequence may have been imposed by the need this structural protein has for multiple contacts in complexes with other histones. Analogies may be seen in the evolutionary stability of amino acid sequence of the structural protein actin in lower forms (e.g. ameba) compared with that in muscle of trout and rabbit[107]; the same is true in a comparison of the tubulin proteins of microtubules in chick brain and sea urchin[108].

Subfractions of the main types of histones represent additional variety which may be important to histone structure and function. These variations include differences in amino acid sequence and an impressive display of posttranslational modifications of the polypeptide chain in the form of methylations, acetylations, and phosphorylations.

In chromatin, as the isolated nucleoprotein is called, the characteristic X-ray diffraction pattern depends on some undetermined complex of histones and DNA. In view of the constancy of histone composition relative to DNA, such complexes are likely to be determined more by DNA secondary structure than by sequence specificity. The identity of the histones and their precise inter-

107. Elzinga, M., Collins, J. H., Kuehl, W. M. and Adelstein, R. S. (1973) *PNAS* **70**, 2687.
108. Luduena, R. F. and Woodward, D. O. (1973) *PNAS* **70**, 3594.

actions with DNA hold essential clues to the organization of the chromosome and the control of its replication. Inasmuch as histones appear to be absent from prokaryotes, the distinctive features of eukaryotic DNA replication (e.g. slower fork movement, smaller replication fragments, and limitation to a small fraction of the cell growth cycle) may be partly associated with the histones.

Recent X-ray and biochemical analyses have led to a novel proposal for how histones organize the chromatin structure[109]. A tetramer, of two each of histones III and IV, and a similar complex of IIb1 and IIb2 form a unit cell containing 200 DNA base pairs and two molecules each of these four histones. Estimated dimensions for such a unit cell, 100 Å along the chromatin fiber axis with a fiber diameter of 100 Å, would be in keeping with electron microscopic data as well. A DNA packing ratio of 6.8 in the fiber (ratio of DNA duplex length to that of the chromatin fiber) suggests an accordian-like folding, or coiling, of the DNA duplex. The fiber, consisting of tightly packed DNA and associated protein alternating with more extended DNA, would resemble a string of beads. Interruptions in the histone coating of DNA at unit-cell intervals would explain the enzymatic cleavage of chromatin into DNA fragments of this size multiple[110] and might account for some of the distinctive features of chromatin replication[111].

Recent studies of histone biosynthesis and its interruption by cycloheximide, an inhibitor of protein synthesis, have suggested a coupling of the rate of DNA chain elongation with the availability of histones[112]. Dependence of histone biosynthesis on DNA synthesis has also been observed in the shutdown of histone synthesis by inhibitors of DNA synthesis[113].

The regulation of DNA replication and expression in chromatin must also depend on the nonhistone chromosomal proteins. They are a complex group of proteins containing, in the rat, 12 to 18 major components and at least 20 to 40 minor ones[114]. Their roles as enzymes, regulators, and structural elements remain to be determined[114a].

9. Proteins that Remove Superhelical Turns[115]

The operation of a swivel is needed to relieve the torque in unwinding a very long DNA duplex. Without it, replication of a fully

109. Kornberg, R. D., (1974) Science 184, 868.
110. Hewish, D. R. and Burgoyne, L. A. (1973) BBA 52, 504.
111. Kriegstein, H. J. and Hogness, D. S. (1974) PNAS 71, 135.
112. Weintraub, H. (1972) Nat. 240, 449.
113. Butler, W. B. and Mueller, G. C. (1973) BBA 294, 481.
114. Wu, F. C., Elgin, S. C. R. and Hood, L. E. (1973) B. 12, 2792.
114a. Stein, G. S., Spelsberg, T. C. and Kleinsmith, L. J. (1974) Science 183, 817.
115. Wang, J. C. (1971) JMB 55, 523; in DNA Synthesis In Vitro (1973) (Wells, R. D. and Inman, R. B. eds.) Steenbock Symposium, University Park Press, Baltimore, p. 163; Champoux, J. J. and Dulbecco, R. (1972) PNAS 69, 143.

covalent circle would be physically impossible. Although an unwinding protein can increase the rate of unwinding of the helix, it cannot alleviate the torque, nor correct the supertwisting, that develops immediately when the strands of an intact circular duplex are taken apart. The function of a swivel can be performed in two ways. One is by the joint action of a nuclease and ligase: nicking one strand of the duplex relieves the torque or twist; sealing the nick by ligase restores the covalent state of the duplex rod or circle. The demonstrable reversibility of ligase (Section 7 of this chapter) might suggest its function in this role but the kinetics seem highly unfavorable. The other alternative is the action of a protein which appears to untwist the duplex without apparent strand breakage and reunion.

An untwisting protein, called ω (no relation to the protein associated with RNA polymerase), was purified from E. coli[116] on the basis of its activity in removing superhelical turns from a highly twisted, circular, λ DNA. Inasmuch as ω causes no other physical damage in the DNA and leaves it as an intact duplex circle, the most attractive mechanism postulates a transient nicking and resealing. Yet, ω operates without an energy source and is unable to catalyze formation of a diester bond at a nick, as ligase does. An alternative mechanism to nucleolytic hydrolysis must therefore exist.

One possible mechanism for an untwisting protein is shown in Fig. 7-27. Instead of creating a nick with free 5′-phosphate and

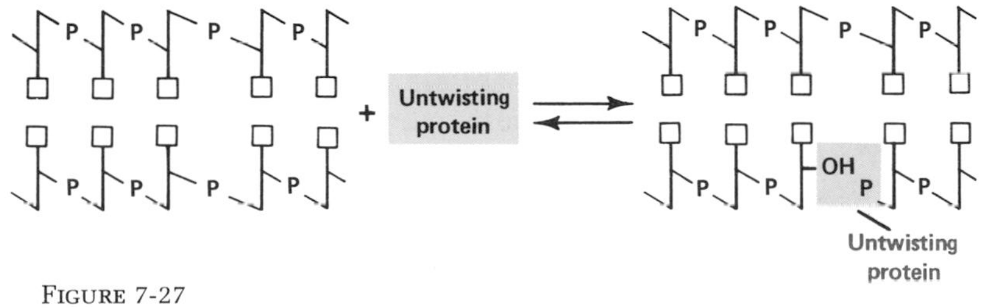

FIGURE 7-27
A possible mechanism for the untwisting of superhelical turns by a protein (see text).

3′-hydroxyl termini as a swivel point, a nick is formed with the untwisting protein attached to one or the other of these terminal groups. The energy of the diester bond is conserved in the attachment and the reaction is reversible. With the protein attached, the two ends of the nick are free to rotate relative to the helix axis. Such rotation takes place when the DNA has a deficiency of helix turns, and is thermodynamically unstable relative to the nicked form of DNA. Following some degree of untwisting of the duplex, the diester bond is reformed at the nick by detachment of the untwisting protein. Assuming that the protein acts catalytically, it can be considered a DNA-untwisting enzyme, a "swivelase".

An enzyme similar to ω has been identified in extracts of nuclei

116. Wang, J. C. (1971) JMB 55, 523.

from mammalian cells[117] by measuring removal of the superhelical turns in polyoma DNA. More recent studies[118] of this activity indicate that a mechanism such as that suggested for a DNA-untwisting enzyme might apply.

10. In vitro Replication[119]

T4 bacteriophage has at least six genes in its chromosome that code for proteins absolutely essential in DNA replication. The more complicated chromosomes of bacteria and larger cells must require at least as many. Which are these DNA replication proteins and what do they do? Some have been discussed already: RNA polymerase for initiation, DNA polymerases for chain extension and completion, unwinding protein for fork movement, and ligase for sealing nicks. The purpose of this section is to evaluate further the physiological contributions of the known replication proteins and to point out some proteins that seem to be missing.

Mutants for enzymes in a vital process such as DNA replication, whose end product, DNA, cannot be furnished in the medium, must of necessity be conditionally lethal. The conditional mutant must be able to produce a protein that sustains DNA replication under a "permissive" condition but is deficient under a "restrictive" condition. A number of such mutants have been obtained that support DNA replication at 30° but not 42°[120]. These temperature-sensitive mutants have provided important clues and guides to the existence and isolation of the enzymes essential to replication.

Among some mutants of *E. coli* that grow at 30° but not at 42°, RNA and protein synthesis are maintained at nearly normal rates at 42° but DNA synthesis is not. In one group of mutants, DNA synthesis stops or slows down (*extension type*), whereas in another it is sustained through one complete round of replication and then stops (*initiation type*).

The many mutants in which DNA synthesis is temperature sensitive, dna_{ts}, have been classified by the several laboratories where they were discovered into seven groups (*dna* A–*dna* G) based on the position of the mutation in the genetic map. Mutants of the *dna* D group now appear to be identified with the same genetic locus as those of the *dna* C group; the *dna* F mutants have since been found to be defective in ribonucleoside diphosphate reductase (Chapter 2, Section 6). Of the five distinctive replication mutants, *dna* A and *dna* C–D mutants are initiation type; *dna* B, *dna* E (identified as the gene for DNA polymerase III), and *dna* G are extension type.

To identify the protein products of genes and their functions in metabolism it is essential, as it is in the study of all metabolic events,

117. Champoux, J. J. and Dulbecco, R. (1972) *PNAS* **69**, 143.
118. Champoux, J. J., personal communication.
119. Richardson, C. C. (1969) *ARB* **38**, 795; Goulian, M. (1971) *ARB* **40**, 855; Klein, A. and Bonhoeffer, F. (1972) *ARB* **41**, 301.
120. Gross, J. D. (1971) **in** "*Current Topics in Microbiology and Immunology*," **57**, 39.

to resolve the system into its parts and then to reconstitute it. In practical terms this means to break the cell barrier and isolate the molecular components. Three experimental systems have been used in the attempts to resolve the enzymes of bacterial chromosome replication: (i) permeable cells accessible to low molecular weight compounds, such as the nucleotide precursors, but not to large proteins or nucleic acids, (ii) partial lysates and membrane complexes that retain many structural elements of the cell but are accessible to both precursors and proteins, and (iii) soluble enzymes in an extract that may be unable to replicate the complete bacterial chromosome but can sustain replication of certain phage chromosomes. Since some phages rely almost entirely on the host replication machinery, the enzymatic components of the host can be probed and isolated and their functions defined.

These three types of systems will be discussed now in more detail.

I PERMEABLE CELLS[121-124]

Rendering E. coli cells permeable to nucleotide precursors, by treatment with tris(hydroxymethyl)aminomethane and ethylenediamine tetraacetate (EDTA)[121], destroyed their capacity for semiconservative DNA replication. A better method uses lipid solvents (ether or toluene) which affect the membrane and appear to leave the cells intact, but unable to multiply and form colonies. Fortified with ATP, the four deoxynucleoside triphosphates, Mg^{2+}, and a K^+ buffer, these cells sustain semiconservative DNA replication at nearly physiological rates (2000 nucleotides per cell per second). ATP is required, not only to maintain deoxynucleoside triphosphate levels, but also for RNA priming, pol III* action, and possibly in additional ways.

In experiments with permeable cells, 10S replication fragments were shown to be the initial DNA products. They could later be joined to the chromosome in the presence of added DPN, the cofactor of E. coli ligase. NMN, which interrupts ligase action by promoting the reverse reaction, prevented joining of 10S fragments in these permeable cells.

DNA synthesis was interrupted in permeable cells of dna_{ts} mutants when the temperature was shifted to 43°. The reaction was also inhibited by drugs such as nalidixic acid and mitomycin, which are known to interrupt in vivo DNA synthesis, and by sulfhydryl-blocking agents such as N-ethylmaleimide. Cyanide and azide, which block DNA synthesis in the intact cell, have no influence in permeable cells.

DNA synthesis in permeable pol A⁻ cells superficially resembled that of wild-type cells growing at a normal rate. However, the

121. Buttin, G. and Kornberg, A. (1966) *JBC* **241**, 5419.
122. Moses, R. E. and Richardson, C. C. (1970) *PNAS* **67**, 674.
123. Moses, R. E., Campbell, J. L., Fleischman, R. A. and Richardson, C. C. (1971) in *Nucleic Acid-Protein Interactions* (Ribbons, D. W., Woessner, J. F. and Schultz, J. eds.) North-Holland-Amsterdam, p. 48.
124. Vosberg, H-P. and Hoffmann-Berling, H. (1971) *JMB* **58**, 739.

pol A⁻ permeable cells were deficient in several ways, one of which has been called a repair reaction. For example, a large amount of DNA synthesis in permeable, wild-type cells attributable to repair of degradation by endonuclease I was not observed in permeable *pol* A⁻ cells. This so-called "repair synthesis" was characterized in three ways: (i) absence in mutants lacking endonuclease I, (ii) lack of dependence on added ATP, and (iii) lack of inhibition by sulfhydryl-blocking agents. From this behavior it has been inferred that the polymerase active in repair of nuclease breaks is pol I, whereas replicative synthesis is attributable largely to another DNA polymerase, presumably pol III*.

Distinctions between DNA synthesis attributable to repair and DNA synthesis in chromosome replication have been noted in the behavior of permeable cells. This takes us a step beyond the inferences possible from studies of intact cells, but it is a step that has inherent risks. Disrupting the cell membrane and the organization of protein complexes that rely on its integrity may give a distorted image of in vivo events without offering in return a clearer resolution of the components of those events. The cells are inaccessible to antibody-size molecules and usually to smaller proteins as well. The next step in analysis of replication clearly requires a more extensive disruption of the permeable cell.

II PARTIAL LYSATES[125]

Extensive lysis of cells by lysozyme or detergents results in a major loss of DNA-replicating activity. However, when lysis is gentle and measures are taken to avoid dilution of the macromolecules in the cell, DNA replication resembling that of permeable cells in rate and character can be observed. Three such systems will be mentioned: (a) porous cells, prepared by toluene treatment and rendered permeable to macromolecules by the presence of a nonionic detergent, (b) cells lysed on a cellophane disk seated on an agar surface; and, (c) membrane complexes prepared by washing gentle lysates and purifying them by sedimentation in sucrose gradients.

(a) Porous cells. DNA replication in toluene-treated *E. coli* was maintained in the presence of one percent Triton X-100 even though pol I diffused freely into and out of the cell[126]. Repair synthesis was inhibited by antibody to pol I but the ATP-dependent, sulfhydryl-sensitive synthesis observed in a *pol* A⁻ strain was unaffected. Free passage of pol I was demonstrated in another way. X-ray-treated cells, in which the damaged DNA was unsuitable as a template for replicative synthesis, recovered capacity for replication when pol I was added to the incubation medium; presumably pol I contributed to the repair of the DNA lesions that had interrupted replication.

125. Schaller, H., Otto, B., Nüsslein, V., Huf, J., Herrmann, R. and Bonhoeffer, F. (1972) *JMB* **63**, 183.
126. Moses, R. E. (1972) *JBC* **247**, 6031.

Replication was blocked at elevated temperature in dna_{ts} mutants, but could be reestablished in the presence of Triton X-100 by replacing the defective protein with an active one supplied from extracts of wild-type cells. Thus, an assay for the purification of the *dna* gene products was made possible. The use of Triton X-100 also revealed an interesting facet in the control of DNA replication. The marked decrease in rate of replicative synthesis of cells entering stationary phase could be relieved by the presence of the detergent. It would seem that the detergent permits toluenized, stationary-phase cells with a suppressed level of DNA synthesis to express the full capacity for synthesis they display in exponential growth.

A system comparable to the toluene-Triton preparation of *E. coli* is the Brij 58 treatment of azide-poisoned *B subtilis*[127]. Such cells are permeable to proteins, lose 95 percent of their DNA-synthesizing activity (pol I) but still sustain their capacity to initiate new rounds of replication at the chromosome origin.

(b) Cellophane disk lysate. Cells may be lysed on a cellophane disk held on an agar surface, and then transferred with the disk onto a droplet of incubation mixture[128]. Small molecules can diffuse through the disk and the macromolecular components of the replication machinery are preserved at high concentration in the lysate on top of the disk. The lysate has proved to be accessible to macromolecules applied directly to the droplet[129]. The most crucial features of this system appear to be retained in cells plasmolyzed by exposure to 2 M sucrose but not actually lysed[130].

DNA synthesis in the gentle lysate on the cellophane disk could be prevented by pretreatment of cells with ultraviolet light, mitomycin, or chloramphenicol, agents that block DNA replication in vitro. Lysed cells had the same precursor requirements for DNA synthesis and the same response to inhibitors as permeable cells, and also were temperature-sensitive when prepared from dna_{ts} mutants, corresponding to the properties of initiation (*dna* A and *dna* C–D) and extension (*dna* B, *dna* E, and *dna* G) mutants. These characteristics have led to the use of this gentle lysate system as an assay for purification of several proteins from wild-type cells that are missing or defective in the lysates from mutant cells. For example, *dna* E and *dna* G proteins were purified from wild-type cells by assaying for the ability of a protein fraction to restore activity to a lysate of *dna* E or *dna* G mutant, respectively. Thus far, only the *dna* E and *dna* G proteins have been obtained in relatively pure form. At this point the limitations of the gentle lysate become apparent. For it is exceedingly difficult to define the function of a

127. Ganesan, A. T. (1971) *PNAS* **68**, 1296.
128. Smith, D. W., Schaller, H. E. and Bonhoeffer, F. J. (1970) *Nat* **226**, 711.
129. Schaller, H., Otto, B., Nüsslein, V., Huf, J., Herrmann, R. and Bonhoeffer, F. (1972) *JMB* **63**, 183.
130. Wickner, R. B. and Hurwitz, J. (1972) *BBRC* **47**, 202.

purified protein with a crude and complex lysate. The identity of the *dna* E product with pol III was established[131], but only because a separate polymerase assay was available and the properties of pol III were already known. By contrast, the precise function of the *dna* G protein, suggested to be in the initiation of small replication fragments[132], has not been resolved as yet.

(c) Membrane complexes[133]. Prepared by washing the gentle lysates and then sedimenting them in sucrose gradients, such complexes have not been as effective in maintaining DNA replication as the concentrated lysate on the disk, nor have such preparations proved successful in resolving the replication proteins.

At a certain stage in attempts to resolve the replication proteins and to understand their functions, advances with permeable cells and partial lysates are limited. Biochemical studies with a soluble enzyme system become essential.

III SOLUBLE ENZYMES[134]

DNA replication with all the characteristics of that obtained in permeable-cell preparations has not yet been attained with soluble enzyme fractions, although the polymerases, ligase, and unwinding proteins are now purified and can be supplied. A soluble extract of *E. coli*[135], developed to perform the first stage in the replication of a small viral chromosome, holds promise for successfully resolving more complex replicative steps[136].

Lysozyme lysis of *E. coli* at low temperature and without detergents yields an extract with soluble components that convert M13 and ϕX174 single strands to their circular, duplex replicative forms. This type of extract can also sustain stages in the replication of T7 and λ phages, and perhaps those of larger chromosomes. Another advantage for these studies is the virtually complete removal of host DNA from the extract.

Although the procedure for obtaining the extract can be described simply and has no exotic operations (Table 7-8), the details are exceedingly important and require close attention. The requirements emphasize the need for young cells of a certain genotype, harvested and extracted in a particular way. Cells prepared and extracted for the conversion of M13 viral strands to replicative form may not be optimal for another phage. The details are likely to differ depending not only on the phage studied and the stage of replication

131. Nüsslein, V., Otto, B., Bonhoeffer, F. and Schaller, H. (1971) *NNB* **234**, 285; Otto, B., Bonhoeffer, F. and Schaller, H. (1973) *EJB* **34**, 440.
132. Lark, K. G. (1972) *NNB* **240**, 237.
133. Knippers, R. and Strätling, W. (1970) *Nat.* **226**, 713; Okazaki, R., Sugimoto, K., Okazaki, T., Imae, Y. and Sugino, A. (1970) *Nat.* **228**, 223.
134. Wickner, W., Brutlag, D., Schekman, R. and Kornberg, A. (1972) *PNAS* **69**, 965; Schekman, R., Wickner, W., Westergaard, O., Brutlag, D., Geider, K., Bertsch, L. L. and Kornberg, A. (1972) *PNAS* **69**, 2691.
135. Op. cit. in footnote 134.
136. Weiner, J. H., Brutlag, D., Geider, K., Schekman, R., Wickner, W. and Kornberg, A. (1974) *In* R. Wickner, *DNA Replication*, p. 187 Marcel Dekker, Inc. N.Y.

TABLE 7-8
Factors in the preparation of active extracts for replicating
M13 and ϕX174 DNA

Variables		Purpose or description
General	Specific	
E. coli strain	Pol A$^-$	Reduce background of DNA replication
	Endo I$^-$	Reduce DNA degradation
	PNP$^-$ (polynucleotide phosphorylase)	Reduce background of RNA incorporation
Selected substrain	Rapid growth	Ease and reproducibility of cultivation
	Rapid lysis	Ease of lysozyme lysis
Conditions for growth	Nutrient medium (Hershey broth)	Rich, to sustain growth rate
	No antifoam	Avoid detergents
	Controlled aeration	Avoid foaming
Harvest	Time of harvest	Early to mid-log phase, $A_{590nm} = 0.5$; later stages inactive
	Temperature of harvest	Avoid chilling; use 23°
	Centrifugation	12,000 × g; 15 minutes
	Cell suspension	5–10 × 10^{10} cells/ml (400-fold concentration)
	Quick freeze cells in liquid N$_2$	Cells stable for months
	Storage of cells at −20°	Cells stable for months
Lysozyme lysis of cells (adjusted for successful lysis at 0°)	1. Thaw frozen cells at 4°	Freezing step required before lysozyme treatment
	2. Buffer: Tris-Cl, 50 mM	Phosphate inhibits at later stages
	3. pH 7.5	Range 7 to 8.5
	4. Sucrose, 10%	May preserve native enzyme conformations
	5. NaCl, 0.1 M	Ionic strength: range 0.05–0.50 M Cl$^-$
	6. Lysozyme to 0.1 mg per ml	
	7. 30 min lysis at 0°, 1.5 min lysis at 30° to 37°	Gentle lysis, avoid shear
	8. No auxiliary detergents	
The extract	Centrifugation: 200,000 × g for 60 minutes at 0°	Active, clear supernatant; free of DNA
	Storage: Quick frozen in liquid N$_2$; stored at −20°	Stable for months

examined but on the type of cell being used. What is most important is that a crude <u>enzyme extract</u> has been obtained in which the complex events of phage DNA replication can be enacted. The soluble enzymes in this extract are now susceptible to fractionation by standard methods.

11. Enzymes of Replication

Given an active extract, the prime requirement for purifying an enzyme is a simple and quick assay. The standard technique for measuring incorporation of labeled deoxyribonucleotides into DNA has made it possible to assay the enzyme in some twenty fractions from a separation procedure in an hour's time. Of course it is crucial in these assays that the nucleotide incorporation observed is dependent on the addition of the particular template-primer, on other agents known to affect the reaction, and on the identity of the DNA product expected. With such an assay in hand, purification of enzymes involved in the replication of M13, ϕX174, T4, T7, and λ DNA has been underway.

M13 replication. Conversion of the M13 viral strand to the duplex replicative form requires RNA polymerase and the *E. coli* unwinding protein for initiation[137] and the DNA polymerase III* system for covalent extension of the chain (Table 7-9; Fig. 7-28). The product of

137. Brutlag, D., Schekman, R. and Kornberg, A. (1971) *PNAS* **68**, 2826; Geider, K. and Kornberg, A. (1974) *JBC* **249**, 3999.

TABLE 7-9
Proteins required for conversion of M13 and ϕX174
viral strands to RF

Replication stage	Proteins required
M13: Single strand to RF II	RNA polymerase DNA polymerase III* ⎫ Pol III Copolymerase III* ⎭ holoenzyme *E. coli* unwinding protein DNA polymerase I Ligase
ϕX174: Single strand to RF II	*dna*A product *dna*B product *dna*C–D product (DNA polymerase III* ⎫ Pol III Copolymerase III* ⎭ holoenzyme *dna*G *E. coli* unwinding protein, and spermidine DNA polymerase I Ligase

ATP required in the pol III* system

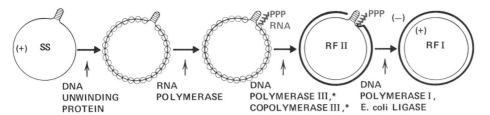

FIGURE 7-28
A scheme for the conversion of M13 viral strands to RF I by action of six purified proteins. In the absence of pol I, the location of the gap in the RF produced by the purified proteins is at the same unique place as that produced by the crude cell extract. The functions of the several proteins, either demonstrated or inferred, are as follows: RNA polymerase, to form a priming fragment; DNA unwinding protein, to mask the single-stranded DNA in all but the "promoter" region for RNA polymerase and to support pol III*-copolymerase III* in carrying out most of the DNA chain growth; pol I, to fill the small remaining gap, including that created by its 5′ → 3′ exonucleolytic excision of the remaining RNA fragment; and ligase, to join the polynucleotide ends to complete the synthesized complementary circle.

A duplex, hairpin region is suggested as the "promoter" to account for its relative freedom from being bound by the unwinding protein and for its resemblance to the standard promoters of transcription in duplex DNA.

replication is RF II, a duplex in which a nearly full-length, complementary strand has been synthesized on an intact, circular viral strand[138]. Removal of the RNA priming fragment and filling the small gap created by the excision can be performed by pol I; RNase H, which removes RNA from an RNA-DNA hybrid, may also serve as excising agent. Ligase is essential for sealing RF II to produce the covalently closed, circular duplex, RF I. It seems likely that the enzymes active in this replication event are packaged in the cell in a hypothetical complex, called a "replisome", analogous to one described below for ϕX174 replication. Based on the RNA polymerase involvement in the in vivo replication of plasmids and the F and colicinogenic E$_1$ factors (see Chapter 8, Section 8), we may suggest that invading M13 DNA competes for and exploits the cellular machinery involved in the ongoing replication of cellular plasmids. If this supposition were correct, replication of M13 would be utilizing the "plasmid replisome".

ϕX174 replication.   Despite the superficial similarity of the ϕX174 chromosome to that of M13, the enzyme system involved in its replication is strikingly different[139]. Indications are clear that ϕX174 unlike M13, utilizes the cellular machinery responsible for replicating the host chromosome. Gene products judged to be essential in vivo for both initiation (dna A and dna C–D) and extension (dna B, dna E, and dna G) are required (Table 7-9). Current efforts

138. Wickner, W., Brutlag, D., Schekman, R. and Kornberg, A. (1972) PNAS **69**, 965.
139. Schekman, R., Wickner, W., Westergaard, O., Brutlag, D., Geider, K., Bertsch, L. L. and Kornberg, A. (1972) PNAS **69**, 2691; Wickner, R. B., Wright, M., Wickner, S. and Hurwitz, J. (1972) PNAS **69**, 3233.

to purify each of these proteins employ two different approaches. One depends on a *complementation assay*. An extract of a dna_{ts} mutant containing a particular gene product inactive at elevated temperature is complemented by extracts of wild-type cells that contain this protein in a form that functions at high temperature. Thus an assay for purification of the protein is provided. In this way the purification of *dna* A, *dna* B, *dna* C–D, *dna* E, and *dna* G products has been pursued. The *dna* E product has been identified with pol III and pol III* (Chapter 5, Section 3); the functions of the others are still under study.

The other approach is a straightforward enzyme fractionation of wild-type cells based on nucleotide incorporation in the conversion of ϕX174 single strands to RF. An observation that replication proteins appear to be organized in a complex form has influenced some of the fractionation work. Assuming that the ϕX174 system depends on a "host-chromosome replisome", the assays in both approaches depend on assimilating the protein correctly into the complex as a prerequisite for its catalyzing the replicative event.

In the current understanding of the enzymatic steps in the conversion of ϕX174 single strand to RF II, it is clear that synthesis of the initiating RNA fragment is carried out by enzymes distinct from the classic RNA polymerase, and that chain growth depends on the pol III* system. Spermidine, a polybasic amine ($H_2N(CH_2)_4$ $NH(CH_2)_3NH_2$), is used for replication studies in vitro but must be supplemented by the *E. coli* unwinding protein. As in the replication of the M13 viral strand, the initial product is a complementary linear strand which requires excision of the 5'-RNA primer end and its replacement by DNA before ligase closure can produce the RF I.

The enzymatic conversions of M13 and ϕX174 viral strands to the duplex form have been achieved by incubating these strands with extracts of uninfected cells. However it is evident, as will be explained in Chapter 8, that free single strands are not intermediates in the conversion in vivo; removal of coat protein from the viral DNA is tightly coupled to its replication. Furthermore, the conversion requires the participation of a protein in the virion, and is an operation oriented in or on the cell membrane[140]. A proposal for how the phage DNA competes successfully with DNA of the host for the plasmid and chromosomal replisomes, and how it exploits the cell membrane for their operation, will be presented in the next chapter.

T4 replication[141]. Insights into the replication of the more complex T4 chromosome may be anticipated from the isolation of products of several genes known to be essential for in vivo replication. The complementation assay approach has been used. The protein products of genes 41, 44, 45, and 62 have been obtained in nearly pure form and their molecular weights determined; gene 44 and 62

140. Jazwinski, S. M., Marco, R. and Kornberg, A. (1973) *PNAS* **70**, 205.
141. Barry, J. and Alberts, B. (1972) *PNAS* **69**, 2717.

proteins form a very tight complex. While each of the proteins stimulates DNA synthesis in crude extracts lacking the specific protein, in no instance has its function been determined as it has in the case of the established replication proteins, DNA polymerase (gene 43), ligase (gene 30), and unwinding protein (gene 32). Based on what is known of the interactions of these proteins with DNA and each other, a hypothetical assembly of them has been proposed (Fig. 7-29). Certainly additional proteins, either phage-induced or

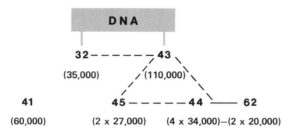

FIGURE 7-29
Possible interactions of T4 replication proteins in their function as a replication assembly unit. (Barry, J. and Alberts, B. (1972) *PNAS* **69**, 2717.)

of host origin, will be implicated in the organization and function of the complete T4 replisome.

DNA replication in general. Enzymes utilized in the replication of the small, single-stranded phage chromosomes are borrowed from the apparatus used by the host in replicating its own larger, duplex chromosome. How some of these enzymes may operate at a replicating fork is suggested in Fig. 7-30. Proteins known to be

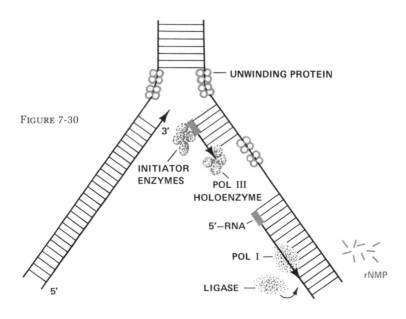

FIGURE 7-30

needed for ϕX174, *E. coli*, and T4 replication, and not included in this illustration, will (when their functions are understood) provide material for correcting or amplifying the illustration.

12. Nucleases in Replication[142]

The importance of nucleases as reagents (Chapter 9, Section 1) in the analysis of DNA structure and functions has been mentioned; just a few examples should suffice as reminders. The 3',5'-phosphodiester backbone was determined by the combined actions of pancreatic DNase and snake venom phosphodiesterase; the 5'-deoxyribonucleotides released by this action provided substrates for the initial studies of enzymatic synthesis of DNA. Micrococcal nuclease was essential for nearest-neighbor analysis of the enzymatic product and with it proof of the antiparallel strand arrangement of the double helix. The 5' $\longrightarrow$ 3' direction of replication in DNA replication fragments was determined with specific exonucleases. Discussions of sequence analysis and synthesis of genes in Chapter 11 will furnish additional examples of the usefulness of nucleases as reagents.

The essential functions that nucleases perform in biosynthetic pathways have not been sufficiently appreciated. Because we have come to reject the notion that biosynthesis occurs by reversal of breakdown reactions and we know that synthetic routes are distinct from degradative pathways, it is important to emphasize that some degradative operations may be required in a biosynthetic sequence. Several instances can be cited: In glycogen synthesis, fructose-1, 6-diphosphatase is a key enzyme; in triglyceride synthesis, phosphatidic acid phosphatase is essential; in protein synthesis, proteolytic cleavage generates enzymes from zymogens, insulin from proinsulin, and viral coat proteins from a larger precursor. In the manufacture of ribosomal, messenger, and transfer RNAs, the active molecules are derived by nucleolytic scissions of larger precursor molecules.

Nucleases may participate in DNA biosynthesis in several ways.

(i) Nicking that might in certain instances create an origin for replication (Chapter 8, Sections 3, 4).

(ii) Coordinate nicking and reunion to produce a swivel for relief of helical torque in replication (see Section 9 of this chapter).

(iii) Excision of the RNA fragment that primes the initiation of DNA chains (see Section 6 of this chapter).

(iv) Maturation or packaging of viral chromosomes, for which specific scissions are required (Chapter 8).

(v) Repair of lesions that interrupt replication (Chapter 9).

142. Lehman, I. R. (1967) *ARB* **36**, 645.

(vi) Salvage of precursors required for synthesis, as in the break-down of host DNA in cells infected with viruses (Chapter 8).

13. Role of the Membrane[143]

The membranes of interest in replication are the plasma membrane in bacteria and nuclear membrane in animal cells, particularly their inner surfaces, on which DNA replication may be oriented. A current conception of membrane organization is that of a lipid bilayer in which proteins are placed among the lipids in a mosaic-like arrangement. Presumably the location of DNA at the membrane surface is governed by interaction with a specific protein fixed on, or partially in, the lipid bilayer. Inasmuch as the organization of lipid and protein components of a membrane depends entirely on non-covalent bonds, it is inherently difficult to preserve and define their specific relationships. Hydrophobic interactions, electrostatic bonding, and varying combinations of these linkages establish the continuity of the membrane. Detergents that discharge DNA from the membrane also disrupt the membrane, and dissociation of electrostatic interactions by salt treatments are no more definitive.

Replication of the bacterial chromosome is geared to cell division during which the genomes must be segregated and partitioned equally between the daughter cells. It seems plausible that this might be achieved by linking the growing chromosome to the cell membrane. Expressing this idea, one model proposes that the growing chromosome is attached to membrane sites[144]. Concurrent growth of new cell membrane separates the developing genomes, and finally septation of the newly synthesized membranes partitions them.

No one has yet succeeded in isolating a unique membrane attachment site for DNA, and the DNA-membrane association remains indecisive. The available lines of evidence for such an association are as follows:

ELECTRON MICROSCOPIC AND
AUTORADIOGRAPHIC STUDIES
Thin sections of B. subtilis appear to show nuclear material touching a complex invagination of the membrane, called the mesosome[145]. As daughter cells separate during cell division, the mesosome of each cell retains contact with the DNA. In protoplasts, where mesosomes are absent, there also seems to be a point of contact between the DNA and membrane. Although the electron microscope image has pointed out the proximity of DNA to membranes, the existence or nature of chemical linkage between them remains to be found.

143. Siegel, P. J. and Schaechter, M. (1973) Ann. Rev. Microb. 27, 261; Dworsky, P. and Schaechter, M. (1973) J. Bact. 116, 1364; Worcel, A. and Burgi, E. (1974) JMB 82, 91.
144. Jacob, F., Ryter, A. and Cuzin, F. (1966) Proc. RS Lond. (B) 164, 167.
145. Ryter, A. (1968) Bact. Rev. 32, 39.

In animal cells, the observations have shown chromosomes attached to the nuclear membrane during interphase and the membrane participating in the arrangement of chromosomes during meiosis. However, recent autoradiographic studies now lead to the conclusion that replication starts throughout the chromosome[146]. Furthermore, some protozoan nuclei and polytene chromosomes in *Diptera* have replication patterns that seem to foreclose any contact with the peripheral nuclear membrane[147].

ASSOCIATION OF NASCENT DNA WITH A SEDIMENTABLE MEMBRANE FRACTION

Nascent DNA was recovered with the rapidly sedimenting cellular fraction which contained, among other things, the plasma membrane. When *B. subtilis*[148] and *E. coli*[149] were given a brief pulse of ^{3}H-thymidine, the membrane fraction was several-fold enriched in labeled DNA relative to total DNA. This enrichment could be reduced by a chase with unlabeled thymidine. The genetic markers at the origin of the *B. subtilis*[150] and the *E. coli*[151] chromosomes also remain associated with the membrane fraction.

Replicating DNAs of infecting phages (P22, λ, T4, φX174, and M13) at several stages of their life cycles were also found in the host cell membrane fraction[152]. These associations may include DNA sites not undergoing replication. The evidence also suggests that maturation of many phages involves the membrane, since components of the phage coat in various stages of its development were found associated with it.

ASSOCIATION OF NASCENT DNA WITH PHOSPHOLIPIDS[153]

The nascent DNA of cell extracts lysed with Sarkosyl (sodium lauryl sarcosinate) was associated with crystals of the detergent that form with Mg^{2+}. These crystals form a sharp band, called the M-band, in sucrose gradients. The phospholipids in membranes band with the Sarkosyl crystals. Attachment of the DNA to the M-band is indirect, and depends on phospholipid in the membrane fraction. Although it had been inferred that DNA was attached to the membrane at one point through the replicative fork, recent evidence indicates instead multiple points of attachment[153a]. Despite a 20-fold fragmentation of the DNA by X-ray irradiation, there was no decrease in its binding to

146. Huberman, J. A., Tsai, A. and Deich, R. A. (1973) *Nat* **241**, 32; Fakan, S., Turner, G. N., Pagano, J. S. and Hancock, R. (1972) *PNAS* **69**, 2300.
147. Prescott, D. M. and Murti, K. G. (1973) *CSHS* **38**, 609; Rudkin, G. T. (1972) **in** *Results and Problems in Cell Differentiation*, (W. Beermann, ed.), Vol. 4 Springer-Verlag, N.Y. p. 59.
148. Ganesan, A. T. and Lederberg, J. (1965) *BBRC* **18**, 824.
149. Fuchs, E. and Hanawalt, P. (1970) *JMB* **52**, 301.
150. Sueoka, N. and Quinn, W. G. (1968) *CSHS* **33**, 695. Yamaguchi, K. and Yoshikawa, H. (1973) *NNB* **244**, 204.
151. Fielding, P. and Fox, C. F. (1970) *BBRC* **41**, 157.
152. Siegel, P. J. and Schaechter, M. (1973) *Ann. Rev. Microb.* **27**, 261.
153. Ballesta, J. P., Cundliffe, E., Daniels, M. J., Silverstein, J. L., Susskind, M. and Schaechter, M. (1972) *J. Bact.* **112**, 195; Tremblay, G. Y., Daniels, M. J. and Schaechter, M. (1969) *JMB* **40**, 65.
153a. Dworsky, P. and Schaechter, M. (1973) *J. Bact.* **116**, 1364.

the M-band. Also, the reduction in binding of nascent DNA by rifampicin, an inhibitor of RNA transcription, suggests that this attachment of nascent DNA depends more on a transcriptional complex than on a replicative one.

An indirect, but suggestive metabolic interrelationship between DNA replication and membranes has come from studies with E. coli mutants bearing a temperature-sensitive defect in an enzyme essential for phospholipid biosynthesis[154]. Upon shifting the cells to the restrictive temperature, DNA synthesis stops abruptly, but so do the syntheses of RNA and proteins. It is not at all clear yet how these macromolecular biosyntheses are tied to phospholipid and membrane production.

In other E. coli mutants, defective in the dna A product essential for the initiation of a new chromosome, alterations in membrane properties have been noted at the high temperature, including abnormal sensitivity to lysis by a detergent[155]. Some changes have also been seen in the polyacrylamide gel patterns of membrane proteins denatured in sodium dodecyl sulfate[156].

ASSOCIATION OF THE FOLDED NASCENT
CHROMOSOME WITH CELL MEMBRANES

Gentle lysis of E. coli spheroplasts with nonionic detergents in high salt or spermidine yields the complete chromosome attached to cell membrane fragments, provided the temperature is kept at 10° or lower (see Chapter 1, Section 6)[157]. At higher temperatures or upon dissolution of the membranes by Sarkosyl, the chromosome is released[158]. Of great importance are the findings[159] that this chromosome attachment is correlated with initiation of a round of replication, and depends upon the synthesis of specific proteins. In amino-acid-starved cells that have completed a round of replication, all the DNA was isolated in an unbound state. Only after the resumption of protein synthesis did isolation of a chromosome-membrane complex become possible. Presumably cycles of chromosome-membrane association and disassociation accompany the replicative phases of initiation and growth of the chromosome.

Despite its plausibility, and the morphologic and biochemical evidence that support it, the DNA-membrane relationship still remains poorly defined. The difficulty is inherent in the structure of membranes. It can be seen that reasons for believing in the attachment of the chromosome to the cell membrane are of several kinds: contiguity in electron microscope pictures, metabolic relationships, and cosedimentation in velocity gradients. None of the evidence provides convincing information about the chemical nature of DNA

154. Glaser, M., Boyer, N., Bell, R. and Vagelos, R. (1973) PNAS **70**, 385.
155. Hirota, Y., Mordoh, J. and Jacob, F. (1970) JMB **53**, 369.
156. Shapiro, B. M., Siccardi, A. G., Hirota, Y. and Jacob, F. (1970) JMB **52**, 75; (1971) JMB **56**, 475.
157. Stonington, O. G. and Pettijohn, D. E. (1971) PNAS **68**, 6; Pettijohn et al. **in** Wells, R. D. and Inman, R. B. (eds.) (1973) DNA Synthesis In Vitro, Steenbock Symposium, Univ. Park Press, Baltimore.
158. Worcel, A. and Burgi, E. (1972) JMB **71**, 127.
159. Worcel, A. and Burgi, E. (1974) JMB **82**, 91.

attachment to the membrane. Nevertheless such an attachment is a very useful working hypothesis. Further and hopefully definitive information may come from *examining specific DNA functions* that are dependent on, or influenced by, a membrane interaction in a cell-free system. Replication of small viral chromosomes (described in the next chapter) may provide a system for studying such sharply defined functions.

14. Inhibitors of DNA Replication[160-163]

Mutants have been useful in elucidating complex biochemical events and in assessing the physiological function of isolated enzymes. However the mutant approach becomes complicated in analyzing a reaction sequence essential for life of the cell. Such mutants can be obtained under only two circumstances: (i) when the mutation can be circumvented by uptake of the reaction product supplied to the medium, or (ii) the mutation is conditional; that is, when production, stability, or function of the gene product occurs under one condition and not another (i.e. temperature-sensitivity). It is only the latter kind of mutant that can be and has been used for analyzing the actual replication of DNA. Even so, conditional mutants suffer from the difficulty that the protein defect may be only partially manifest under the stressful conditions or may be slow in developing.

The limitations of mutants in analysis of biosynthesis of essential macromolecules encourages the use of antibiotic inhibitors. It should be possible to obtain an almost limitless arsenal of antibiotics. The large variety of antibiotics which bind and inhibit the ribosome suggest that selective pressures have led to the development in nature of antibiotics for virtually every essential bacterial protein. With diligent searching and chemical modifications, it should be possible to match the usefulness of mutants with antibiotic agents directed against every protein and every reaction used by a cell. Just as the application of antibiotic inhibitors has permitted dissection of the intricate mechanisms of protein biosynthesis, so can analysis of nucleic acid biosynthesis by inhibitors be uniquely useful. Many examples are already at hand but much more can be done by expanding the number of agents, and by a more assiduous application of those available.

A discussion of inhibitors of DNA replication must include inhibitors of RNA synthesis too, because of the direct role of RNA transcription in initiation of DNA chains as well as the instructive analogies between both processes. Inhibitors of nucleic acid synthesis can be categorized by their effects: (i) template binding, (ii)

160. Goldberg, I. H. and Friedman, P. A. (1971) *ARB* **40**, 775.
161. Suhadolnik, R. J. *Nucleoside Antibiotics.* (1970) John Wiley & Sons, N.Y.
162. Roy-Burman, P. (1970) *Analogues of Nucleic Acid Components.* Springer-Verlag, N.Y.
163. Sober, H. A. (Ed.) (1970) *Handbook of Biochemistry, Selected Data for Molecular Biology,* 2nd edition, the Chemical Rubber Co., Cleveland, Ohio (Section G, p. 115).

nucleotide analogs, (iii) enzyme binding, and (iv) unknown mechanism (Table 7-10).

TABLE 7-10
Inhibitors of DNA and RNA Synthesis.

Type of Inhibition	Inhibitor	Mechanism	Inhibition or effect
I. Template-binding A. Noncovalent			
	Actinomycin D	Binds to and intercalates between dG-dC pairs	RNA and DNA chain elongation
	Anthracyclines (e.g. Nogalamycin)	Binds and intercalates alternating A–T sequence	RNA and DNA chain elongation
	Acridine dyes (e.g. Acriflavin)	Intercalates	Frameshift mutation; RNA chain initiation
	Ethidium bromide	Intercalates	DNA replication and mutagenesis
	Kanchanomycin	Strong Mg^{2+} complex with template	E. coli DNA polymerase I
	8-Aminoquinolines		E. coli and M. luteus DNA polymerase I; RNA polymerase only partially
B. Covalent			
	Mitomycin (reduced)	Cross-links	DNA replication
	Bleomycin, Phleomycin	Chain breaks	DNA replication
	Anthramycin		DNA and RNA synthesis
II. Nucleotide analogs A. Chain terminator			
	Dideoxynucleoside triphosphates	Incorporated into DNA	DNA chain growth and $3' \rightarrow 5'$ degradation
	Arabinosyl nucleoside triphosphates	Incorporated into DNA	DNA chain growth and $3' \rightarrow 5'$ degradation; bacterial and animal cells
	Cordycepin triphosphate (3'-deoxy ATP)	Incorporated into DNA and RNA	DNA and RNA chain growth
	3'-Amino ATP	Incorporated into DNA and RNA	DNA chain growth
B. Defective DNA			
	dUTP	Incorporated into DNA	Template is degraded
	5-Hydroxyuridine, or 5-aminouridine	Incorporated into RNA	RNA and DNA synthesis
	5-Bromouracil	Incorporated into DNA	Mutagenic, causes replication errors
	Tubericidin "ATP", Formycin "ATP"	Incorporated into RNA and DNA	RNA and DNA synthesis and functions
C. Enzyme binding			
	Hydroxyphenyl-hydrazinouracil	Ternary complex with template and enzyme	B. subtilis polymerase III; reversed by dGTP
	Hydroxyphenylhy-drazinoisocytosine	Ternary complex with template and enzyme	B. subtilis polymerase III; reversed by dATP
D. Nucleotide synthesis[a]			

[a] See Chapter 2

(continued)

TABLE 7-10 (continued)
Inhibitors of DNA and RNA Synthesis

Type of Inhibition	Inhibitor	Mechanism	Inhibition or effect
III. Enzyme binding			
	Rifampicin, Streptovaricin	β-subunit of bacterial RNA polymerase	RNA initiation
	Streptolydigin	β-subunit of bacterial RNA polymerase	RNA chain growth
	α-Amanatin		Mammalian nuclear RNA polymerase
	Kanchanomycin	} See IA	E. coli RNA polymerase
	8-Aminoquinolines		See IA
IV. Unknown mechanism			
	Edeine		DNA replication in bacteria
	Nalidixic acid		DNA replication in bacteria

I. Template Binding

Most antibiotics that inhibit DNA replication do so by binding the template. Binding may be through ionic, hydrophobic or other noncovalent interactions, or through covalent attachments.

(A) Noncovalent binding. *Actinomycin D* (Fig. 7-31) is one of the most extensively used and studied inhibitors of nucleic acid synthesis, particularly in blocking RNA synthesis. It has a lesser effect on DNA synthesis. An attractive model, based on X-ray diffraction patterns of nucleotide-drug crystals, pictures intercalation of the planar phenoxazone ring system between alternating dG–dC base

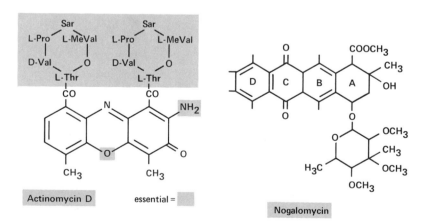

FIGURE 7-31
Actinomycin D and nogalomycin. Elements of the actinomycin structure essential to its action are indicated in color. MeVal, methylvaline; Pro, proline; Sar, sarcosine; Thr, threonine; Val, valine.

pairs and hydrogen-bond linkage of the cyclic peptide portion with the 2-amino group of guanine in the minor groove[164].

Nogalamycin, an intercalative anthracycline, prevents the de novo synthesis by *E. coli* DNA polymerase I of the alternating copolymer poly d(A–T), thus enabling the homopolymer pair poly (dA)·(dT) to be formed[165]. This result suggests strong binding between the alternating dA–dT base pairs.

Acriflavin and *ethidium bromide* (Fig. 7-32) have received much attention because of their use in buoyant density separation of nicked (relaxed) DNA from closed (twisted) circular DNA[166]. The

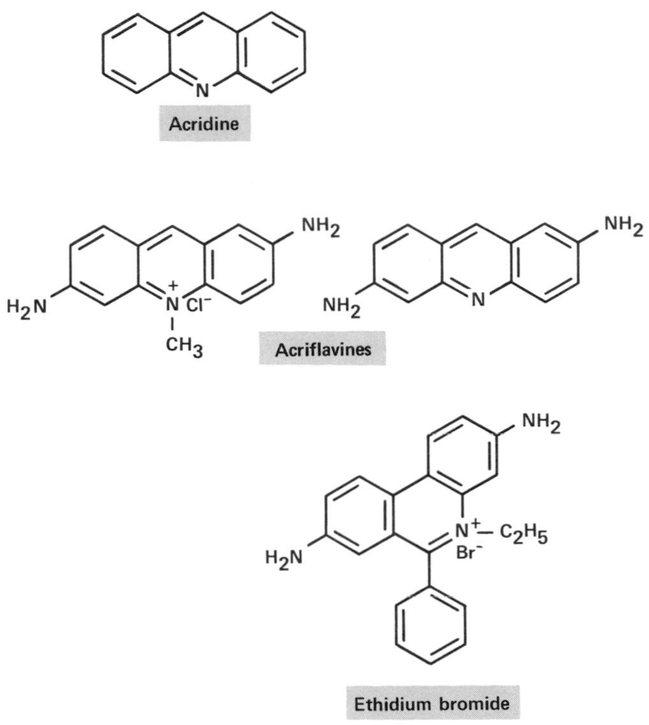

FIGURE 7-32
Acridine, two acriflavines, and ethidium bromide.

buoyant density of the DNA decreases in proportion to the amount of dye intercalated; more dye molecules bind to the open than to the constrained closed duplexes. Acridine dyes "cure" bacterial cells of their episomes, an effect that may be explained by removal of these closed circles from the pool of replicating molecules, causing their disappearance by dilution among the multiplying cell population. Intercalating dyes produce frameshift mutations, by causing the deletion or addition of a base upon replication. Mutagenic effects

164. Sobell, H. M. and Jain, S. C. (1972) *JMB* **68**, 21.
165. Olson, K., Luk, D. and Harvey, C. L. (1972) *BBA* **227**, 269.
166. Radloff, R., Bauer, W. and Vinograd, J. (1967) *PNAS* **57**, 1514.

on the twisted, circular mitochondrial genes of yeast are greater
than on nuclear genes[167]. Proflavin, another acridine, inhibits RNA
synthesis at initiation rather than in elongation, perhaps by blocking
formation of the DNA-RNA polymerase complex[168].

Kanchanomycin inhibits DNA polymerase I on duplex DNA by
blocking template function, and as such, the inhibition can be overcome by increasing the template concentration. However, RNA
polymerase inhibition may involve enzyme inactivation. Inhibition
is not antagonized by increased template concentration, but rather
by increased enzyme levels[169]. Likewise the antimalarial *8-amino-
quinolines* (Fig. 7-33) bind to DNA and inhibit DNA polymerase

FIGURE 7-33

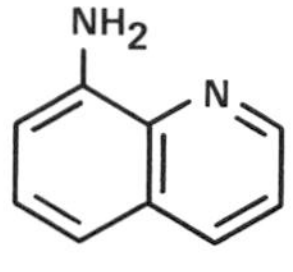

NH2

8—Aminoquinoline

activity; the enzyme from *M. luteus* is affected more than DNA
polymerase I of *E. coli*[170]. As with kanchanomycin there is evidence
for enzyme inactivation as well as template binding.

(B) Covalent binding. Mitomycin, bleomycin, phelomycin, and
anthramycin (Fig. 7-34) inhibit DNA synthesis by covalent attachment to the template[171]. Mitomycin selectively inhibits DNA synthe-

167. Slonimski, P. P., Perrodin, G. and Croft, J. H. (1968) *BBRC* **30**, 232.
168. Richardson, J. P. (1966) *JMB* **21**, 83.
169. Goldberg, I. H. and Friedman, P. A. (1971) *ARB* **40**, 775.
170. Whichard, L. P., Washington, M. E. and Holbrook, J. (1972) *BBA* **287**, 52.
171. Goldberg, I. H. and Friedman, P. A. (1971) *ARB* **40**, 775.

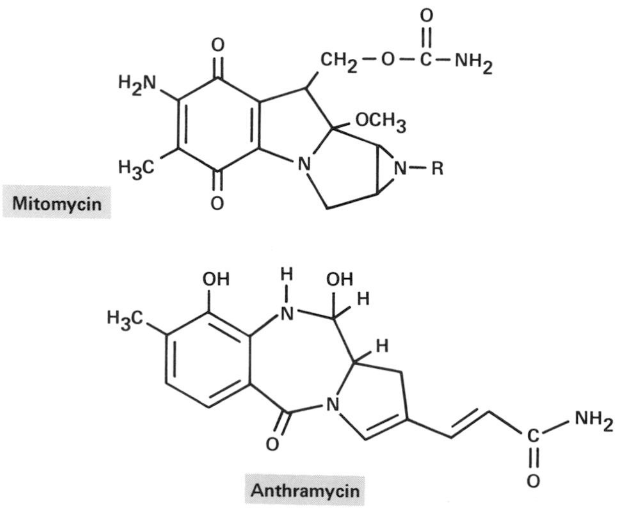

FIGURE 7-34
Mitomycin and anthramycin.

sis in cells by interstrand cross-linking of the DNA. The drug must first be reduced to generate the active bifunctional alkylating agent, and it is suggested the keto group of guanine is involved in the cross-linking reaction[171]. In mutants defective in repair of ultraviolet light damage, synthesis of host DNA is blocked, but that of small phage DNA is not[172]. Inasmuch as a single hit inactivates the chromosome, the host DNA offers a far larger target than the viral DNA.

Bleomycin and *phleomycin* seem to attach covalently to DNA with an apparent specificity for the 2-carbonyl position on thymine[173]. Phleomycin specifically inhibits DNA polymerase and $3' \rightarrow 5'$ exonucleolytic degradation of DNA[174]. Although the binding of both antibiotics results in single-strand breaks in the DNA in vivo and in vitro[175], their binding may be different since bleomycin decreases and phleomycin increases the melting temperature[176]. Also bleomycin stimulates the initial rate of DNA synthesis by DNA polymerase[177], presumably by providing more primer termini for the initiation of synthesis.

Anthramycin binds to native DNA and remains associated with the denatured strands at alkaline pH. RNA and DNA syntheses are inhibited, although inhibition of RNA synthesis requires a helical template. Binding of the drug depends on unknown parameters of structure and composition. It binds the homopolymer pair poly dG·poly dC but not poly dA·poly dT[178].

II. Nucleotide Analogs

These compounds interfere with replication by altering the template-primer after incorporation, by competitive binding at the active site of the enzyme, or by preventing the synthesis of the nucleotide precursors. The latter group has been discussed in Chapter 2.

(A) Chain terminators. The *3'-deoxy* and the *arabinosyl* analogs lack the 3'-hydroxyribo configuration and are incorporated at a very slow rate. Once incorporated, they represent primer termini that cannot be extended by DNA polymerases[179]. To the extent that they are not removed by nuclease action they maintain their block of chain growth.

Cordycepin triphosphate (3'-deoxy ATP) inhibits both RNA and DNA polymerase chain elongation by terminating the primer. The nucleoside does not inhibit growth of bacterial cells, probably because it is phosphorylated very poorly. The triphosphate strongly inhibits RNA synthesis in ascites tumor and HeLa cell lines and has

172. Lindqvist, B. H. and Sinsheimer, R. L. (1967) *JMB* **30**, 69.
173. Pietsch, P., Garrett, H. (1968) *Nat.* **219**, 488.
174. Falaschi, A. and Kornberg, A. (1964) *Fed. Proc.* **23**, 940.
175. Stern, R., Rose, J. A. and Friedman, R. M. (1974) *B.* **13**, 307.
176. Op. cit. in footnote 171.
177. Yamaki, H., Suzuki, H., Nagai, K., Tanaka, N. and Umezawa, H. (1971) *J. Antibiot.* **24**, 178.
178. Op. cit. in footnote 171.
179. Suhadolnik, R. J. *Nucleoside Antibiotics.* (1970) John Wiley & Sons, N.Y.; Roy-Burman, P. (1970) *Analogues of Nucleic Acid Components.* Springer-Verlag, N.Y.; Atkinson, M. R., Deutscher, M. P., Kornberg, A., Russell, A. F. and Moffatt, J. G. (1969) **B.** 8, 4897.

recently been shown to inhibit addition of poly A segments to the 3' end of completed heterogeneous nuclear mRNA[180]. Like 3'-deoxy ATP, *3'-amino ATP* inhibits RNA synthesis in isolated nuclei and extracts of ascites tumor cells by terminating the primer. Although the nucleoside inhibits DNA synthesis in these cells, extracts show no inhibition of DNA synthesis by the triphosphate; ribonucleotide reductase may be the target of in vivo inhibition[180].

(B) Defective DNA. dUTP, the uracil analog of dTTP, is incorporated by *E. coli* DNA polymerase I at a good initial rate into DNA but the extent of reaction is sharply limited[181]. In permeable cells, incorporation of dUTP into replicating ϕX174 DNA is followed by rapid degradation of the replicating molecules, presumably by $5' \rightarrow 3'$ exonuclease action[182]. Helical DNA with A–U base pairs appears to have structural features which make it susceptible to nuclease degradation, both in vitro (Chapter 4, Section 16) and in vivo.

Analogs of uridine and deoxyuridine with *5-hydroxy* or *amino* substituents inhibit DNA, RNA, and also protein synthesis in *E. coli*, and interfere in undetermined ways with the functions of the RNA and DNA molecules into which they are incorporated[183].

Uridine or *deoxyuridine nucleotides* containing *bromine* in the 5 position can also cause replication errors when incorporated into DNA, and light-induced bromouracil dimers are strongly mutagenic[183]. Phage λ DNA carrying the *lac* operator is an order of magnitude more effective in *lac* repressor binding when it contains bromouracil[184]. Thus the presence of bromouracil nucleotides in DNA may cause altered recognition of specific replication signals.

Tubercidin[185] and *formycin*[186] (Fig. 7-35) are cytotoxic analogs of

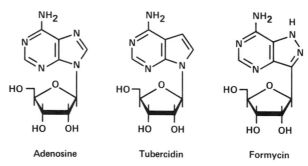

Adenosine   Tubercidin   Formycin

FIGURE 7-35
Tubercidin and formycin, analogs of adenosine.

180. Roy-Burman, P. (1970) *Analogues of Nucleic Acid Components.* Springer-Verlag, N.Y.
181. Bessman, M. J., Lehman, I. R., Adler, J., Zimmerman, S. B., Simms, E. S. and Kornberg, A. (1958) *PNAS* **44**, 633.
182. Wovcha, M. G. and Warner, H. R. (1973) *JBC* **248**, 1746. Geider, K. (1972) *EJB* **27**, 554.
183. Op. cit. in footnote 180.
184. Lin, S.-Y. and Riggs, A. D. (1972) *PNAS* **69**, 2574.
185. Nishimura, S., Harada, F. and Ikehara, M. (1966) *BBA* **129**, 301.
186. Ikehara, M., Murao, K., Harada, F. and Nishimura, S. (1968) *BBA* **155**, 82; Ward, D. C., Cerami, A., Reich, E., Acs, G. and Altwerger, L. (1969) *JBC* **244**, 3243.

adenosine. According to in vitro studies with RNA polymerase they serve, in the form of nucleoside triphosphates, as substitutes for ATP in RNA synthesis. Their inhibitory properties may be due to the formation of nonfunctional RNA or even in other reactions in which precursors and products of ATP are involved.

(C) *Enzyme binding*. *Hydroxyphenylhydrazinopyrimidines* produce unusually strong and specific inhibition of B. subtilis DNA polymerase III in vivo and in vitro[187]. They bind the enzyme in a ternary complex despite the lack of sugar or phosphate groups. As discussed earlier (Chapter 5, Section 4), the mechanism is judged to involve base-pairing with template pyrimidines[188]. This analog has been most useful in identifying the replicative patterns and enzymology of B. subtilis and illustrates the potential benefit that a discovery of comparable inhibitors of other replicative systems would provide.

(D) *Nucleotide synthesis*.   A variety of compounds which interfere at various levels with nucleotide biosynthesis have been used as inhibitors of RNA and DNA biosynthesis, and were discussed in Chapter 2.

III. Enzyme Binding

Rifampicin (Fig. 7-36) is an intensively studied drug; it specifically inhibits RNA synthesis in prokaryotes[189]. Some systems (M13, colicin factor replication) are inhibited by rifampicin because RNA polymerase provides a primer for DNA replication. The drug blocks initiation and not propagation of RNA chains, but it does not prevent binding of RNA polymerase to the DNA template[189]. The drug binds to the β subunit of E. coli RNA polymerase; mutants resistant to the drug possess a β subunit of altered electrophoretic mobility[190]. Rifampicin action is discussed further in Chapter 10, Section 2.

Yeast mitochondrial, and virtually all eukaryotic, RNA synthesis is insensitive to rifampicin but not to a related drug rifampicin AF-013. Resistance to the latter can be restored by addition of an undefined yeast protein factor which stimulates all RNA polymerases[191]. Hydrophobic derivatives of rifampicin (at the position in the structure indicated by color in Fig. 7-36) can inhibit the RNA-directed DNA synthesis by tumor viral polymerases[192]. At high concentrations these derivatives are capable of inhibiting certain mammalian RNA polymerases, and the initiation of RNA synthesis is specifically affected[193]. However, at these concentrations the drug

187. Mackenzie, J. M., Neville, M. M., Wright, G. E. and Brown, N. C. (1973) *PNAS* **70**, 512.
188. Gass, K. B., Low, R. L. and Cozzarelli, N. R. (1973) *PNAS* **70**, 103.
189. Wehrli, W. and Staehelin, M. (1971) *Bact. Rev.* **35**, 290.
190. Rabussay, D. and Zillig, W. (1969) *Febs. Lett.* **5**, 104.
191. DiMauro, E., Hollenberg, C. P. and Hall, B. D. (1972) *PNAS* **69**, 2818.
192. Gerard, G. F., Gurgo, C., Grandgenett, D. P. and Green, M. (1973) *BBRC* **53**, 194; Thompson, F. M., Tischler, A. N., Adams, J. and Calvin, M. (1974) *PNAS* **71**, 107.
193. Meilhac, M., Tysper, Z. and Chambon, P. (1972) *EJB* **28**, 291.

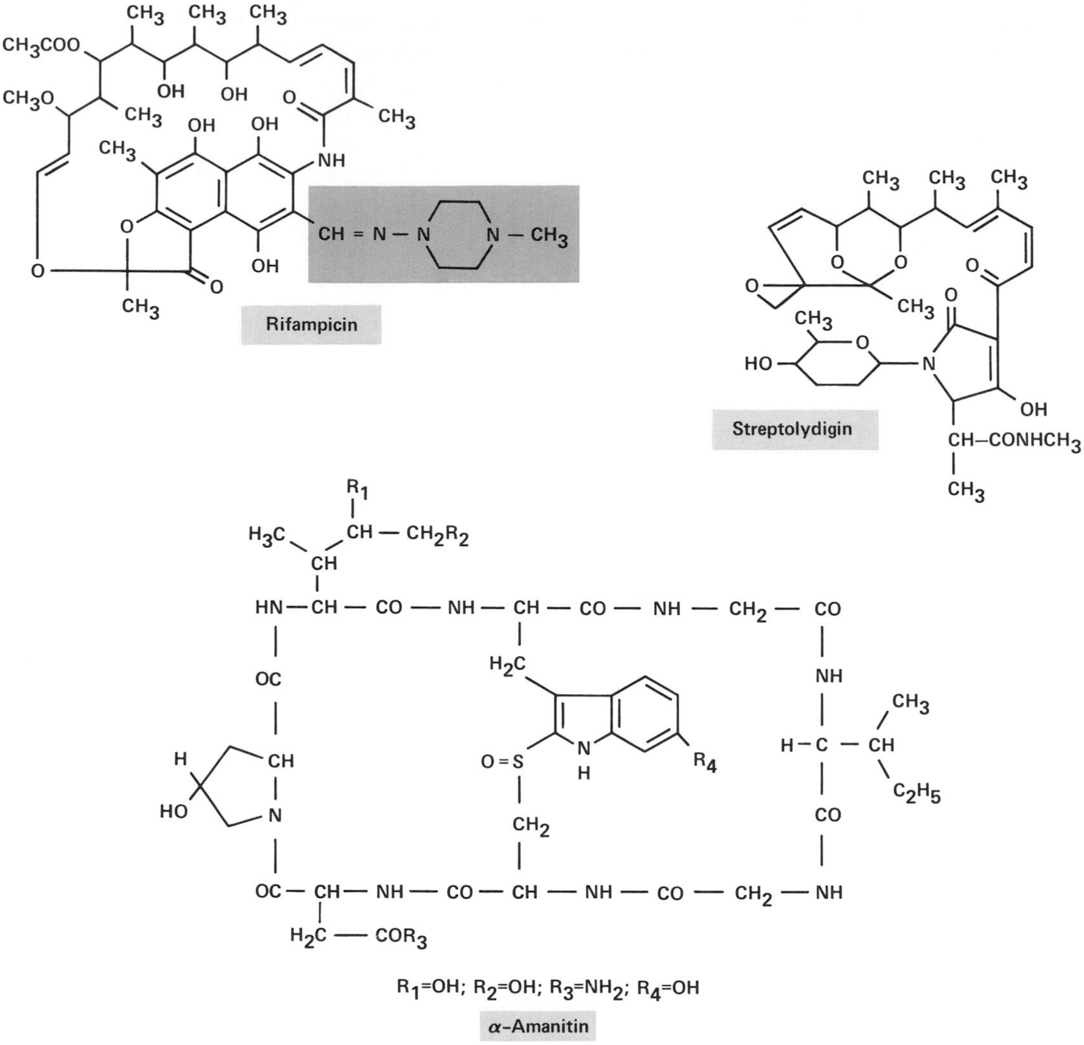

FIGURE 7-36
Rifampicin, streptolydigin, and α-amanitin.

also inhibits several other related enzymes, and its value as a specific reagent is therefore questionable.

Streptovaricin, a drug with similar origin to that of rifampicin, shows the same specificity and mode of action. Streptolydigin (Fig. 7-36), unlike rifampicin, preferentially blocks RNA chain elongation[194]. Yet the site of action is still the β subunit of RNA polymerase.

The sulfur-containing bicyclic polypeptide α-amanitin (Fig. 7-36) from the mushroom Amanita phalloides specifically inhibits mammalian nuclear RNA synthesis[194]. The purified nuclear RNA polymerase, activated by Mg^{2+} and ammonium sulfate, presumably

194. Goldberg, I. H. and Friedman, P. A. (1971) ARB **40**, 775.

the target in vivo, is inhibited. The drug does not inhibit nucleolar RNA polymerase nor the incorporation of ^{3}H-uridine into RNA of KB cells. Thus the drug does exhibit some selectivity among eukaryotic RNA polymerases. Recently, a cell line of Chinese hamster ovary resistant to α-amanitin has been isolated and shown to contain a correspondingly resistant RNA polymerase[195]. The drug is stably bound to the enzyme with a molar stoichiometry near one.

IV. Unknown Mechanism

Edeine (Fig. 7-37) inhibits bacterial DNA synthesis rather specifically[196]. The presence of a spermidine moiety in the molecule suggests a possible noncovalent association with the phosphate backbone of DNA. Although *E. coli* DNA synthesis is interrupted by edeine, conversion of ϕX174 single strand to replicative form by

195. Chan, V. L., Whitmore, G. F. and Siminovitch, L. (1972) *PNAS* **69**, 3119.
196. Kurylo-Borowska, A. and Szer, W. (1972) *BBA* **287**, 236.

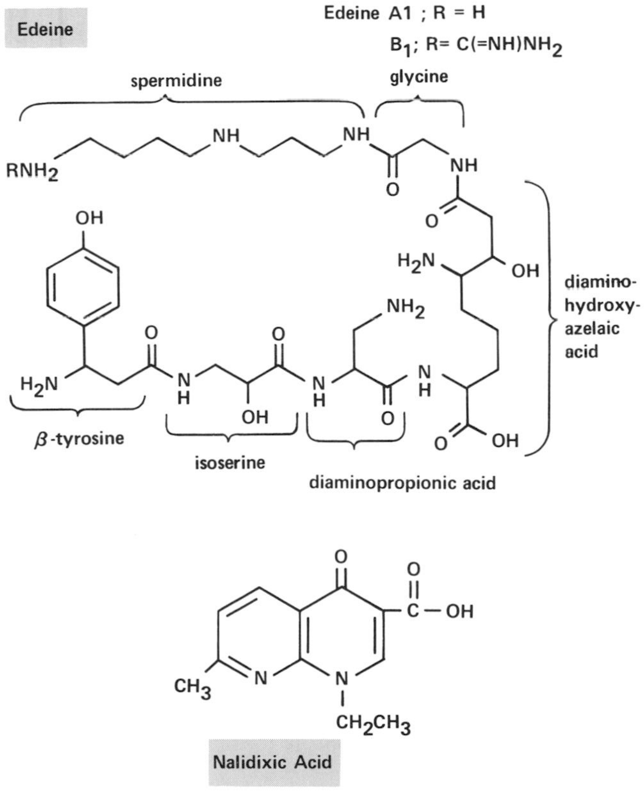

FIGURE 7-37
Edeine and nalidixic acid.

enzymes in extracts is stimulated[197]. This stimulation can be largely replaced by spermidine alone, indicating that structures favorable for replicating single-stranded DNA may drastically differ from structures suitable for double-stranded replication.

How _nalidixic acid_ (Fig. 7-37) inhibits bacterial DNA synthesis is still unknown. The drug does not interfere with conversion of ϕX174 or M13 single-stranded DNA to the double-stranded replicative form, but does inhibit multiplication of the replicative form[198].

197. Bertsch, L. L. and Kornberg, A., unpublished results.
198. Fidanian, H. M. and Ray, D. S. (1974) _JMB_ **83**, 63; Ray, D. S., personal communication.

8

Replication of DNA Viruses

1. Viral Windows on Replication[1,2]

Depending on a host cell for its development, a virus must either use the cell's machinery for its replication or bring in the genetic information to tailormake its own machinery — or do some of both. In any case, understanding the replication of the relatively small, well-defined viral chromosome illuminates the replication process of the cell itself. It may be arguable whether molecular biology originated solely in bacteriophage research[1]. But no vigorous objections should be raised to dramatizing the contributions phages have made to our understanding of DNA synthesis. Perhaps many of the still missing parts of the replication puzzle in both animal and bacterial cells will be seen through windows opened by research with viruses.

1. Cairns, J., Stent, G. S. and Watson, J. D. (Eds.) (1966) *Phage and the Origins of Molecular Biology* CSHL.
2. Stent, G. S. (1971) *Molecular Genetics*, W. H. Freeman and Co., San Francisco.

The degree to which the virus relies on the host's replicative apparatus depends on its size (Table 8-1). In the smallest DNA bacteriophages, M13 and ϕX174 with only eight to ten genes, most of the chromosomal information is devoted to defining the viral coat and its assembly, and the viruses rely almost completely on the machinery of the host cell to replicate their DNA. The small chromosome, only one-thousandth the size of the host chromosome, is then a promising instrument with which to identify the host replicative enzymes and their functions. The two phages M13 and ϕX174, furthermore, appear to use different replicative assemblies in the host cell.

TABLE 8-1
Size of some viruses

Virus	Related viruses	Particle shape	DNA		Shape	Comments
			Mol. wt. $\times 10^{-6}$	Number of base pairs, in kb		
SV40	Polyoma	Polyhedron	3.4	5.1	Duplex, circular	Animal cell host
ϕX174	S13	Polyhedron	1.8	5.4	Single strand, circular	Duplex replicative form
M13	fd, f1	Filament	1.9	5.7	Single strand, circular	Duplex replicative form
T7	T3	Head, short tail	23	35.4	Duplex, linear	Terminal redundancy
λ	ϕ80, 434, P2, 186	Head, tail	32	49	Duplex, linear	Cohesive ends form replicative circles, lysogenic
T5		Head, tail	76	115	Duplex, linear	Nicked
T4	T2, T6	Head, tail	120	180	Duplex, linear	Terminal redundancy; permuted
R17	MS2, f2 Qβ	Polyhedron	1.0	3.0	Single strand, linear	RNA, not DNA

Phages of intermediate size such as T7 induce a few enzymes for their own replication but depend also on host enzymes, whereas large phages like T4 contain twenty or more genes that define synthesis of precursors and a relatively complete and independent replicative apparatus. Even so, it is easier to grapple with the T4 genome than the *E. coli* chromosome, which is twenty times larger.

Lysogenic phages like λ, whose chromosome can be inserted into that of the host, can in addition be used to discover how such recombination takes place, and what factors determine whether the course of infection will be temperate or virulent. Infection of *E. coli* with phage λ can take one of two courses. One is favored in metabolically active cells. DNA and phage multiplication and cell lysis proceed as with T phages; this is a *virulent* infection (*lytic response*). The other course is favored when glucose, the primary energy source, is exhausted, cyclic AMP levels become elevated, and alternative energy sources are called upon. A *temperate* infection (*lysogenic response*) ensues in which the phage DNA is integrated into the host chromosome and replicated as part of it without detectable effect on the cell. The phage genome, called _prophage_, remains dormant for many cell generations until an external agent, such as ultraviolet irradiation or some metabolic alteration, supervenes. The result is an excision of the prophage from the host chromosome followed by multiplication of the phage DNA in a typically virulent infection.

Replication of virulent as well as temperate phages has been studied by means of host mutants with defects that impede development of the phage but do not interfere with growth of the host. New insights into replication have been obtained by identifying these defects and by discovering, in turn, phage mutations that overcome these defects (see Section 7 of this chapter).

As bacteriophages can be employed to clarify DNA replication in the bacterial cell, so can animal cell viruses be applied toward understanding animal cell DNA replication. Reference has already been made (Chapter 6, Section 6) to the DNA polymerases induced by large animal viruses; attention will be given in Section 9 of this chapter to the smallest of the DNA animal viruses, polyoma and SV40. Although these viruses are similar to the phages M13 and ϕX174 in size, they produce temperate as well as virulent infections. Of great interest is their ability to induce intensive and uncontrolled proliferative growth in the host cells and tissues.

In addition to characterization of the enzymes of DNA synthesis, there remain the difficult questions of their localization in the cell and the control of their activities. Once again it may prove that elucidation of spatial and temporal factors in the phage life cycle will reveal basic facts about the regulation of replication and expression of the host DNA.

2. Stages in the Viral Life Cycle[3]

To understand how the viral DNA is synthesized, we must know how the DNA is organized within the virus, delivered into the cell, multiplied, packaged, and finally released from the cell. These stages, each containing several steps, may be designated as follows.

(i) _Adsorption_ of the virus to a specific receptor on the cell surface.

(ii) _Penetration_ of the DNA through the cell envelopes and membrane layers to the interior of the cell. Uncoating of the DNA from its protein shell may be tightly coupled to its replication.

(iii) _Recognition_ and _expression_ of the DNA by _replication_ to a duplex form (in the single-stranded DNA viruses, M13 and ϕX174), or by _transcription_ of a particular region of the DNA (in the double-stranded viruses, T phages and λ), or by _translation_ of a special region of the RNA (in RNA phages and animal viruses).

---

3. Luria, S. and Darnell, J. E., Jr. (1968) _General Virology_, John Wiley & Sons, N.Y.

(iv) *Multiplication* of the DNA through various intermediates leading ultimately to a mature form appropriate for packaging.

(v) *Assembly* into the protein shell of the virus.

(vi) *Release* of virus particles as they are packaged, by progressive budding through the cell surface without major damage to the cell, or by rupture of the cell after the particles have accumulated inside the cell in large numbers.

In reviewing the little that is known of the chemistry of the viral life cycle, one question remains most central. How is the viral DNA recognized and handled with a specificity that insures its rapid and error-free duplication in competition with an extraordinary excess of cellular DNA? Based on the data and evidence described in preceding chapters, it appears most likely that one or more specific proteins will prove responsible for recognizing and guiding the viral DNA through the many stages of development. It is helpful to pursue the simple hypothesis that a protein molecule does some or most of these jobs.

According to this hypothesis, an adsorption protein, now renamed *pilot protein*, guides the virus DNA to the specific cell surface receptor. By its recognition and strong interaction with DNA in the virus particle, the pilot protein orients a particular DNA region toward the cell membrane. In order for a polyelectrolyte such as DNA to penetrate the lipid bilayer of the membrane, a pore or sheath is essential. Passage of even small ions through the plasma membrane requires a protein channel. The pilot protein may provide the hydrophobic exterior required for interaction with membrane lipids and the hydrophilic inner surface needed for passage of the DNA. Spheroplasts, which lack cell walls and outer envelopes but are surrounded by a plasma membrane, can be infected by free DNA, but are less efficient by several orders of magnitude than a phage infection of intact cells; the infection with DNA may depend on porelike structures present in the plasma membrane.

Perhaps interaction with surface receptors and entry into the cell through the membrane cause conformational changes in the virus that expose part of the DNA. How is this small stretch of viral DNA recognized inside the cell and how does it compete with a million times more cellular DNA? In the case of M13, the competition would be for the cellular enzyme unit or replisome responsible for replication of the extrachromosomal plasmids, whereas with ϕX174 evidence suggests that the viral DNA must appropriate the replicative assembly used for the duplication of the cell chromosome itself. When a duplex viral DNA such as T7 invades the cell, certain gene loci must be promptly transcribed by the host RNA polymerase and translated so that early proteins, including phage-specific polymerases (RNA and DNA) are made available.

If a pilot protein of a single-stranded DNA phage is designed to direct it to a replicative assembly, and that of a duplex DNA phage to RNA polymerase, then a pilot protein of an RNA phage should direct the entering RNA to a ribosome for expression of its messenger function in the synthesis of a phage-specific protein. In each instance, the invading nucleic acid, needing to be expressed by replication, transcription, or translation, finds itself in a medium in which the endogenous DNA and RNA would appear to hold an overwhelming advantage.

It is proposed that the pilot protein has several faces, one recognizing a region of the viral nucleic acid and another the key protein element of a replicative, transcriptional, or ribosomal unit. This specific recognition of a gene expression unit might provide the invading DNA or RNA with the competitive advantage it needs. In the same vein, the operation of analogous host-cell pilot proteins would be presumed. These would differ only in recognizing a specific nucleotide sequence of the host rather than the viral nucleic acid. Assuming that phage development, including the early events, is situated at the membrane, an additional hydrophobic face or tail on the pilot protein—for attachment to the lipid bilayer—may be present.

Further speculations about particular functions of the pilot protein in subsequent stages of DNA multiplication, and phage morphogenesis and release are premature at this point. However, as phage life cycles and related cellular functions are described in succeeding sections of this chapter, evidence or clues for the existence or operation of a pilot protein should be considered (Table 8-2). The gene 3 protein of M13 and gene H protein of ϕX174 are discussed in succeeding sections.

TABLE 8-2
Proteins with pilot functions

Phage, plasmid, or DNA	Protein	Molecular weight	Comments
M13	gene 3	70,000	Adsorption, replication
ϕX174	gene H	37,000	Adsorption, replication
R17(MS2)[a]	gene A	40,000	Adsorption, maturation
Col E1, col E2[b]			Complex with DNA
F-factor[c]	copilin(?)	60,000	Complex with DNA; transfer of DNA
ϕ29[d]			Transfection

[a]Krahn, P. M., O'Callaghan, R. J. and Paranchych, W. (1972) Virol. **47**, 628.
[b]Blair, D. G., Clewell, D. B., Sheratt, D. J. and Helinski, D. R. (1971) PNAS **68**, 210.
[c]Kline, B. C. and Helinski, D. R. (1971) B. **10**, 4975.
[d]Ortin, J., Vinuela, E. and Salas, M. (1971) NNB **234**, 275; Hirokawa, H. (1972) PNAS **69**, 1555.

A most remarkable feature in an overview of viral multiplication is the fact that so many viruses are produced in so short a time, for example, several hundred *E. coli* phages in twenty minutes. The awesome speed and efficiency of the process do not necessarily make it simpler for the biochemist to fathom, but by being elevated above the background of normal cellular activity, the viral operations become more evident. Despite the extraordinary complexity of the viral life cycle, its dissection and eventual reconstruction with chemically defined components in a cell-free system from start to finish remains an attractive horizon.

3. Small Filamentous Viruses: M13 (fd, f1)[4]

E. coli with hairlike structures, called F pili, are infected by a threadlike virus that goes under several names, depending on the laboratory of its formal origin. The viruses most commonly referred to are: f1 (New York)[5], fd (Heidelberg)[6] and M13 (Munich)[7]. They are virtually indistinguishable. For convenience, they will be referred to here as a single entity, M13. The virus is about one micron long and only 60 Å wide (Fig. 8-1 and Table 8-3). It contains the

4. Marvin, D. A. and Hohn, B. (1969) *Bact. Rev.* **33**, 172.
5. Loeb, T. (1960) *Science* **131**, 932; Zinder, N. D., Valentine, R. C., Roger, M. and Stoeckenius, W. (1963) *Virol.* **20**, 638.
6. Marvin, D. A. and Hoffmann-Berling, H. (1963) *Nat.* **197**, 517.
7. Hofschneider, P. H. (1963) *Zeitschrift für Naturforschung*, **186**, 203.

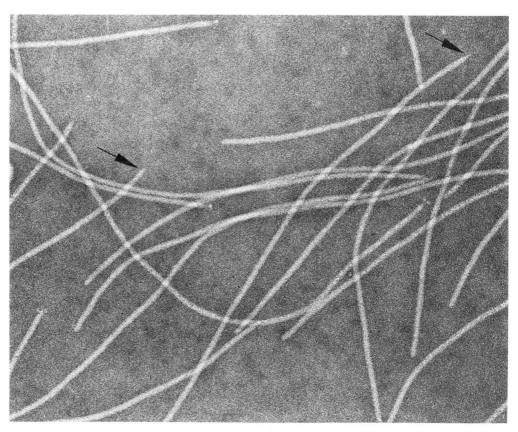

FIGURE 8-1
Electron micrograph of M13 (× 200,000). Some filaments are pointed at one end (arrows). (Courtesy of Professor R. C. Williams.)

TABLE 8-3
Characteristics of M13 and ϕX174

243

SECTION 3:
Small Filamentous
Virus: M13 (fd, f1)

	M13	ϕX174
Shape	Filament	Polyhedron with spikes
Length	10,000 Å (1 μ)	250 Å
Diameter	60 Å	250 Å
Particle weight	11×10^6	6×10^6
Weight/10^{14} particles	2 mg	1 mg
Major coat protein	gene 8 product (B protein)	gene F protein
Adsorption protein	gene 3 product (A protein)	gene H protein
Spike proteins	None	2
DNA, molecular weight	2.0×10^6	1.8×10^6
DNA/10^{14} particles	330 μg	300 μg
Base composition: A (in percent)	23	25
T	36	33
G	21	24
C	20	18

DNA as a single-stranded circle elongated into a loop extending the axial length of the virus. The DNA is ensheathed in a protein tube made up of a helical strand of 3000 protein subunits in a shinglelike arrangement[8]. The virus attaches to the cell, at or in the vicinity of the F pilus (Fig. 8-2). Infection results in the extrusion of one thousand or more particles in each cell generation without serious impairment of cellular growth. Many unusual features of the virus and its life cycle (Fig. 8-3) recommend it for an intensive study of the broad aspects of DNA synthesis.

M13 can be prepared in large quantities, about one gram per ten liters of culture, and is the only bacterial virus thus far crystallized in a form suitable for X-ray diffraction structure analysis. The coat protein, coded by gene 8, contains 50 amino acid residues (about 5200 daltons)[9]; histidine, cysteine, and arginine are lacking. The sequence (Fig. 8-4) is divisible into an acidic region at one end (residues 1-20), a basic region at the other (residues 40–50), and a hydrophobic interior (residues 21–39). Based on the amino acid composition and sequence, an α-helical conformation has been suggested. The asymmetric distribution of lysine residues and the adjacent line of hydrophobic residues are reminiscent of lysine-rich histones. This simple nucleohistone, which could be changed systematically in amino acid composition by genetic means, is an appealing model for probing deoxyribonucleoprotein organization.

In addition to the major coat protein, M13 contains a few molecules of a much larger protein, near 70,000 daltons, coded by

8. Marvin, D. A., Wiseman, R. L. and Wachtel, E. J. (1974) *JMB* **82**, 121.
9. Asbeck, F., Beyreuther, K., Köhler, H., Von Wettstein, G. and Braunitzer, G. (1969) *Zeitschrift für Physiologische Chemie*, **350**, 1047; Snell, D. T. and Offord, R. E. (1972) *BJ* **127**, 167.

244

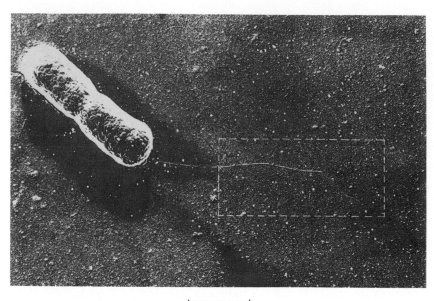

1.0 μm

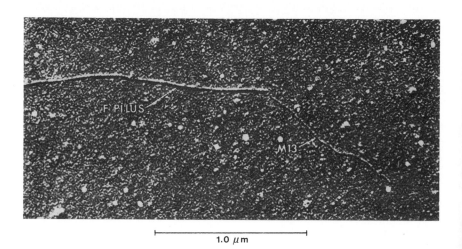

1.0 μm

FIGURE 8-2
Electron micrograph of an F pilus of a male *E. coli* with an M13 phage filament
attached to its tip. Despite this demonstrable interaction between pilus and phage,
the direct role of the pilus in adsorption is uncertain (see text below). (Courtesy of
Dr. Jack Griffith.)

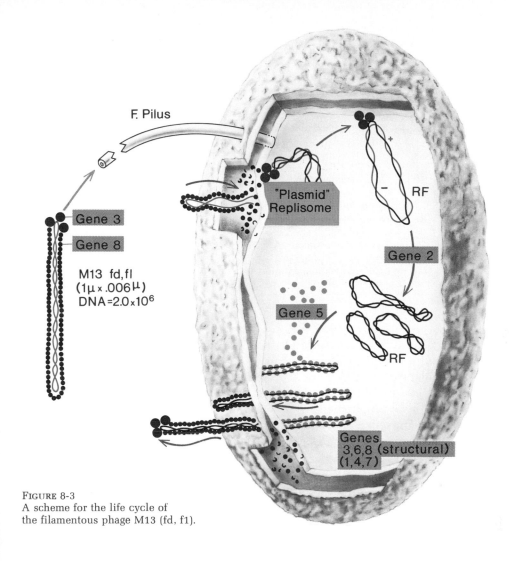

F. Pilus

Gene 3

Gene 8

M13 fd, fl
(1μ × .006μ)
DNA = 2.0 × 10⁶

"Plasmid" Replisome

RF

Gene 2

Gene 5

RF

Genes 3,6,8 (structural) (1,4,7)

FIGURE 8-3
A scheme for the life cycle of
the filamentous phage M13 (fd, f1).

1   5   10   15   20
H Ala-Glu-Gly-Asp-Asp-Pro-Ala-Lys-Ala-Ala-Phe-Asp-Ser-Leu-Gln-Ala-Ser-Ala-Thr-Glu-   **ACIDIC**

39   35   30   25   21
Ile-Gly-Ile-Thr-Ala-Gly-Val-Ile-Val-Val-Val-Met-Ala-Trp-Ala-Tyr-Gly-Ile-Tyr   HYDROPHOBIC

          40   45   50
          Lys-Leu-Phe-Lys-Lys-Phe-Thr-Ser-Lys-Ala-Ser-OH   **BASIC**

FIGURE 8-4
Amino acid sequence of the gene 8 protein. The original determination (Asbeck, F.,
Beyreuther, K., Köhler, H., Von Wettstein, G. and Braunitzer, G. (1969) *Zeitschrift
für Physiologische Chemie* **350**, 1047) has been modifed to include an additional
alanine residue at position 27. (Snell, D. T. and Offord, R. E. (1972) *BJ* **127**, 167).

TABLE 8-4
M13 genes and function

Gene	Function	Size of protein, in daltons	Number of copies per virion, or infected cell
1	Unknown; (assembly?)	~30,000	
2	RF multiplication; nicking	~40,000	
3	Minor coat protein (A protein); adsorption, parental RF formation	70,000	2–3
4	Unknown; (assembly ?)	~50,000	
5	Viral strand synthesis; unwinding protein	10,000	800,000
6	Minor coat protein (?); (terminator?)	~3000	30
7	Unknown; (assembly ?)		
8	Main capsid protein (B protein); lacks histidine, arginine, cysteine	5200	3000

gene 3[10] (Table 8-4) and presumed to contain a full complement of amino acids. Generally considered to be the protein responsible for adsorption, the gene 3 protein will be seen below to qualify as a pilot protein. There is recent evidence for another minor coat protein, distinguished by its very small size (about 3000 daltons) and the presence of cysteine and arginine[11]. It seems possible from earlier evidence[10] that this may be the gene 6 protein and that about 30 molecules are used to terminate the filament.

The small size of the M13 genome speaks for its relative simplicity. The DNA circle of 2 μ contains 6000 nucleotides representing eight genes, of which only four (so far) have assigned functions (Table 8-4). The map of the genome[12], shown in Fig. 8-5, is not as complete as the ϕX174 map (see below). Like ϕX174, this small virus is easily prepared and handled biochemically and genetically, inviting attempts to study its replication and its impact on the cellular host.

Adsorption. Because M13 requires male *E. coli* strains bearing F pili and has been photographed moored to the end of an F pilus (Fig. 8-2) this structure has been implicated as the means by which the M13 phage delivers its DNA into the cell[13]. The F pilus has also been assigned the function of transferring male *E. coli* chromosomal DNA to female cells that lack F pili, and for getting RNA from RNA phages into the cell. How the pilus might provide for the passage of circular M13 DNA into the cell and of chromosomal DNA out of the cell is still an unsolved problem.

One model for M13 adsorption proposes that the filamentous DNA nucleoprotein is conducted along a groove of the F pilus to

10. Henry, T. J. and Pratt, D. (1969) *PNAS* **62**, 800.
11. Bertsch, L. L., Marco, R. and Kornberg, A., unpublished observations.
12. Lyons, L. B. and Zinder, N. D. (1972) *Virol.* **49**, 45.
13. Caro, L. G. and Schnös, M. (1966) *PNAS* **56**, 126; Jacobson, A., (1972) *J. Virol.* **10**, 835.

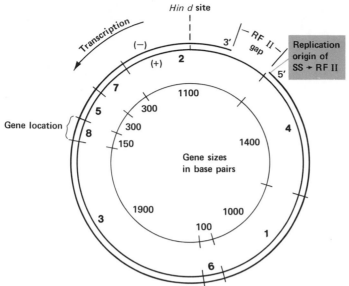

FIGURE 8-5
Genetic map of M13 (f1, fd) showing a tentative ordering of gene
locations, an estimate of their sizes, the restriction site of Hin
d endonuclease, and the position of the gap in the enzymatically
synthesized parental RF II. (Lyons, L. B. and Zinder, N. B. (1972)
Virol. **49**, 45; Schaller, H., personal communication; see text below.)

the cell surface[14] while another favors a progressive depolymerization of the pilus that retracts the tip, with the phage attached, down
to the base of the pilus at the cell surface[15]. Experimental proof for
neither model is persuasive[16]. Yet there is recent evidence, with
filamentous phages specific for piliated *E. coli* and a related organism[17], which is most readily explained by a retraction mechanism. For lack of understanding the chemistry of F-pilin and the
polymerization and depolymerization of this 11,800-dalton building block of F-pili, the important functions of the pilus remain
controversial.

Mutants in the gene 3 protein are defective in adsorption[18]. The
defect in temperature-sensitive mutants at the elevated temperature
appears to be a reduction in the rate of adsorption rather than an
absolute block. The gene 3 protein has not been isolated and characterized, nor has its location in the phage particle been determined.

Penetration. M13 DNA is not injected into the cell but must be
replicated to the duplex form in order to be drawn in. In experiments with rifampicin, an agent that blocks the initiation of replication, the phage were permitted to adsorb to the cell but not to
undergo removal of coat protein from the DNA[19] (Fig. 8-6). Such

14. Brinton, C. C. (1971) *Critical Reviews of Microbiology,* **1**, 105.
15. Marvin, D. A. and Hohn, B. (1969) *Bact. Rev.* **33**, 172.
16. O'Callaghan, R., Bradley, R. and Paranchych, W. (1973) *Virol.* **54**, 220.
17. Novotny, C. P. and Fives-Taylor, P. (1974) *J. Bact.* **117**, 1306; Bradley, D. E. (1974) *Virol.*
    58, 149.
18. Henry, T. J. and Pratt, D. (1969) *PNAS* **62**, 800.
19. Brutlag, D., Schekman, R. and Kornberg, A. (1971) *PNAS* **68**, 2826.

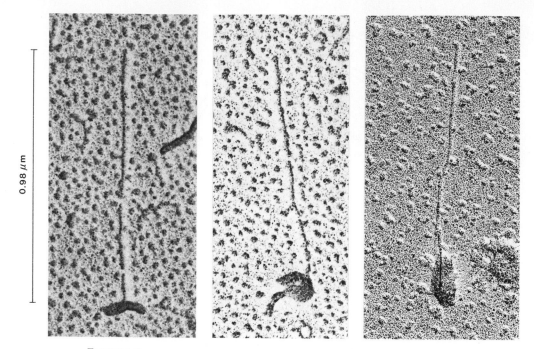

0.98 μm

FIGURE 8-6
Electron micrograph of M13 phages interrupted in penetration by rifampicin.
Filaments attached to membrane fragments are full-length particles (average 0.98 μm).

aborted phages showed no change in buoyant density upon sedimentation to equilibrium in density gradients. The phages had suffered conformational changes in the nucleoprotein, as judged by susceptibility of the DNA to nicking by nucleases that fail to attack DNA in intact phages. Ultraviolet irradiation of M13 phages producing an average of one lesion per DNA circle led to only partial penetration by the phage particle[20]. These and related experiments all indicate that decapsidation of the phage nucleoprotein is tightly coupled to replication of the DNA.

Since replication must take place inside the cell, the virus is presumed to penetrate the several cell envelopes as a relatively intact nucleoprotein. The M13 capsid protein, unlike that of most other phages, is not completely discarded into the medium. Instead, a considerable fraction of the protein is deposited in the inner cell membrane and is salvaged for coating the progeny DNA[21]. The notable experiment[22] with T2 phage that indicated DNA to be the genetic substance because so much of the DNA and so little of the protein entered the cell would have been less persuasive had M13 been used instead.

Inasmuch as the gene 3 pilot protein is expected to be at the

20. Griffith, J. and Kornberg, A. (unpublished).
21. Trenkner, E., Bonhoeffer, F. and Gierer, A. (1967) *BBRC* **28**, 932; Henry, T. J. and Brinton, C. C., Jr. (1971) *Virol.* **46**, 754; Smilowitz, H. (1974) *J. Virol.* **13**, 94.
22. Hershey, A. D. (1953) *CSHS* **18**, 135.

advancing tip of the M13 filament and, as will be indicated next, is found tightly bound to the replicated viral DNA and associated with the inner membrane, it may serve in guiding the DNA through the lipid bilayer of the membrane.

Conversion of viral DNA to replicative form (RF).   This is the first of three stages in the replication of a single-stranded viral DNA. It requires no synthesis of phage-directed proteins and relies on a host replicative system. It does require, as will be indicated presently, the gene 3 pilot protein.

Studies of conversion of the single-strand circle to the circular duplex form brought into focus once again the old question of how a DNA chain is initiated. It was these studies that provided the first insight into RNA priming (Chapter 7, Section 5). A short RNA segment produced by RNA polymerase in the presence of an *E. coli* unwinding protein is extended in vitro by the DNA polymerase III* system[23] (Chapter 7, Section 11). The product synthesized by a crude enzyme extract (from pol A⁻ cells[24]) or by purified enzymes[25] is a duplex circle with a small gap (Fig. 7-28). The product contains the intact viral DNA and nearly full-length synthetic complementary strands with a persistent portion of the RNA priming segment at the 5′ end[26].

The gap in the enzymatically synthesized RF II is located at a unique region in relation to the viral DNA (Fig. 8-7)[27]. This conclusion is based on electrophoretic analysis of well-defined DNA fragments produced by restriction enzymes (Chapter 9, Section 4)

23. Wickner, W., Schekman, R., Geider, K. and Kornberg, A. (1973) *PNAS* **70**, 1764; Geider, K. and Kornberg, A. (1974) *JBC* **249**, 3999.
24. Wickner, W., Brutlag, D., Schekman, R. and Kornberg, A. (1972) *PNAS* **69**, 965.
25. Geider, K. and Kornberg, A. (1974) *JBC* **249**, 3999.
26. Westergaard, O., Brutlag, D. and Kornberg, A. (1973) *JBC* **248**, 1361.
27. Tabak, H. F., Griffith, J., Geider, K., Schaller, H. and Kornberg, A. (1974) *JBC* **249**, 3049.

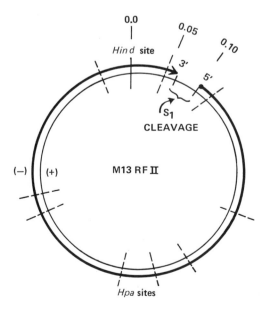

FIGURE 8-7
Unique location of the gap in the enzymatically synthesized RF II of M13 determined through distinctive cleavages by restriction endonucleases of *H. influenzae* (Hin d II, III) and *H. parainfluenzae* (Hpa II) and by single-strand specific S₁ endonuclease.

and the observation that, when the RF II is cleaved at the gap by an endonuclease to produce a linear duplex, subsequent melting and reannealing restores linear duplexes rather than generating circles. Were the gaps located at several regions, reannealing would lead to overlaps and circles would be seen microscopically. The unique location of the RF II gap suggests a unique origin of replication at or near that point.

A highly significant feature of the RF produced in vivo is its association with the gene 3 pilot protein[28]. Incorporation of histidine, which is absent from the capsid protein, labels only the gene 3 protein. Parental RF was isolated from cells infected with [³H]-histidine phage and treated with chloramphenicol to interrupt further utilization of parental RF. Histidine-labeled protein was found bound to the DNA, and the radioactivity was quantitatively conserved relative to the viral DNA. Fractionation of the infected cell components showed the complex of RF with protein of gene 3 origin to be associated exclusively with the inner cell membranes.

Indications[28] that the gene 3 protein performs a function essential to the conversion of the viral strand to RF come from the behavior of M13 containing a thermosensitive mutation in gene 3. Cells infected with this mutant failed to form any RF at an elevated temperature. Nevertheless, adsorption of the phage must have occurred because such infections were complemented at the high temperature by simultaneous infections with mutants carrying a nonsense mutation in another gene (e.g. gene 2). Presumably production of gene 3 protein directed by the nonsense gene 2 mutant permitted entry and replication of the temperature-sensitive gene 3 mutant which, in turn, directed production of the gene 2 protein needed by the nonsense mutant.

The transfer of gene 3 protein with the DNA and evidence for its role as pilot protein in the first stage of replication prompted a successful search for an analogous protein in φX174 infections (see Section 4 of this chapter). These findings with DNA phages also suggest a similar role for the adsorption protein of RNA phages. The gene A protein of MS2, labeled with (³H)-histidine, and transferred into the cell with the invading RNA, may facilitate translation of the RNA replicase gene and serve subsequently in maturation of phage progeny[29].

Utilization of a rifampicin-sensitive system for the conversion of M13 viral DNA to the parental RF is a property also displayed by some DNA plasmids in their replication. The system is distinct from that utilized by φX174 in its replication (Table 8-5). A fertility plasmid (F factor)[30] and a colicinogenic factor (Col E_I, see Section 8 of this chapter) may use a common replicative system shared by M13 DNA. The large, twisted F factor has been seen anchored in the inner cell membrane at the base of the F pilus. Both the F factor

28. Jazwinski, S. M., Marco, R. and Kornberg, A. (1973) PNAS **70**, 205.
29. Nathans, D. and Kozak, M. (1971) NNB **234**, 209; Paranchych, W., Ainsworth, S. K., Dick, A. J. and Krahn, P. M. (1971) Virol. **45**, 615; Krahn, P. M., O'Callaghan, R. J. and Paranchych, W. (1972) Virol. **47**, 628.
30. Bazzicalupo, P. and Tocchini-Valentini, G. P. (1972) PNAS **69**, 298.

TABLE 8-5

251
Distinctions between M13 and ϕX174[a]

	M13	ϕX174
Structure of virion	Filamentous, simple	Polyhedral, complex
Adsorption receptor	F pilus (?); inner membrane	Lipopolysaccharide; outer membrane
Host strain	Male	Rough
Salvage of coat protein by progeny	Yes	No
Replication system: SS $\rightarrow$ RF	Rifampicin-sensitive ("plasmid")	Rifampicin-resistant ("chromosomal")
Viral strand synthesis: Unwinding protein	Gene 5 protein	Unknown
Coupled to encapsidation	No	Yes
Pool of uncoated progeny DNA	Yes	No
New DNA in parental coat	Yes	No
Intact parental DNA in new coat	Yes	No
Coat assembly in the membrane	Yes	No
Mode of release	Budding	Cell lysis
Infection	Parasitic; cells continue to multiply	Virulent

[a]There is no (<10%) demonstrable homology (hybridization) between M13 and ϕX174 DNA.

and colicinogenic factor DNAs have been isolated in complexes with an unusual protein[31]. During certain denaturing treatments, the protein nicked and relaxed the twisted circle, and became covalently linked to the 5' end of the nicked chain. In conjugal transfer of DNA from a male to female cell, it is the nicked strand of the F factor led by its 5' end that has been identified in genetic studies.

Perhaps the proteins complexed to DNA plasmids will be found to resemble the phage pilot proteins. Knowledge gained about the contribution of these proteins to M13 DNA replication may prove to be important in understanding the plasmid replicative systems.

Multiplication of RF.   This is the second stage of DNA multiplication (Fig. 8-8). It requires the action of gene 2 protein, presumably to nick the parental RF, as an early event in replication of the duplex[32]. Inhibition of this stage by rifampicin indicates an action by RNA polymerase[33], but the inhibitory action of nalidixic acid is not explained. Characteristics of the structures and labeling patterns of intermediates are consistent with the rolling-circle model proposed for ϕX174 replication (see Section 4 of this chapter), but the alternative θ structure mechanism, observed in *E. coli* or phage λ (Chapter 7, Section 3) has not been excluded. Perhaps knowledge about the mechanism of the first replication stage will be helpful in resolving the components responsible for this more complex event.

31. Kline, B. C. and Helinski, D. R. (1971) *B.* **10**, 4975; Blair, D. G., Clewell, D. B., Sheratt, D. J. and Helinski, D. R. (1971) *PNAS* **68**, 210.
32. Francke, B. and Ray. D. S. (1972) *PNAS* **69**, 475; Fidanian, H. M. and Ray, D. S. (1972) *JMB* **72**, 51; (1974) *JMB* **83**, 63; Tseng, B. Y. and Marvin, D. A. (1972) *J. Virol.* **10**, 384.
33. O'Callaghan, R., Bradley, R. and Paranchych, W. (1973) *Virol.* **54**, 220.

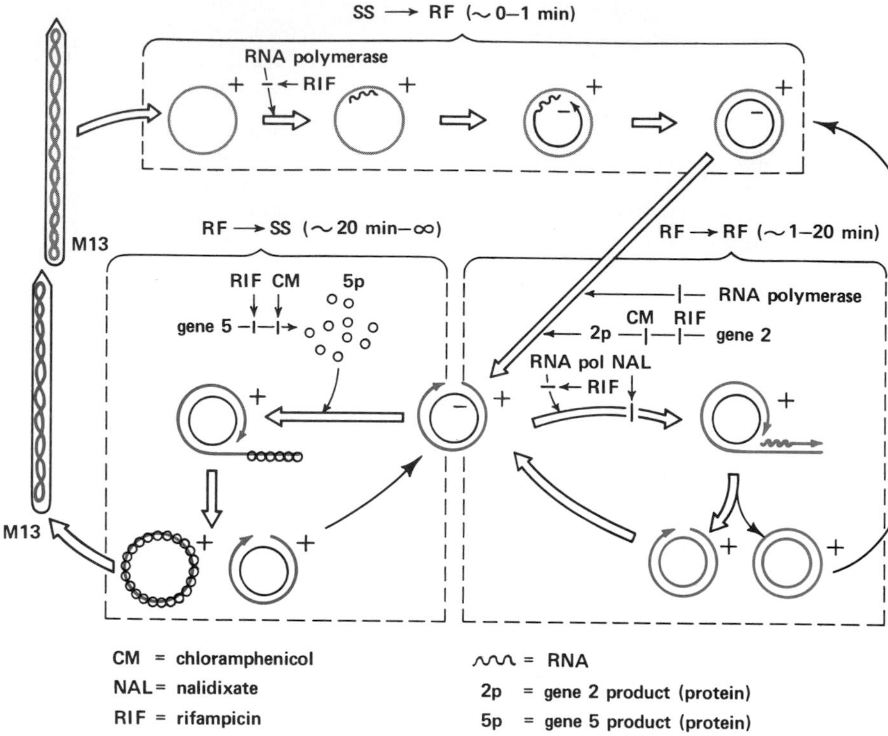

FIGURE 8-8
A scheme proposed for M13 DNA replication in three successive stages:
SS → RF → RF → SS. Functions of the phage gene 2 and gene 5 products
(2p and 5p) and actions of inhibitors are indicated. (Ray, D. S., personal
communication; Fidanian, H. M. and Ray, D. S. (1972) *JMB* **72**, 51; (1974) *JMB*
83, 63. (Courtesy of Professor D. S. Ray.)

Synthesis of progeny single strands.   The final stage in M13 DNA
replication is the production of viral strands from a replicative
intermediate. Available evidence suggests a rolling circle mechanism, as described for the same stage of ϕX174 replication[34] (see
Section 4 of this chapter). However there is at least one major distinction. Synthesis and accumulation of viral M13 strands depend
on their conversion by gene 5 protein to a distinctive nucleoprotein[35] whereas, with ϕX174, complexing with coat and morphogenetic proteins is required.

The amount of gene 5 protein available determines whether the
pathway of replication (Fig. 8-8) will produce single strands or
RF[36]. Cells infected with a mutant, temperature-sensitive in gene 2
and held at a high temperature, accumulated enough gene 5 protein
so that when the temperature was shifted down to a permissive
level only single-stranded DNA was synthesized. Thus the accumulation of an RF pool is not an obligatory stage preceding single-
strand production.

34. Ray, D. S., Fidanian, H. M. and Francke, B. *Virol.* (in press); Fidanian, H. M. and Ray, D. S. (1972) *JMB* **72**, 51; (1974) *JMB* **83**, 63.
35. Pratt, D., Laws, P. and Griffith, J. (1974) *JMB* **82**, 425.
36. Mazur, B. J. and Model, P. (1973) *JMB* **78**, 285.

Gene 5 protein has the properties of an unwinding protein[37] (Chapter 7, Section 7). It has a mass of 10,000 daltons and binds four nucleotides in the M13 chain. Thus 1500 molecules are required to complex an M13 circle; if 100 viral strands were present at one time in the cell during infection, they would require 150,000 molecules. Gene 5 protein behaves like an unwinding protein in binding tightly and cooperatively to single-stranded DNA, and in facilitating the unwinding of helical DNA.

The function performed by the gene 5 protein can only be surmised. It may facilitate the replicative operation that generates the viral strand, prevent the viral strand from being used as template and diverted to RF, and protect it from nucleases. RF accumulates in cells infected with gene 5 mutants. In the absence of coat proteins or other gene products essential for phage assembly, complexes of the viral strand with gene 5 protein were found to accumulate in large numbers.

The structure of the gene 5—DNA complex as a well-defined intracellular nucleoprotein[38] deserves careful analysis, among other reasons, in order to understand how the coated viral strand orients itself at the inner surface of the membrane for efficient assembly into mature phage.

Is RF I a key intermediate in replication? Inasmuch as RF II, the nicked form, appears to be the key intermediate in currently favored schemes (Fig. 8-8), the question arises whether RF I is only a spur from the main pathway, rather than directly on it. If only a spur, why is it produced directly from the parental strand and then multiplied many times over? The answer may prove to be simple if the supercoiled RF I, by virtue of its special structure, is the form required for transcription[39]. Many copies of RF I would be needed for rapid and effective synthesis of phage gene products. Expression of gene 2 is essential at the outset and the level of gene 5 protein may determine directly the extent of phage DNA synthesis and assembly.

Phage assembly. The intracellular location of the coat protein is exclusively in the inner cell (plasma) membrane[40]. How this unusual polypeptide is deposited in the membrane, and disposed in some particular arrangement, are intriguing questions whose solution may have implications not only for the budding of a filamentous virus but for protein secretion in general. The coat polypeptide, in α-helical conformation, is a 75 Å rod which could extend across the lipid bilayer. Its lysine-rich end may face toward the cell interior to react with the viral DNA and displace gene 5 protein; the hydrophobic center would provide a lipophilic interface with lipid hydrocarbons and the acidic end would face outside toward the periplasmic layer. Formation of the viral nucleohistone is part of a budding process that enables the micron-long viral filament to

37. Alberts, B., Frey, L. and Delius, H. (1972) *JMB* **68**, 139; Oey, J. L. and Knippers, R. (1972) · *JMB* **68**, 125.
38. Pratt, D., Laws, P. and Griffith, J. (1974) *JMB* **82**, 425.
39. Puga, A. and Tessman, I. (1973) *JMB* **75**, 83.
40. Smilowitz, H., Carson, J. and Robbins, P. W. (1972) *Journal of Supramolecular Structure*, **1**, 8.

be extruded through all layers of the cell envelope without materially interfering with cellular growth.

Mini M13 phage. Filaments only about one-third the standard length were discovered in a stock of phages obtained after many passages beyond an initial single-plaque isolation[41]. The mini particles varied in length from 0.2 to 0.5 that of M13 and contain DNA circles corresponding in size. The particles are reproduced only upon coinfection with normal M13, and were favored in multiplication. Analysis of mini phage DNA by heteroduplex electron microscopy showed them to be fragments of the M13 circle containing the unique site for the *Hin* d restriction endonuclease (Fig. 8-7) located near the replication origin and extending for a variable distance in one direction from this site (Fig. 8-9).

FIGURE 8-9
A scheme for the position of mini M13 sequences of varying length relative to the M13 map. The sequences are shown beginning at the *Hin* d cleavage site, which they possess, and extending in the direction of complementary strand synthesis to points from 0.2-to 0.5-length of M13 before being circularized.

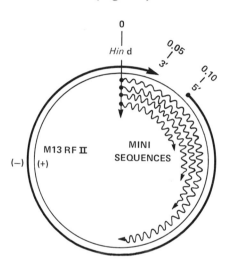

One possible mechanism for the first appearance of mini M13 DNA is an uncommon event in replication in which a circular fragment is closed prematurely. Possessing the site for the origin of replication, the mini DNA is readily replicated. Mini-sized RF has also been discovered in cells infected with amber mutants defective in gene 5 protein[42], and this observation may offer some clues about the production of mini M13. Since assembly of the viral DNA requires its enclosure in a tube of capsid protein, there is no apparent problem in completing a filament of virtually any mini- or poly-length. The mini phages should prove useful in studying both the site and structure of the replication origin of M13, and many questions about its life cycle.

M13 as a parasite rather than a phage. M13 alone among the bacterial viruses does not produce a virulent infection[43]. Rather its

41. Griffith, J. and Kornberg, A. (1974) *Virol.* **59**, 139.
42. Grandis, A. S. and Webster, R. E. (1973) *Virol.* **55**, 14.
43. Marvin, D. A. and Hohn, B. (1969) *Bact. Rev.* **33**, 172.

behavior is similar to an early misconception of a lysogenic phage infection, in which the cell produces and secretes virus particles without undergoing lysis. M13 more closely resembles the myxovirus group of animal viruses, which include influenza, mumps, sendai, simian virus 5 (SV5), and others. These filamentous viruses, composed of nucleoprotein units, are expelled through the cell surface with relatively few pathological effects. However, with any block in maturation of viral particles at the cell membrane, the nucleoprotein units accumulate, the cells fuse with one another, disintegrate, and die. Similarly, interruption in development of M13 at any stage past RF replication results in accumulation of the aborted intermediate form, development of large intracellular whorls of the plasma membrane, and loss of cell viability[44]. The reasons for cellular pathology found in abortive infections, and the relatively commensal relationship when the cell is supporting a rapidly reproducing virus, are unknown and are likely to be based on the biosynthesis and assembly of proteins and lipids in the cell membranes.

4. Small Polyhedral Virus: φX174

The earliest reward from the study of the smallest of the bacteriophages came from work on φX174[45]. It was the discovery of a chromosome made of a single strand of DNA and continuous or circular in outline. Since then, genetic, physical, and biochemical studies of φX174 and its life cycle (Fig. 8-10), and of the very closely related phage S13[46] have done much to advance our understanding of the molecular biology related to DNA replication.

The φX174 particle is an icosahedral polyhedron with a mass of 6.2×10^6 daltons and a diameter of 250 Å[47] (Fig. 8-11). The phage capsid is composed of 60 molecules of F protein (50,000 daltons). There are twelve vertices of fivefold symmetry (Fig. 8-12). Fitted into the capsid at each vertex is a knob (spike).

The φX174 chromosome is a DNA molecule containing about 5400 nucleotides, shared by nine genes arranged as shown on the map[48] in Fig. 8-13. Assuming that all the DNA is used to code for amino acids and that their average weight is 120, then the DNA is sufficient to specify approximately 216,000 daltons of protein. It is evident from Table 8-6 that the combined molecular weights assigned to the known gene products already exceeds that limit, if in fact each product is coded by a separate gene and the above values are correct.

---

44. Schwartz, F. M. and Zinder, N. D. (1968) *Virol.* **34**, 352.
45. Sinsheimer, R. L. (1968–1969) *The Harvey Lectures*, **64**, 69.
46. Baker, R. and Tessman, E. (1967) *PNAS* **58**, 1438; Godson, G. N. (1973) *JMB* **77**, 467.
47. Edgell, M. H., Hutchison, C. A. III. and Sinsheimer, R. L. (1969) *JMB* **42**, 547.
48. Benbow, R. M., Mayol, R. F., Picchi, J. C. and Sinsheimer, R. L., (1972) *J. Virol.* **10**, 99; Baas, P. D. and Jansz, H. S. (1972) *JMB* **63**, 569.

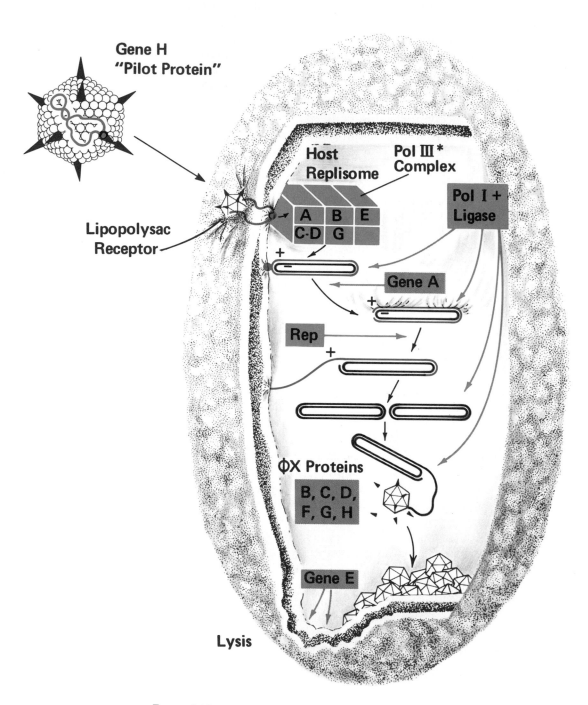

Gene H
"Pilot Protein"

Lipopolysac
Receptor

Host
Replisome

Pol III*
Complex

A B E
C-D G

Pol I +
Ligase

Gene A

Rep

ΦX Proteins

B, C, D,
F, G, H

Gene E

Lysis

FIGURE 8-10
A scheme for the φX174 life cycle. See Fig. 8-3 and Table 8-5 for comparison
with phage M13.

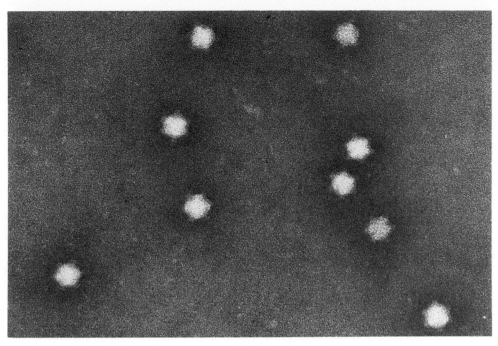

FIGURE 8-11
Electron micrograph of ϕX174 ($\times$ 200,000). (Courtesy of Professor R. C. Williams.)

FIGURE 8-12
Schematic representation of the ϕX174 subunits
in the polyhedron (left) and an enlargement of a
spike region (right) suggesting orientations of gene
H, gene G, and the major coat (gene F) proteins.
(Edgell, M. H., Hutchison, C. A. and Sinsheimer,
R. L. (1969) *JMB* **42**, 547.)

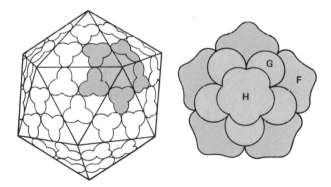

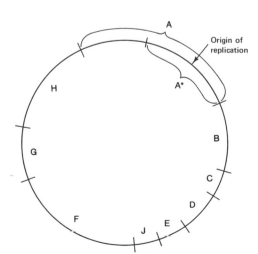

FIGURE 8-13
Genetic map of ϕX174
(see Table 8-6 for details).

TABLE 8-6
ϕX174 Genes and Functions[a]

Gene		Function	Size of protein[b]	Number of copies
New	Old			
[c]A	VI	RF multiplication—nicking	62,000	
B	IV	Morphogenesis	25,000	
C	VIII	Viral strand synthesis	10,000[d]	
D	V	Viral strand synthesis	14,500	
E	I	Cell lysis	17,500(?)	
F	VII	Major coat protein	50,000	60
G	III	Spike protein	18,500	60
H	II	Spike protein, adsorption	37,000	12
J		Unknown	9,000	60
			244,000	

[a]Source: Benbow, R. M., Mayol, R. F., Picchi, J. C. and Sinsheimer, R. L., (1972) J. Virol. **10**, 99; Baas, P. D. and Jansz, H. S. (1972) JMB **63**, 569.
[b]Baker, R. and Tessman, E. (1967) PNAS **58**, 1438; Godson, G. N. (1973) JMB **77**, 467.
[c]Intragenic regulation of protein synthesis produces a protein, called A*, of 35,000 daltons; the complete protein is A. Both forms are found in infected cells. (Source: Linney, E. and Hayashi, M. (1974) Nat. **249**, 345.
[d]Borras, M.-T., Vanderbilt, A. S. and Tessman, E. S. (1971) Virol. **45**, 802.

Adsorption and penetration. The spikes, representing tiny tail-like structures, may function as primitive attachment organs. Each of the twelve spikes contains five molecules of G protein (18,500 daltons each) and one molecule of H protein (37,000 daltons). It seems likely from current studies[49] that H protein is transferred with the DNA into the cell and is serving as the pilot protein for ϕX174.

The phage adsorbs irreversibly to lipopolysaccharide in the outer membrane layer of rough strains of *E. coli* and *Salmonella typhimurium*[50]. These strains lack all the O-antigen outer chains but contain intact core and lipid *A* structures (Fig. 8-14). Adsorption includes a reversible stage in which the detached phage has not been detectably altered and remains infectious, followed by an irreversible stage called eclipse. Eclipsed phage have lost infectivity for fresh cells and their DNA has become susceptible to attack by pancreatic DNase. Intact lipid *A* is required for progress of adsorption to the eclipse stage, but not for the earlier reversible one[51]. As with M13, removal of the coat protein from the DNA is tightly coupled to replication of the viral DNA.

Not only is the ϕX174 particle attached to the outer cell membrane but so is the parental RF[52]. This localization of ϕX174 is in sharp contrast to that of M13, which is associated with the inner cell membrane. In fact, this distinction can be demonstrated in membrane fractions of *E. coli* simultaneously infected with ϕX174 and M13 (Fig. 8-15). An intriguing observation in such membrane

49. Jazwinski, S. M., Marco, R. and Kornberg, A., unpublished observations.
50. Incardona, N. L. and Selvidge, L. (1973) J. Virol. **11**, 775.
51. Jazwinski, S. M. and Lindberg, A., personal communication.
52. Op. cit. in footnote 49.

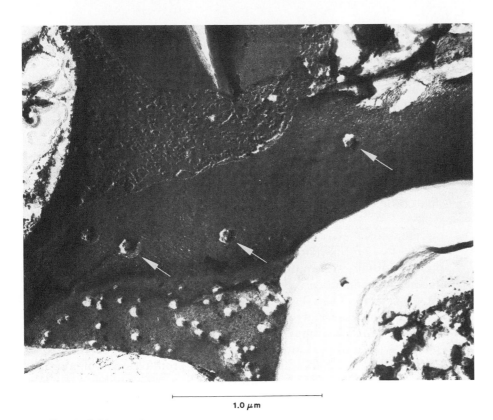

1.0 μm

FIGURE 8-14
Adsorption of φX174 to outer membrane of *E. coli* as seen in a freeze-etch electron micrograph. (Courtesy of Dr. Jack Griffith.)

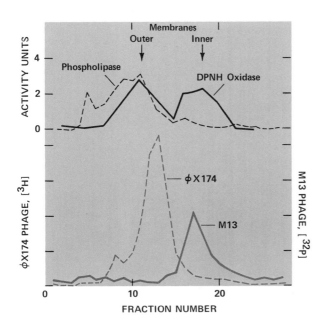

FIGURE 8-15
φX174 and M13 phages and their replicative forms attach to distinctive membrane fractions in a simultaneous infection of *E. coli*. The less dense fraction marked by a peak of DPNH oxidase contains the inner plasma membrane; the denser fraction marked by phospholipase A contains the lipopolysaccharide outer membrane. Presence of DPNH oxidase in the outer membrane fraction suggests regions of adhesion between inner and outer membranes in these cells.

fractionations are indications, based on an inner-membrane enzyme marker, that the outer membrane fraction to which ϕX174 attaches and to which ϕX174 RF is bound has some inner membrane adhering to it. Points of adhesion between inner and outer membrane have been seen in the electron microscope and suggested as preferred sites for ϕX174 adsorption[53] and may be identified by the presence of a distinctive phospholipase A1[54]. Since the folded *E. coli* chromosome also appears to be firmly attached to a membranous site characterized by outer membrane markers[55], there is an intriguing possibility that the ϕX174 receptor, by its location on the external surface of this site, gives the infecting DNA direct access to the chromosomal replicative apparatus fixed at the inner surface.

REPLICATION

In 1959 when the discovery of a single-stranded chromosome was first reported[56], its replication appeared to pose an exception to models for semiconservative replication. Of course, the mechanism of replication of ϕX174 DNA proved instead to be a potent confirmation of the semiconservative model. The viral strand, called the (+) strand, has a base composition (Table 8-3) in which the purines are not equivalent to pyrimidines, nor does A match T, nor G match C. In the cell, replication of the (+) strand to produce a complementary (−) strand yields a duplex with the base-pairing characteristics of all natural DNAs.

The replication cycle of ϕX174 DNA[57] includes three stages outlined in Table 8-7 and discussed below. Details of each stage can only be surmised from available data, so what follows are hypothetical models rather than firm demonstrations.

53. Bayer, M. E. and Starkey, T. W. (1972) *Virol.* **49**, 236.
54. Scandella, C. J. and Kornberg, A. (1971) *B.* **10**, 4447.
55. Worcel, A. and Burgi, E. (1974) *JMB* **82**, 91.
56. Sinsheimer, R. L. (1959) *JMB* **1**, 43.
57. Sinsheimer, R. L. (1968–1969) *The Harvey Lectures*, **64**, 69.

TABLE 8-7
Replication cycle of ϕX174

Stage	Time min.	Events
I	0–1	Adsorption and penetration
		viral SS → parental RF II
		Transcription of RF I
II	1–10	parental RF II → 20 progeny RF
	15	RF multiplication stops
		Host DNA synthesis stops
III	15–20	~ 20 progeny RF → ~ 200 viral SS in phage particles
	20–30	Cell lysis

Replicative stage I (SS → RF). In the eclipsed particle the DNA is partially exposed, and so is made available as a template to the host replicating system responsible for duplication of the host chromosome. A protein of 37,000 daltons, presumably the H protein, may direct the initiation of replication after serving in the adsorption-penetration process[58]. Parental RF isolated from in vivo infections had several molecules of this protein but no other phage proteins.

Initiation of the complementary strand depends on an RNA priming segment synthesized by a rifampicin resistant system (Tables 7-9; 8-5)[59]. The *dna* A, *dna* B, *dna* C–D, *dna* G, and additional host replication proteins are involved. Growth of the complementary strand depends on the pol III* system[60], and completion of the chain by pol I. Excision of RNA, gap-filling, and ligase action have been discussed in Chapter 7. The duplex product of single-strand conversion by a soluble enzyme extract from *pol* A⁻ cells contains a small gap in the complementary strand.

Replicative stage II (RF → RF)[61]. This operation requires the gene A product, which is thought to nick the (+) strand. Unlike the counterpart gene 2 product of M13, the gene A product cannot be shared in the same cell with other φX174 genomes mutant in this gene. Thus, the gene A product is said to act only in a "cis" fashion whereas the M13 gene 2 product acts also in a "trans" fashion. The gene A region may be the locus for the origin of replication in this and other stages[62].

These indications, that the gene A protein initiates RF replication, have been confirmed with its recent isolation[63]. It is an endonuclease which breaks one phosphodiester bond in the (+) strand of φX174 RF I but fails to act on all other DNAs tested. Inasmuch as the enzyme also cleaves φX174 SS, the site of its action may be the hairpin loop of the (+) strand, recognized in the superhelical RF I as well as in the SS. The "cis" action, observed in vivo but not in vitro, may be explained by tight binding of mutant protein fragments to the cleavage site which prevent the nuclease from acting.

A host gene[64] called *rep* is also required for RF multiplication. Mutants of *rep⁻* genotype appear to show no defect in the replication of their own DNA but this conclusion is based only on the gross criterion of cellular growth rate. When the function performed by the *rep* gene product in φX174 replication becomes clear, so may its role in the replication of cellular DNA.

58. Jazwinski, S. M., Marco, R. and Kornberg, A., unpublished observations.
59. Schekman, R., Wickner, W., Westergaard, O., Brutlag, D., Geider, K., Bertsch, L. L. and Kornberg, A. (1972) *PNAS* **69**, 2691.
60. Wickner, W., Schekman, R., Geider, K. and Kornberg, A. (1973) *PNAS* **70**, 1764.
61. Francke, B. and Ray, D. S. (1972) *PNAS* **69**, 475; Fidanian, H. M. and Ray, D. S. (1972) *JMB* **72**, 51; (1974) *JMB* **83**, 63; Greenlee, L. L. (1973) *PNAS* **70**, 1757; Schröder, C. H. and Kaerner, H-C. (1972) *JMB* **71**, 351.
62. Benbow, R. M., Hutchison, C. A., III, Fabricant, J. D. and Sinsheimer, R. L. (1971) *J. Virol.* **7**, 549; Baas, P. D. and Jansz, H. S. (1972) *JMB* **63**, 569; Johnson, P. H. and Sinsheimer, R. L. (1974) *JMB* **83**, 47.
63. Henry, T. J. and Knippers, R. (1974) *PNAS* **71**, 1549.
64. Denhardt, D. T., Dressler, D. H. and Hathaway, A. (1967) *PNAS* **57**, 813.

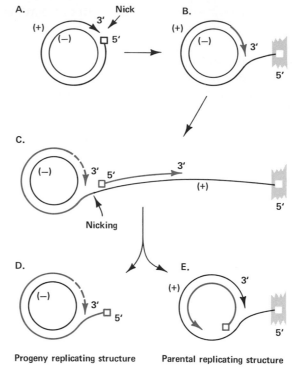

A.   Nick   B.

FIGURE 8-16
Rolling-circle mechanism for ϕX174 replication
(see text).

    The rolling circle is a currently favored, but not proven, model for the replication of RF. It proposes a covalent extension at the 3' primer terminus of the nicked (+) strand (Fig. 8-16A) using the (−) strand as template[65]. The origin is at a fixed point on the (+) strand; replication is unidirectional and clockwise with respect to the genetic map of ϕX174. The 5' terminus, fixed and protected in some manner, is displaced as the advancing replicating fork "rolls" the template circle clockwise (Fig. 8-16B).

    The displaced (+) strand at some time becomes a template itself when a new (−) strand is initiated upon it (Fig. 8-16C). Multiple initiations of the (−) strand[66] representing discontinuous synthesis, as in larger chromosomes, would distinguish this replication from the continuous mode of the (+) strand. Synthesis of (+) and (−) strands now proceed simultaneously, including a second length of (+) strand (Fig. 8-16C). A nicking event at the origin of replication generates two replicating structures. One contains the fixed parental (+) strand (Fig. 8-16E), while the other has the rolling (−) strand (Fig. 8-16D).

65. Dressler, D. (1970) PNAS 67, 1934; Knippers, R., Whalley, J. M. and Sinsheimer, R. L. (1969) PNAS 64, 275.
66. Eisenberg, S. and Denhardt, D. T., (1974) PNAS 71, 984.

Based on the linear kinetics of RF replication it is assumed that there are only a few master templates that function as "stamping machines" fixed to membrane sites. The progeny RF are released to the cytoplasm, to be used for transcription and, in the next stage, for viral strand synthesis.

Evidence for the rolling circle model has been obtained from biochemical experiments and electron microscope observations. Open RF were found to be broken only in the (+) strand, and that strand was commonly greater than genome length. Structures with σ (lariat) shapes were observed in which a single-stranded limb, presumably (+) strand, protruded from a duplex circle. These and similar results have led to the proposal of asymmetric RF replication with the (−) strand serving as master template.

The significance of RF I as an obligatory intermediate in replication was questioned when the M13 pathway was considered (see Section 3 of this chapter). Whereas the nicked form, RF II, occupies the key position in a favored replication scheme (Fig. 8-8), the prime function of RF I may be as a template for transcription. It is the preferred form for in vitro transcription[67] and, in gene A mutants that accumulate the superhelical RF, the template efficiency for mRNA synthesis was increased manyfold[68].

Replicative stage III (RF $\rightarrow$ SS)[69]. Progeny RF function as templates for synthesis of viral single strands. Assuming a rolling-circle mechanism as proposed for RF multiplication, an RF nicked at a specific point in the (+) strand would be extended by copying the (−) strand as template (Fig. 8-17). There are several major differences from RF replication.

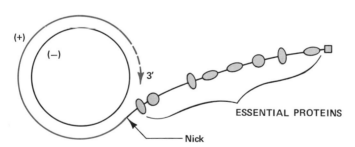

FIGURE 8-17
Dependence of viral single-strand synthesis on phage capsid and other proteins.

(i) Synthesis of (+) strands uniquely requires that the coat proteins F and G and additional proteins coded by genes C and D, be present to bind and envelop the viral strand as it is displaced.

67. Hayashi, Y. and Hayashi, M. (1971) B. **10**, 4212.
68. Mazur, B. J. and Model, P. (1973) JMB **78**, 285.
69. Francke, B. and Ray, D. S. (1972) PNAS **69**, 475; Fidanian, H. M. and Ray, D. S. (1972) JMB **72**, 51; (1974) JMB **83**, 63. Greenlee, L. L. (1973) PNAS **70**, 1757.

(ii) Synthesis of (−) strand on the displaced (+) strand is precluded.

(iii) A single cytosine residue in the (+) strand is methylated[70]. Finally, a full-length (+) strand must be cut and then closed by ligase to form a circle.

How closure of the viral strand is accomplished is not immediately obvious. A clue has been provided by the observation that open viral strands accumulating in ligase-deficient cells can infect wild-type spheroplasts[71]. One possibility is that such open strands utilize a duplex, hairpinlike region[72] to furnish a nick which ligase can seal. It would follow then that ligase is provided with a similar substrate in the normal closure of open viral strands (Fig. 8-18). With the

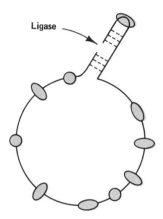

FIGURE 8-18
A scheme suggested for utilizing the duplex, hairpin-like region in the viral strand for its closure.

completion of the DNA packaging some twenty minutes after the start of infection, several hundred particles fill the cell and are released when an endolysin induced by gene E produces cell lysis.

5. Medium-size Virus[73]: T7

T7, a virulent *E. coli* bacteriophage with a short and primitive cone-like tail, is more complex than M13 and φX174 but smaller and simpler than T4 (Fig. 8-19). The chromosome is 23×10^6 daltons, corresponding to 35,400 base pairs. T7 DNA molecules, unlike T4 (see Section 6 of this chapter), are all the same. Sequence determinations, which have penetrated only a few residues at each end[74] (Fig. 8-19), verify that all T7 duplexes are complete to the ends and are terminated by a 5′-phosphate and a 3′-hydroxyl at each end. Structural, genetic, and biochemical analyses confirm that one T7 duplex molecule is not distinguishable from another.

70. Razin, A., Sedat, J. W. and Sinsheimer, R. L., (1973) *JMB* **78**, 417; Razin, A. (1973) *PNAS* **70**, 3773.
71. Schekman, R. W. and Ray, D. S. (1971) *NNB* **231**, 170.
72. Schaller, H., Voss, H. and Gucker, S. (1969) *JMB* **44**, 445.
73. Studier, F. W. (1972) *Science* **176**, 367.
74. Weiss, B. and Richardson, C. C. (1967) *JMB* **23**, 405; Price, S. S., Schwing, J. M. and Englund, P. T. (1973) *JBC* **248**, 7001.

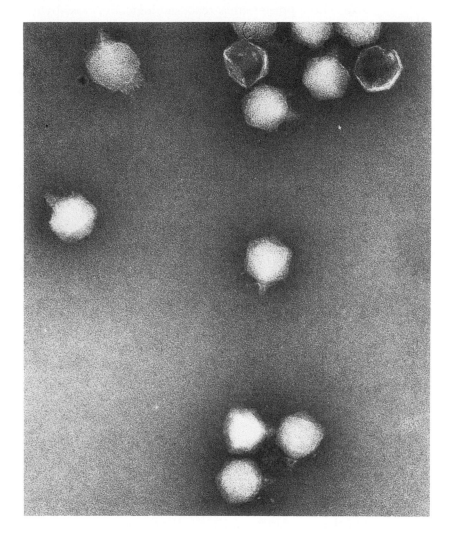

FIGURE 8-19
Electron micrograph of phage T7 (× 200,000) and the terminal sequences of T7 molecules showing that the duplexes are complete and that the residues contain 5′-phosphate and 3′-hydroxyl groups. (T7 micrograph courtesy of Professor R. C. Williams.)

T3 is a phage closely related to T7, resembling it in shape and size. These phages cross-react serologically, cross-hybridize in DNA annealing, recombine genetically, and yet induce distinctive enough enzyme patterns to preclude successful complementation in a mixed infection.

T7 infection is especially interesting for two reasons: (i) as a linear chromosome, its replication poses special problems whose solution should have general significance, and (ii) its chromosome is nearly completely mapped and achieving a full understanding of its replication and transcription is an inviting challenge.

Terminal redundancy in the T7 chromosome. In common with other T phages, but unlike bacterial and many viral chromosomes, the T7 chromosome is not circular. Nor does it have a circular stage during replication like phage λ, which will be described in Section 7 of this chapter. As a linear duplex, T7 DNA might be susceptible to exonuclease action. It requires also a mechanism for fully replicating the 3′ end of each template strand, a problem that a circular template does not face. At its 5′ end, the initial part of the replica, as in other DNA chain initiations, should be RNA. This RNA segment must later be excised and that terminal section filled by DNA. Since no mechanism is known for 3′ → 5′ DNA chain growth, the solution to this problem is not obvious. However, a remarkable structural feature of the T7 chromosome suggests an answer: it contains a stretch of several hundred nucleotides repeated at both ends of the duplex, known as "terminal redundancy".

Terminal redundancy can be demonstrated following exonucleolytic removal of terminal segments[75]. Limited exonuclease III treatment generates 5′ single-strand tails (Fig. 8-20). These tails behave as cohesive or "sticky" ends, and link to form circles readily observable in the electron microscope. At higher DNA concentrations, intermolecular linkages form chromosome dimers and longer repeating chains, called _concatemers_. The manner in which terminal redundancy and concatemers enable replication to include the 3′ ends will be discussed presently.

Genetic and biochemical map. The genetic map of T7 appears to be nearly saturated (Fig. 8-21). By gel electrophoretic analysis, the induction of some 30 new polypeptides was observed after T7 infection, 23 of them identified with specific mutations on a linear map. The size of the polypeptides accounts for nearly 90 percent of the coding capacity of T7 DNA[76].

The genes of T7, like those of T4 and λ, are clustered according to function and order of expression. Genes that code for early functions in phage development, including DNA replication, are situated at the left end of the map and are transcribed first, whereas the genes that serve phage morphogenesis are transcribed later and are

75. Ritchie, D. A., Thomas, C. A., Jr., MacHattie, L. A. and Wensink, P. C. (1967) _JMB_ **23**, 365.
76. Studier, F. W. (1972) _Science_ **176**, 367.

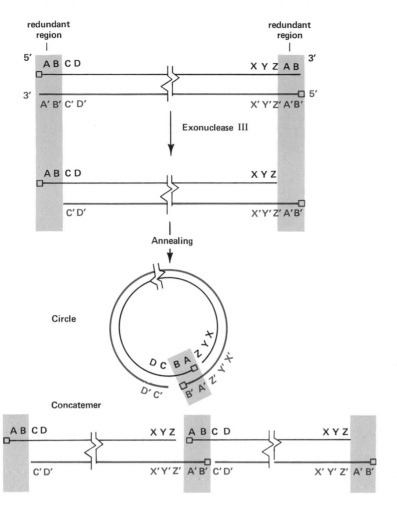

FIGURE 8-20
A model illustrating how the presence of redundant regions at ends
of phage T7 DNA can be detected by conversion of linear to circular and
concatemeric forms (see text).

located in the right half. Transcription is exclusively from one strand
of the duplex.

Nineteen of the T7 genes are known to be essential: of these, six
(genes 1, 2, 3, 4, 5, and 6) are required for replication; gene products
1, 3, 5, and 6 have been purified and characterized. A specific RNA
polymerase (gene 1 product)[77] is required for transcription of the
next five genes (2, 3, 4, 5, 6) but might, in addition, be directly involved in initiating replication. The DNA polymerase (gene 5 product)[78] is similar to the more intensively studied T4 enzyme, in
requiring single-stranded DNA and a consequent failure to utilize
helical DNA as template-primer (see Chapter 5, Section 5). The

77. Chamberlin, M. and Ring, J. (1973) *JBC* **248**, 2235, 2245.
78. Grippo, P. and Richardson, C. C. (1971) *JBC* **246**, 6867.

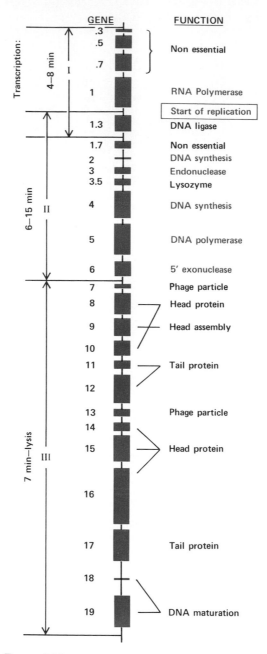

FIGURE 8-21
Genetic map of phage T7. The conventional
orientation of the map from left to right is shown
here from top to bottom. Indicated are
the gene numbers, their estimated sizes and
functions; those essential for DNA replication
are in color. The region transcribed early, the
subsequent stages of the phage cycle, and
their duration are also shown.

enzyme lacks a $5' \rightarrow 3'$ exonuclease but its gene location is followed on the map by gene 6, which codes for an enzyme[79] with this function. This nuclease resembles phage λ exonuclease in action (Chapter 9, Section 1). Its behavior also recalls the $5' \rightarrow 3'$ exonuclease functions of pol I, in requiring a double strand, in attacking a 5' terminus whether it contains a hydroxyl or a phosphate group, and in releasing oligonucleotides as well as 5'-mononucleotides.

Both the $5' \rightarrow 3'$ exonuclease and an endonuclease I coded[80] by gene 3 contribute to the extensive breakdown of host DNA which furnishes most of the nucleotides for T7 DNA synthesis. The endonuclease, despite a strong preference for single-stranded DNA, converts duplex DNA also to short oligonucleotides with 3'-hydroxyl and 5'-phosphate termini. Whether the essential functions of these nucleases are confined to salvage of host DNA for precursors in T7 DNA replication or are more directly involved in phage replication, as will be suggested presently, remains to be determined.

The protein products of genes 2 and 4, the remaining two genes essential for replication[81], have not been identified. Their purification and characterization should provide added insight into the mechanism of T7 DNA replication. T7 also induces a DNA ligase[82] and a DNA-binding protein[83]. Ligase, the gene 1.3 product, is an ATP-dependent enzyme; it can be replaced by host ligase but is clearly required in cells of low ligase content. The DNA-binding protein resembles the gene 32 unwinding protein induced by T4 (Chapter 7, Section 7) in stimulating T7 DNA polymerase replication of the single-stranded template specific for the T7 enzyme; the gene locus for this binding protein has not been determined.

Recent studies[84] of T7 DNA replication with a soluble enzyme system (Chapter 7, Section 11) have disclosed a remarkable capacity of the T7 DNA polymerase to replicate duplex T7 DNA in the presence of a novel host cofactor, relatively small in size and tightly bound to the enzyme. This functional form of T7 polymerase, reminiscent of the holoenzyme complex of polymerase III and copolymerase III* (Chapter 5, Section 3), will add to our understanding of both T7 replication and *E. coli* DNA replication.

DNA Replication: A Suggested Role for Concatemeric DNA

Host DNA continues to be synthesized for the first five minutes after infection and then begins to be degraded with the onset of T7

79. Kerr, C. and Sadowski, P. D. (1972) *JBC* **247**, 305, 311.
80. Center, M. S., Studier, F. W. and Richardson, C. C. (1970) *PNAS* **65**, 242; Sadowski, P. D. and Kerr, C. (1970) *J. Virol.* **6**, 149; Center, M. S. and Richardson, C. C. (1970) *JBC* **245**, 6285, 6292; Sadowski, P. D. (1971) *JBC* **246**, 209.
81. Studier, F. W. (1972) *Science* **176**, 367; Strätling, W. and Knippers, R. (1973) *Nat.* **245**, 195.
82. Masamune, Y., Frenkel, G. D. and Richardson, C. C. (1971) *JBC* **246**, 6874.
83. Reuben, R. C. and Gefter, M. L. (1973) *PNAS* **70**, 1846.
84. Modrich, P. and Richardson, C. C., personal communication.

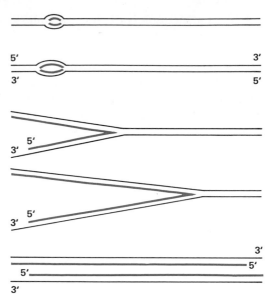

FIGURE 8-22
Origin and direction of replication on the T7
chromosome. Bidirectional replication starts at a
point about 17 percent from the left end to generate
"eye" forms. With equal rates of fork movement
the left end is completed first to generate a "Y"
form. Note that a region at the 3' end of each
template strand remains unreplicated.

DNA replication. Degradation continues for 10 to 15 minutes, shut-
ting off at about the time of lysis, 25 to 30 minutes after infection.

Electron microscope studies indicate T7 DNA replication is started
at a point 17 percent from one end of the molecule, designated as
the left end, and proceeds in both directions. The bubble formed
gives the replicating intermediate an eye-shaped appearance[85]
(Chapter 7, Section 3, and Fig. 8-22). When replication has reached
the left end, a Y-shaped structure is produced and then expands.
Second rounds may be initiated before the first is completed giving
rise to multiple eyes and forks. Most remarkable and informative
are: (i) the development of single-strand regions on one of the two
growing arms of the replicating chromosome (see Fig. 7-10), and (ii)
the appearance of concatemers containing two, three, and even
longer unit lengths of T7 DNA.

Single-stranded regions were seen in nearly all molecules in a
characteristic distribution[85] (Fig. 8-23); they were asymmetrically
positioned on one of the growing arms of the replicating fork in a
manner consistent with discontinuous growth (see Chapter 7, Section 3). These gaps in the duplex must eventually be filled, and the

85. Wolfson, J., Dressler, D. and Magazin, M. (1972) PNAS **69**, 499; Dressler, D., Wolfson, J. and
Magazin, M. (1972) PNAS **69**, 998; Wolfson, J. and Dressler, D. (1972) PNAS **69**, 2682.

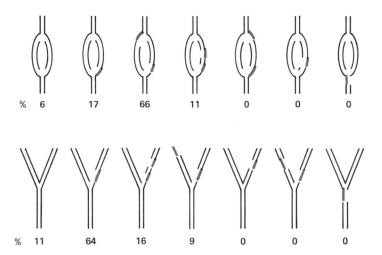

%   6   17   66   11   0   0   0

%   11   64   16   9   0   0   0

FIGURE 8-23
Growing forks of the T7 chromosome contain single-stranded
regions, in both the "eye" and "Y" forms, in characteristic
locations. (Wolfson, J. and Dressler, D. (1972) *PNAS* **69**, 2682.)

combined action of pol I and ligase has already been cited as an
example of how this may be accomplished for the host chromosome
(Chapter 7, Section 6). However, two gaps will remain at the 5′ ends
of the replicated strands opposite the 3′ ends of the parental template
strands (Fig. 8-22). None of the actions of available enzymes can
easily explain how gaps in these terminal locations can be filled.

More specifically, if initiation of DNA replication opposite the
3′-terminal regions of the template strands involves RNA priming,
then replacement of the RNA fragment would be impossible except
by 3′ → 5′ chain growth, and for this no DNA polymerase is known.
An unknown DNA polymerase capable of starting a DNA chain
cannot be excluded but, once again, there is no precedent for such
an enzyme. A hypothesis, proposed to solve the dilemma, suggests
that concatemers developed by a typical recombinational mechanism are initiated at the redundant regions at each end of the
T7 molecule; known nucleases and polymerases then cut and fill
to complete them.

Concatemer formation might proceed in this way[86]. The complementary sequence of the 3′-ended tail regions of two T7 molecules can base-pair to form a dimer linked by hydrogen-bonding
of these tail regions (Fig. 8-24). If both single-strand regions (tails)
are longer than the redundant zone (Fig. 8-24a), then the dimer is
readily completed by gap filling; if the tails do not include the entire
redundant zone (Fig. 8-24b), then the extra lengths of single-strand
regions might be accommodated by exonuclease action. As an example, the gene 6 product acting from the 5′ terminus might remove

86. Watson, J. D. (1972) *NNB* **239**, 197.

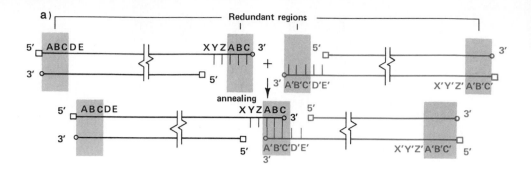

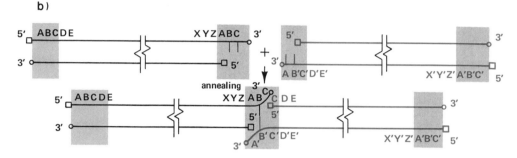

FIGURE 8-24

A scheme utilizing concatemers as intermediates in replication of T7 DNA. When the unreplicated 3'-tail region extends to a point *beyond* the redundant region as in (a), the annealing of two such molecules at their redundant regions leaves gaps to be filled. When the unreplicated 3'-tail region extends to a point *within* the redundant region, as in (b), annealing of two such molecules at their redundant regions leaves single-stranded lengths of DNA which are unmatched and require excision. (Watson, J. D. (1972) *NNB* **239**, 197.)

a length of chain that would enable the extra 3' terminated single strand to take its place. Or, the single-strand spur might be erased by 3' → 5' exonuclease action. For any gaps that remain, filling and sealing by polymerase and ligase would assure that the double-length concatemer is complete.

Concatemers of varying lengths have been observed during the course of T7 replication. Buildup of multimeric sizes can be explained in two ways. Concatemers can double in size by the scheme just described for duplication of the unit-length duplex. Alternatively, creation of 3' tails on the concatemer by 5'-exonuclease action (i.e. gene 6 product), followed by annealing and nuclease and polymerase action, leads to the joining of duplexes of all sizes. The experimental evidence also suggests that interference with final assembly of the virus leads to the development and accumulation of long concatemers. Examples are infections with mutants in genes 8, 9, or 10, which code for head proteins, and mutants in genes 18 or 19, whose products are still unknown. Head proteins are known to be essential for φX174 viral DNA synthesis and for maturation of T4 and λ DNA; the gene 18 and 19 products may be more directly involved in carving up the concatemer, as suggested below.

Maturation of the genome. Incorporation into the phage head requires that a concatemer be cut to the precise sequence of the T7 molecule. Although the steps involved are still unknown, there is enough experience with other systems to suggest how this might happen (Fig. 8-25). Nicking of the concatemer at specific, staggered points would be analogous to the conversion of λ concatemers

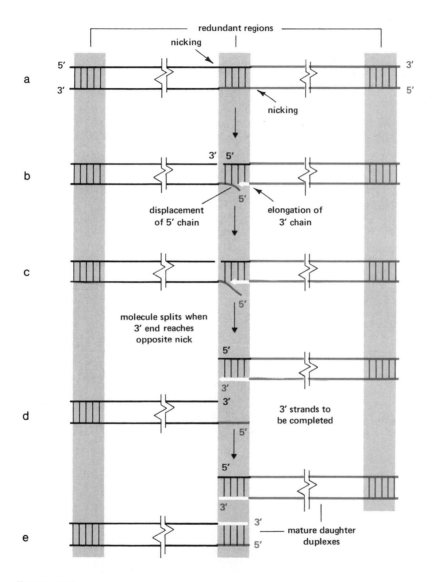

FIGURE 8-25
A scheme for maturation of T7 DNA from the concatemeric intermediate.
(a) The concatemer is nicked at specific, staggered points. (b) The 3′ hydroxyl ends created by nicking are elongated by polymerase action displacing the 5′ ends. (c) The concatemer comes apart when extension of the 3′ hydroxyl end reaches the nicked region on the template strand. (d) Molecules with incomplete 3′ ends are available for elongation by polymerase action. (e) Unit-length T7 molecules with redundant terminal regions are ready for packaging in phage heads. (Watson, J. D. (1972) *NNB* **239**, 197.) (See also Fig. 8-29.)

to linear forms. The gene 18 and 19 products could conceivably be responsible for such specific cleavages. In order to separate unit-length T7 molecules from the concatemer, the overlapping ends in the redundant region must be dissociated by some enzymatic action. This might be accomplished by replicative extension of the 3'-hydroxyl-terminated chain at one of the nicks. Displacement of the 5' terminus would then replace the hydrogen-bonded linkage when the advancing 3'-ended chain came close to the nick on the template (opposite) chain. Polymerase filling of incomplete regions on both duplexes would finish them for packaging into the phage particle.

6. Large Virus[87]: T4

The structure and life cycle of T-even phages (T2, T4, and T6) have become so familiar as to suggest that they are well-understood. Despite considerable knowledge about these phages, most of the major questions about replication, transcription, and assembly of the chromosome have not been answered. Certain features of DNA biosynthesis that have been revealed or are best illustrated by replication of T-even phages will be selected for discussion. Although attention will focus on T4, the facts also apply to the closely related T2 and T6 phages.

The anatomy of the long-tailed T4 phage and the syringe-like injection of its DNA through the E. coli surface have been widely

87. Cohen, S. S. (1968) *Virus-Induced Enzymes*, Columbia University Press, N.Y.; Edgar, R. S. and Epstein, R. H., (1965) *Sci. Amer.* **212**, 70; Wood, W. B. and Edgar, R. S., (1967) *Sci. Amer.* **217**, 60.

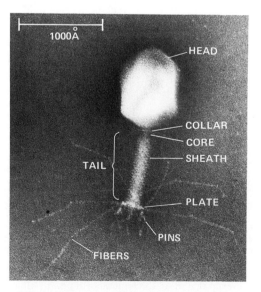

FIGURE 8-26
Electron micrograph of phage T4 ($\times$ 270,000).
(Courtesy of Professor R. C. Williams.)

pictorialized (Fig. 8-26). The organization of the T4 chromosome in the phage head and the mechanism of its penetration through the plasma membrane are hardly understood at all. T4 DNA is a linear duplex of 120×10^6 daltons, a mass sufficient to code for about 200 genes. To date nearly 100 of the T4 genes have been identified and mapped[88] (Fig. 8-27). Of these, nearly 30 are in some way involved

88. Stahl, F. W., Craseman, J. M., Yegian, C., Stahl, M. M. and Nakata, A. (1970) *Gen.* **64**, 157; Wood, W. B. (1973) in *Survey of Genetics* (R. C. King, ed.) Plenum Publishing Corp., N. Y. (in press); Mosig, G. (1968) *Gen.* **59**, 137.

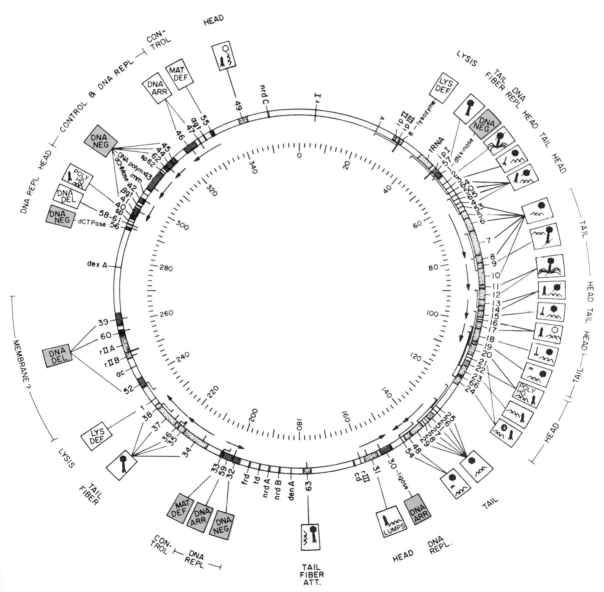

FIGURE 8-27
Genetic map of phage T4. Genes involved in DNA replication are in color (see Table 8-8 for details). (Wood, W. B. (1973) in *Survey of Genetics* (R. C. King, ed.) Plenum Publishing Corp., N.Y., in press.)

TABLE 8-8
T4 genes involved in DNA replication

	T4 replication genes		
	Gene	Map location degrees	Function
DNA negative	1	50	Deoxynucleotide kinase
	32	200	Unwinding protein
	41	296	
	42	300	dCMP hydroxymethylase
	43	308	DNA polymerase
	44	316	
	45	317	
	56	290	dCTPase, dUTPase
	62	314	
DNA arrest	30	152	DNA ligase
	46⎱ 47⎰	325	Cytosine specific DNase (?)
	59	204	
DNA delay	39	262	
	52	238	
	58⎱ 61⎰	292	
	60	256	
Maturation defective	33	204	RNA polymerase modifier
	βgt	298	HMC β-glucosyl transferase
	αgt	330	HMC α-glucosyl transferase
	55	332	RNA polymerase modifier
Nucleotide biosynthesis	cd	162	dCMP deaminase
	nrd A⎱ nrd B⎰	188	Nucleoside diphosphate reductase
	nrd C	350	Thioredoxin
	td	194	dTMP synthetase
	frd	196	Dihydrofolate reductase

in DNA replication (Table 8-8). Defects in these genes either block DNA replication entirely (DNA *neg*), delay it (DNA *del*), prematurely arrest it (DNA *arr*), or prevent its maturation to a packageable form (mat *def*).

Shutoff of host syntheses[89]. Within one minute after T4 phage adsorption, synthesis of host macromolecules has virtually ceased, transcription of certain phage genes has been initiated, and within four minutes replication of the phage DNA is underway. The abrupt and complete switch in metabolic directions from host to phage is unusual even for a virulent infection. This fact together with the distinctive base composition of the T4 DNA made this phage infection an excellent subject for studies of control of chromosome

89. Duckworth, D. H. (1970) *Virol.* **40**, 673; Snyder, L. (1973) *NNB* **243**, 131.

replication and expression. From these investigations came the discovery of messenger RNA, virus-induced enzyme functions[90], early and late transcriptional control[91], and the patterns of organelle morphogenesis[92, 93].

Despite all that is known about T4 infection, there is still no explanation why host macromolecular synthesis stops so promptly and completely. This remains one of the major unsolved questions of molecular biology. One suggestion is that an effect on the membrane by phage penetration secondarily influences the replication and transcription of the host chromosome tied to the membrane. Other evidence invokes a phage-induced protein whose effect is not defined. The attraction of this problem is not so much a compulsive need to complete the catalogue of events surrounding a phage infection, but more the promise of clarifying the spatial organization and control of the *E. coli* chromosome and with it another major insight into molecular biology.

Circular permutation of the T4 chromosome. The T4, like the T7, DNA molecule has terminally redundant regions. They comprise a few percent of the DNA length, and so contain several thousand base pairs. T4 DNA molecules are all of the same size and genetic composition, yet are heterogeneous in sequence. They are circularly permuted as shown in Fig. 8-28a, b. This structural feature, first suggested by genetic studies[94], can be easily demonstrated by denaturing and then reannealing the DNA[95]. Hydrogen-bonded circles with single-stranded tails are generated in a high percentage of the molecules (Fig. 8-28c). The importance of terminal redundancy in replication has already been mentioned for T7 replication, and the additional circular permutation of T4 DNA must be a feature in its replication and maturation.

Replication and maturation[96]. The many observations about T4 replication are not easily connected into a coherent scheme. Not only is the process complex, but attempts to analyze it by various methods have been repeatedly confused by the high level of recombinational activity in which T4 molecules indulge.

Replication starts at a point between gene 42 and gene 43 and proceeds bidirectionally, forming an expanding loop[97]. The 10S replication fragments, first discovered in studies of T4 replication[98],

90. Cohen, S. S. (1968) *Virus-Induced Enzymes*, Columbia University Press, N.Y.
91. Milanesi, G., Brody, E. N., Grau, O. and Geiduschek, E. P., (1970) *PNAS* **66**, 181; Black, L. W. and Gold, L. M. (1971) *JMB* **60**, 365; O'Farrell, P. Z. and Gold, L. M. (1973) *JBC* **248**, 5502.
92. Edgar, R. S. and Epstein, R. H., (1965) *Sci. Amer.* **212**, 70; Wood, W. B. and Edgar, R. S., (1967) *Sci. Amer.* **217**, 60.
93. Wood, W. B. (1973) in *Genetic Mechanisms in Development*, Symp. Soc. Develop. Biol. p. 29; Eiserling, F. A. and Dickson, R. C. (1972) *ARB* **41**, 467.
94. Streisinger, G., Emrich, J. and Stahl, M. M. (1967) *PNAS* **57**, 292.
95. MacHattie, L. A., Ritchie, D. A., Jr., Thomas, C. A. and Richardson, C. C. (1967) *JMB* **23**, 355.
96. Frankel, F. R. (1968) *CSHS* **33**, 485; Huberman, J. A. (1968) *CSHS* **33**, 509; Bernstein, H. and Bernstein, C. (1973) *JMB* **77**, 355.
97. Mosig, G., Bowden, D. W. and Bock, S. (1972) *NNB* **240**, 12.
98. Okazaki, R., Okazaki, T., Sakabe, K., Sugimoto, K., Kainuma, R., Sugino, A. and Iwatsuki, N. (1968) *CSHS* **33**, 129.

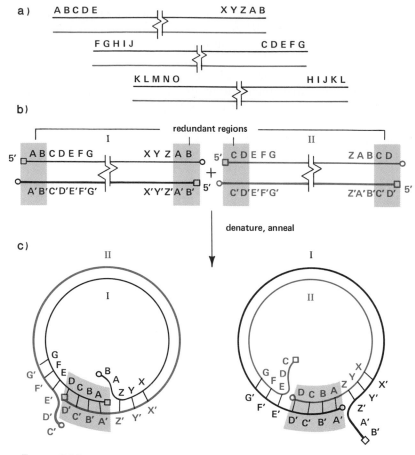

FIGURE 8-28
A model illustrating how circularly permuted, redundant regions at ends of
phage T4 DNA (a,b) can be detected by conversion of the linear to a circular
form (c).

anneal equally to both strands. Numerous electron microscopic
observations and sedimentation studies attest to the large size of
the replicating intermediate. Multiple initiations before completion
result in complex, pleated replicating structures, with as many as
60 replicating forks. Discovery of an RNA-DNA copolymer that
hybridizes to one strand of T4 DNA suggests RNA priming may also
be utilized in T4 replication[99].

There is some evidence for a circular form in T4 replication[100].
Although the observations are compatible with the development of
long concatemers as in λ replication (see Section 7 of this chapter),
a T7-like mechanism has not been excluded. Single-stranded 3'
tails that remain unreplicated can anneal, assisted where necessary
by the trimming and tailoring actions of exonucleases, possibly
those induced by the DNA-*arr* genes 46 and 47.

99. Buckley, P. J., Kosturko, L. D. and Kozinski, A. W. (1972) *PNAS* **69**, 3165.
100. Bernstein, H. and Bernstein, C. (1973) *JMB* **77**, 355.

The enzymes required for chain growth include the products of at least five DNA *neg* genes, some of which were mentioned in Chapter 7. DNA polymerase (gene 43 product), unwinding protein (gene 32 product), the protein complex coded by genes 44 and 62, and the product of gene 45 are all essential and are likely to operate as an integrated unit. Perhaps in such an assembly, DNA polymerase can utilize a duplex template-primer, which the pure enzyme cannot. The role of ligase (gene 30 product) is not essential for chain growth but is, of course, required to connect the 10S fragments. The extent to which the host ligase and other replicative enzymes such as pol I may contribute has been mentioned in, or may be inferred from, earlier discussions (Chapter 4, Section 12 and Chapter 7, Section 6).

Release of DNA from the multimeric complex in mature form ready for packaging into phage heads requires only that scissions produce molecules slightly larger than one genome in length. A double-stranded scission (Fig. 8-29) that provides a "headful"-size piece requires no sequence-specific nicking, as with T7, but rather a mechanism that measures the length desired. Mini T4 phages which

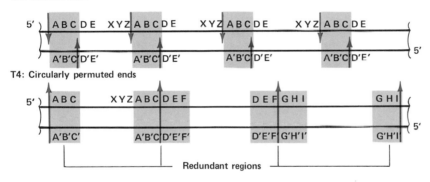

FIGURE 8-29
A scheme suggesting how staggered, double-strand scissions of a concatemer can produce a T7 molecule with redundant and uniform ends, assuming their completion (repair) by polymerase action; and a scheme for producing T4 molecules with redundant and circularly permuted ends by directly opposite double-strand scissions. (See Fig. 8-25 for the T7 mechanism.)

contain a random fraction of the genome evidently must package a length of DNA that fits the size of the mini head.

Scissions of the duplex produce molecules with terminal repetitions and permuted arrangements of the nucleotide sequence. Provided the genetic content is fully represented, some variation in length of the terminal region can be tolerated. Genes 49, 16, and 17, known to be essential for head formation, contribute to the maturation of DNA in ways that have yet to be determined.

Incorporation of a unique base. It was puzzling, when first observed, how *all* the cytosine residues were replaced by hydroxy-

methylcytosine (HMC) in the synthesis of T2, T4, and T6 phage
DNAs. How could dCTP produced in the host cell be prevented from
base-pairing with G in the template and from being incorporated
into the DNA? The ingenious solution of this problem by the infecting phage resides in a set of novel enzymes induced by its genome
(Fig. 8-30)[101].

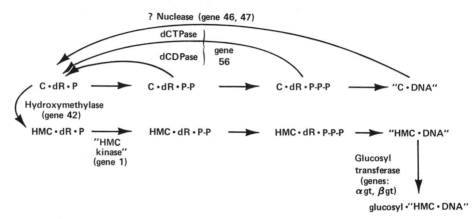

FIGURE 8-30
Pathway of glucosyl hydroxymethylcytosine (HMC)- DNA synthesis and breakdown
of cytosine (C)-containing DNA (see text).

(i) A hydroxymethylase[102] (gene 42 product) to convert C·dR·P
(dCMP) to HMC·dR·P.

(ii) A nucleoside monophosphate kinase[103] to phosphorylate the
latter to the diphosphate, which in turn becomes a substrate for
nucleoside diphosphate kinase to produce the final triphosphate
product, HMC·dR·P·P·P, for incorporation into DNA.

(iii) A family of induced glucosyl transferases[104] to glucosylate
the HMC groups in the DNA in patterns characteristic for the different
T-even phage DNAs (see below).

(iv) A dCDP-dCTP pyrophosphatase (gene 56 product)[105] to remove di- and triphosphates of dCMP as substrates for DNA synthesis,
thereby regenerating dCMP as a substrate for hydroxymethylase and
also for deamination to dUMP, the precursor of dTMP.

(v) Gene 46 and gene 47 products, thought to be a new nuclease[106]
to hydrolyze C-containing but not HMC-containing DNA. Should
such a nuclease exist, it would degrade phage DNA that may have
incorporated C residues, as well as the preexisting host DNA, to
generate dCMP and other deoxynucleotide precursors needed for
phage synthesis.

101. Cohen, S. S. (1968) *Virus-Induced Enzymes*, Columbia University Press, N.Y.
102. Mathews, C. K., Brown, F. and Cohen, S. S. (1964) *JBC* **239**, 2957.
103. Duckworth, D. H. and Bessman, M. J. (1967) *JBC* **242**, 2877.
104. Josse, J. and Kornberg, A. (1962) *JBC* **237**, 1968.
105. Zimmerman, S. B. and Kornberg, A. (1961) *JBC* **236**, 1480.
106. Wiberg, J. S. (1966) *PNAS* **55**, 614; Prashad, N. and Hosoda, J. (1972) *JMB* **70**, 617.

Modification of a macromolecule. The presence of novel bases in DNA in unique patterns that could not be managed by base-pairing is achieved by specific modifications of the polynucleotide after it is synthesized. For example, the genetically determined patterns of glycosylation of HMC residues in T2, T4, and T6 DNA were shown to be derived by specific glycosyl transfers to the DNA within a few minutes after its polymerization (Chapter 9, Section 6). These T-even phage studies were the first demonstration of how macromolecules may be modified at a stage beyond polymerization[107]. It is now also known that the natural occurrence in protein of an amino acid that does not belong to the codon-specified group of twenty amino acids is brought about by modifying the polypeptide after it is assembled. In this manner, proline residues are hydroxylated, lysine methylated, and serine and histidine phosphorylated.

7. Lysogenic Virus[108]: λ

λ is a phage of medium size (Table 8-1) whose capacity to convert *E. coli* to a lysogenic state (temperate infection) makes it especially interesting. An infection with λ seems poised at several stages in its development to pursue either the usual virulent course, of multiplying the λ DNA leading to cell lysis, or the temperate course, of incorporating the λ DNA into the host chromosome where it remains latent for an average of 1000 cell divisions[109]. λ is representative of a number of related lysogenic phages of *E. coli* (ϕ80, 434, P2 and 186) and *Salmonella typhimurium* (P22).

λ DNA is unusual in that it has single-stranded tails at each 5' end which are complementary (Fig. 8-31a)[110]. Lambdoid phages (λ, ϕ80, 82, 21 and 424) have identical 12-nucleotide-long tails; these differ from the 19-nucleotide-long tails of the nonlambdoid phages (186, P2, and 299)[111] as shown in Fig. 8-31b. Cohesion of the complementary sequences by hydrogen bonding converts the linear duplex to a circular form (Fig. 8-31c). Upon entering the cell, the linear molecule is converted by ligase to a covalently closed circle ready for integration or replication.

Integration and excision. Circular λ molecules are integrated by union of a specific region of the phage DNA, called "*att*" (Fig. 8-32), at a precise location between the *gal* and *bio* genes on the *E. coli* chromosome. This site-specific recombination between phage and host DNA takes place in a unique way[112] (Fig. 8-32) and requires λ gene products defined by the *int* locus. The *int* locus protein has

107. Kornberg, A., Zimmerman, S. B., Kornberg, S. R. and Josse, J. (1959) *PNAS* **45**, 772.
108. Hershey, A. D. (ed.) (1971) *The Bacteriophage Lambda,* CSHL; Kaiser, A. D., in Hershey, A. D. (ed.) (1971) *The Bacteriophage Lambda,* CSHL, p. 195.
109. Echols, H. (1972) *Ann. Rev. Gen.* **6**, 157.
110. Wu, R. and Taylor, E. (1971) *JMB* **57**, 491.
111. Murray, K. and Murray, N. E. (1973) *NNB* **243**, 134.
112. Campbell, A., in Hershey, A. D. (ed.) (1971) *The Bacteriophage Lambda,* CSHL, p. 13.

a)

λ DNA

```
                  3′                                                              5′
                  GCGCCCA                                            CCCGCCGCTGGA
                  ────────────────────                              ─────────────P
  P ────────────────────                              ────────────────
    GGGCGGCGACCT                                           GTTACG
    5′                                                              3′
```

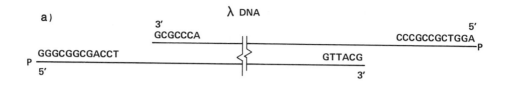

b)

186 DNA

```
                         3′                                                    5′
                         A                                TACGAAAGGGGCGGTGCGG
                         ──────────────                   ──────────────────────P
    ATGCTTTCCCCGCCACGCC                          CA
  P ──────────────────────                       ──────────────
    5′                                                3′
```

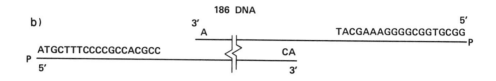

c)

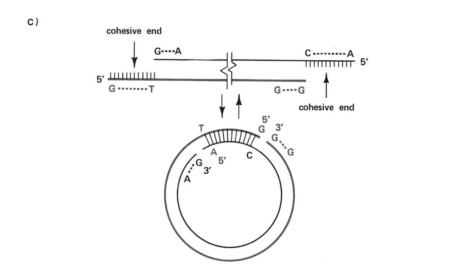

FIGURE 8-31

Structures of linear and circular forms of λ DNA. Sequences of the complementary overlapping 5′-ends of lambdoid phages (e.g. phage λ) (a) are different from those of the nonlambdoid phages (e.g. phage 186) (b). Sequences of 3′-terminal regions of phage λ, recently determined (Weigel, P. H., Englund, P. T., Murray, K. and Old, R. W. (1973) *PNAS* **70**, 1151), are shown in (a). Interconversion of the linear and circular forms by annealing of the overlapping cohesive 5′-ends is demonstrated in (c).

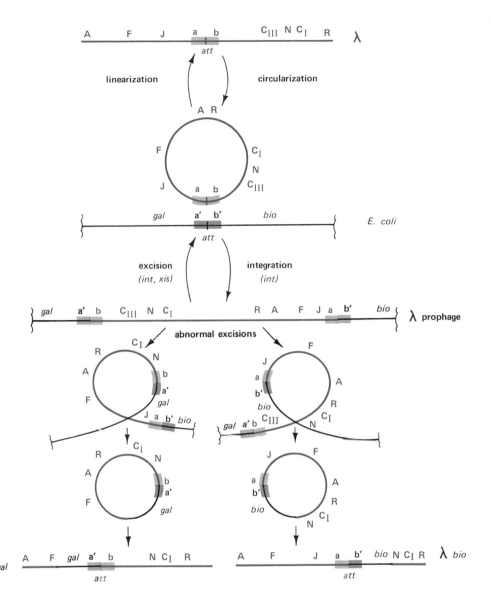

FIGURE 8-32
Mechanism of integration and excision of phage λ DNA into and out of the *E. coli* chromosome. Cleavages of the circularized vegetative phage DNA at locus "ab", and of the *E. coli* chromosome at a corresponding locus "a'b'", lead to integration as indicated; excision of the prophage results from a reversal of these events. Rarely, an excision as shown includes adjacent *E. coli* DNA (*gal*) genes at one end of the prophage or *bio* genes at the other to produce a phage carrying these host genes.

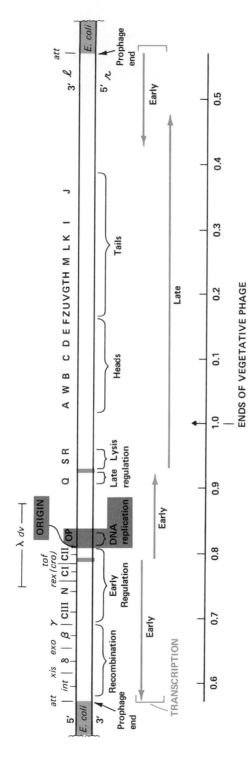

FIGURE 8-33

Genetic map of the prophage form of phage λ. The linear vegetative map reads from A to R; the prophage map is presented because it simplifies the representation of transcription. The genetic content of the plasmid λdv (a defective phage) is shown because it explains how possession of the replication origin and the O and P genes enables λdv to direct its own multiplication in the absence of λ. (See text and references for details.) (Szybalski, W. in *The Bacteriophage Lambda*, (A. D. Hershey, ed.) CSHL, p. 779; Berg, D. (1974) *JMB*, **86**, 59.

been isolated as a polypeptide of 42,000 daltons[113], and studies in vitro of its function are underway[114]. The *int* protein, called <u>integrase</u> is distinct from the proteins, called *red*, specified by λ genes which are responsible for general recombination. (The term *red* has the same significance for λ as the term *rec*, used to symbolize the recombinational genes of the host.) There are two *red* genes[115], *exo* whose product is λ exonuclease (Table 9-1) and *β* whose product, called *β* protein, has no known enzymatic function.

Excision of the phage DNA from the host chromosome is not simply a reversal of integration. It depends on a separate gene, *xis*, in addition to *int*. However, little more can be said about molecular details. One remarkable aspect of excision is the possibility of inclusion of part of the host chromosome along with viral DNA (Fig. 8-32). Although an exceedingly rare event, this inclusion of cell DNA into the viral chromosome can be clearly selected and measured. Genes for galactose utilization (*gal*) or biotin synthesis (*bio*) can become part of the DNA in the phage particle. Transfer of these genes by the phage to a new host, a process called transduction, has been one of the major tools in the modern development of bacterial genetics.

EARLY REPLICATION

Replication of a λ circle starts at a unique point between cII and OP (Fig. 8-33) from which it moves bidirectionally around the circle. The experimental evidence for this conclusion and for discontinuous growth of chains on both parental template strands has been presented earlier (Chapter 7, Section 3). λ replication requires the products of λ genes O and P in addition to host replicative enzymes, but before replication can start, a short stretch of DNA near the origin must first be transcribed. Indications that some RNA synthesis is required to initiate DNA replication came from the genetic and biochemical studies described below.

When λ resides in the *E. coli* chromosome it is replicated only when the chromosome itself is undergoing replication. It was not at first clear what kept the λ unit from being replicated independently starting at its own origin. It had been assumed that λ repressor, synthesized under directions of the integrated phage, prevented transcription of the O and P genes. However, even when those O and P gene products were supplied to the cell by a related phage, replication of λ was still inhibited. What did overcome this inhibition of λ replication were mutations[116] in a region just beyond and to the right of cI, the structural gene for λ repressor, and to the right of the place where the λ repressor is bound.

The most plausible interpretation of these mutations is that a

116. Dove, W. F., Inokuchi, H. and Stevens, W. F., in Hershey, A. D. (ed.) (1971) *The Bacteriophage Lambda*, CSHL, p. 747.
113. Ausubel, S. M. (1974) *Nat.* **247**, 152; Nash, H. A. (1974) *Nat.* **247**, 543.
114. Syvanen, M., (1974) *PNAS* **71**, 2496.
115. Radding, C. M., Rosenzweig, J., Richards, F. and Cassuto, E. (1971) *JBC* **246**, 2510; also preceding and succeeding papers, pp. 2502, 2513; Wackernagel, W. and Radding, C. M. (1973) *Virol.* **52**, 425.

new promoter site not subject to repressor inhibition has been created, and that operation of this promoter site for RNA synthesis is an essential preliminary to replication. The function performed by this brief transcriptional operation is very likely an RNA synthesis that serves to prime the initiation of DNA chain growth (see Chapter 7, Section 5). The RNA transcript, called *oop* RNA, has been isolated recently and found to be 81 nucleotides long with a 5′ terminal sequence[117]: pAUUUUUU · · · . Its synthesis requires the activity of λ genes O and P, the *dna* B and *dna* G replication proteins (Chapter 7, Section 6–9) known to be essential for replication of the *E. coli* and λ chromosomes, and a region on the "l" strand of λ which corresponds to the promoter for replication.

Products of genes N, O, and P are required early in replication. The function of gene N product in replication is indirect. It appears to enhance transcription through some influence on RNA polymerase. Even in the absence of gene N function, λ can undergo two cycles of semiconservative replication. However, without gene O and gene P products, replication is terminated when less than 10 percent of the λ length has been copied. The specific functions of these proteins, still unknown in detail, are directed at initiation of replication. Endonuclease activity has been conjectured. Further studies with a cell-free cellophane disk system (Chapter 7, Section 10), in which a dependence of λ DNA replication on O and P gene products has been observed[118], may lead to a better understanding of these questions.

Multiplication of circles continues from 5 to 15 minutes after infection but these circles are not incorporated into phage particles. Instead they give rise to an elongated, multimeric intermediate which is the major precursor of DNA for phage progeny.

LATE REPLICATION[119]

DNA produced during the latter half of the latent period is distinguished from "early DNA circles" by release, upon denaturation, of single strands many times greater than monomer length. Yet, there is no evidence of the rolling-circle seen in ϕX174 replication. Instead there are indications of long concatemers. Isotopic labeling experiments have suggested that concatemers develop from the nicked circular form, and genetic studies have identified certain loci that appear to be important for this transition.

Are the concatemers of late DNA generated by recombination (see Chapter 9, Section 3), by replication, or by some combination of these processes? An interesting mutant called *gam* identifies the gamma locus whose contribution to late DNA replication may yield important clues about the mechanism. *Gam⁻* mutants produce circles rather than concatemers late in infection and DNA synthesis is reduced in rate. This defect is aggravated in host cells mutant in

117. Hayes, S. and Szybalski, W. (1973) *Fed. Proc.* **32**, 529 abs; Dahlberg, J. E. and Blattner, F. R. (1973) *Fed. Proc.* **32**, 664 abs.
118. Klein, A. and Powling, A. (1972) *NNB* **239**, 71.
119. Enquist, L. W. and Skalka, A. (1973) *JMB* **75**, 185; McClure, S. C. C., MacHattie, L. and Gold, M. (1973) *Virol.* **54**, 1; and succeeding paper.

rec A or pol A (Chapter 9, Section 3) and thus deficient in recombination or gap filling. However, the gam⁻ defect is completely relieved in rec A⁻ cells, by an additional loss of the rec BC (Chapter 9, Section 3) gene function. These results suggest that a precursor of late λ DNA (Fig. 8-34) sensitive to the rec BC nuclease, may be destroyed unless the nuclease is controlled by the gam gene product[120]. Evidence to support this suggestion is the inhibition in vitro of the rec BC nuclease by the isolated gam protein[121].

The hypothetical scheme[122] in Fig. 8-34 indicates alternative

120. Enquist, L. W. and Skalka, A. (1973) *JMB* **75**, 185; Pilarski, L. M. and Egan, J. B. (1973) *JMB* **76**, 257.
121. Sakaki, Y., Karu, A. E., Linn, S. and Echols, H. (1973) *PNAS* **70**, 2215.
122. Enquist, L. W. and Skalka, A. (1973) *JMB* **75**, 185.

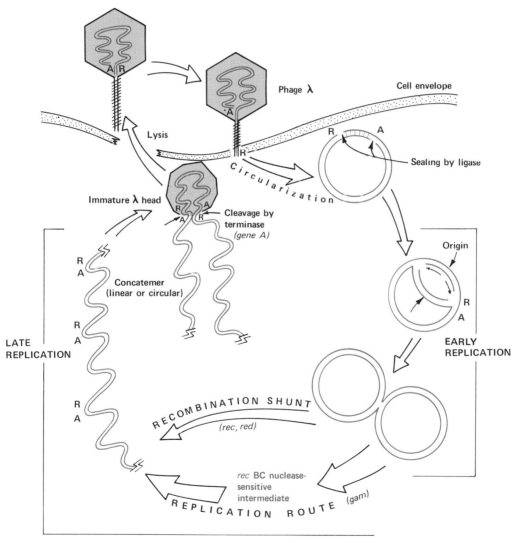

FIGURE 8-34
A scheme for replication of phage λ. Early stages involve doubly forked circular forms. In later stages concatemeric forms support DNA maturation. (After Enquist, L. W. and Skalka, A. (1973) *JMB* **75**, 185.)

pathways to late DNA synthesis: a replication route that relies on the *gam* gene product and a recombinational shunt that utilizes the host *rec* A system and possibly the phage-induced *red* system.

Maturation of late DNA to λ-length linear duplexes with cohesive ends requires that staggered double-stranded scissions be made at precise sequences. A terminase[123] thought to be specified by gene A, and assayed as a nuclease generating λ cohesive ends, is now regarded as responsible for this function. Another absolute require-ment for DNA maturation is the presence of four head proteins (products of genes B, C, D, and E) which define the framework of the phage head. In the absence of any one of these gene products, heads are not assembled and DNA without cohesive ends and several times the unit length accumulates. Unlike T4, there is no evidence for the measuring out of a "headful" of λ DNA. In fact deletion mutants with as much as 25 percent of the DNA lacking are packaged as λ particles. Presumably a specific sequence recognition directs the gene A nuclease to make the scissions that define the length of a λ chromosome.

BACTERIAL FUNCTIONS IN REPLICATION

All phages rely on their host cells for energy and biosyntheses which are also essential to the life of the cell. For its DNA replication, λ depends on enzymes that *E. coli* uses for replication of its own chromosome. There are, in addition, cellular functions a phage may require for its rapid replication that do not appear to be essential for the cell, as judged by its growth rate. Identifying these functions advances not only our understanding of phage replication but pro-vides the first insight into where these functions fit in the economy of the uninfected cell.

To discover a function required for λ, but not cellular growth, a certain mutant of *E. coli* was selected that failed to support λ but showed no other detectable deficiency. A phage mutant was selected in turn that could tolerate this deficiency and multiply in this *E. coli* mutant. When, for example, such a phage mutation was found in gene N, the corresponding *E. coli* mutation was named *gro* N. Thus far, three bacterial *gro* mutants for λ replication have been identified: *gro* N⁻, *gro* P⁻, and *gro* E⁻.

The *gro* N mutation[124] is in RNA polymerase. It is recognized in the isolated enzyme by increased rifampicin sensitivity and dimin-ished salt activation. Presumably the gene N product, whose function is to enhance transcription of certain operons by interacting with RNA polymerase, fails to do so with the *gro* N-altered enzyme. The phage mutants that overcome this defect are of two kinds: (i) those that have lost the requirement for gene N product, and (ii) those whose gene N product has been modified to compensate for the alteration in the RNA polymerase.

123. Wang, J. C. and Kaiser, A. D. (1973) *NNB* **241**, 16.
124. Georgopoulos, C. P. in Hershey, A. D. (ed.) (1971) *The Bacteriophage Lambda*, CSHL, p. 639; (1971) *PNAS* **68**, 2977.

Gro P^{-}[125] bacteria are temperature-sensitive for their own DNA synthesis; they stop making their own DNA after about an hour at 40°. However, synthesis of λ DNA is blocked even at low temperatures. Mutants in gene P are thought to compensate for the gro P^{-} defect, as suggested for the λ gene N mutations, by modifying the gene P product to accommodate the altered gro P product. The latter appears to be the same as the *dna* B product (Chapter 7, Section 10) but how it contributes to DNA replication is still uncertain.

Gro E^{-} mutants[126] fail to support λ growth at the stage of head assembly, and DNA maturation is blocked as a consequence. Mutations in λ gene E, the gene that codes for the major structural protein of the head, can suppress gro E defects. In many gro E mutants, the head assembly of other phages, including T4, T5, ϕ80, and 186, is also blocked. Although the molecular nature of the gro E product is unknown, it appears to satisfy a general requirement for phage morphogenesis. Inasmuch as some gro E mutant cells show filament formation, clumping and other abnormal growth features, the gro E product may be used by the cell in its own morphogenesis.

8. Plasmids[127]

Plasmids are extrachromosomal DNAs found in most bacterial cells. They are duplex, circular, and supercoiled. They resemble bacteriophage chromosomes in size range, and are either very small (about 5×10^6 daltons) or rather large (near 10^8 daltons) (Table 8-9). The small plasmids are maintained at a level of about 20 copies per cell, the large ones at one or two. The remarkable aspects of plasmids are:

(i) The control of their replication that keeps their number fixed from one cell generation to the next.

(ii) The capacity of some, called episomes, to integrate into the host chromosomes.

(iii) The transfer of some (sex and resistance plasmids) to other cells by viral mediated transduction, or by mating via pili ("conjugal transfer").

(iv) The promotion by sex plasmids, upon integration into the host chromosome, of conjugal transfer of the host chromosome.

(v) The variety of plasmid gene products which confer distinctive traits on the cells in which they are established (Table 8-9).

The molecular details of the replication, integration, and transfer of plasmids are largely unknown. However some interesting features about plasmids may offer insights into their replication and perhaps that of viral and cellular DNA as well. One is the complex of closed,

125. Georgopoulos, C. P. and Herskowitz, I. in Hershey, A. D. (ed.) (1971) *The Bacteriophage Lambda*, CSHL, p. 553.
126. Georgopoulos, C. P., Hendrix, R. W., Kaiser, A. D. and Wood, W. B. (1972) *NNB* **239**, 38; also preceding paper.
127. Helinski, D. R. and Clewell, D. B. (1971) *ARB* **40**, 899; Helinski, et al. (1972) *Sixth Miles International Symposium on Molecular Biology*. (R. F. Beers, ed.)

TABLE 8-9
Bacterial plasmids

Plasmid	Size (daltons $\times 10^{-6}$)	No. copies[a] per cell	Product
Colicinogenic factors			
Col E_1	4.2	20	Colicin E_1, membrane changes
Col E_2 (Shigella)	5.0	20	Colicin E_2, "DNase"
Col E_3	5.0	20	Colicin E_3, ribosomal RNase
Col V	94	2	Colicin V; F pilus
Col Ib (Salmonella)	62	2	Colicin Ib; I pilus
Fertility (sex) factors			
F_1	62	2	F pilus
F lac	135	2	F pilus; lac operon
Resistance factors			
R_{64}	76	2	Resistance to drugs; I pilus
R_1	60	2	Resistance to drugs; F pilus
R-pSC101[b]	5.8		Resistance to tetracycline only; no fertility functions
Defective phages			
λdv	4.2	50[c]	Self-reproduction
λdv gal	6.1	(50)	Self-reproduction; gal operon
Unknown			
15T⁻ minicircles	1.4	20	None

[a]Approximate.
[b]This plasmid was generated by shearing a standard-size R factor. When taken up by E. coli, it recircularized to form a new, autonomously replicating plasmid only about one-tenth the size of the original (Source: Cohen, S. S., Chang, A. C. Y., Boyer, H. W. and Helling, R. B. (1973) PNAS **70**, 3240.)
[c]Stated for monomeric units; also present are oligomers.

supercoiled plasmid with a molecule of protein[128]. The DNA is converted to an open, relaxed form upon treatment of the complex with an ionic detergent (sodium dodecyl sulfate), protease (pronase), or alkali (pH 12.5). The relaxed plasmid DNA was found to be nicked only in the "heavy" strand, and the protein appeared to be covalently linked to the 5′ end of that strand. Such relaxation complexes have been described for col E_1, col E_2, and F factor. Their significance can only be surmised:

(i) A signal for the origin of replication, as in a rolling-circle mechanism in which the protein molecule anchors the 5′ end of the strand (Fig. 8-35).

(ii) A device to guide the conjugal transfer of F factor DNA, known to take place by movement of the "heavy" strand led by the 5′ end (Fig. 8-35).

128. Kline, B. C. and Helinski, D. R. (1971) B. **10**, 4975; Blair, D. G., Clewell, D. B., Sheratt, D. J. and Helinski, D. R. (1971) PNAS **68**, 210; Clewell, D. B. and Helinski, D. R. (1972) J. Bact. **110**, 1135.

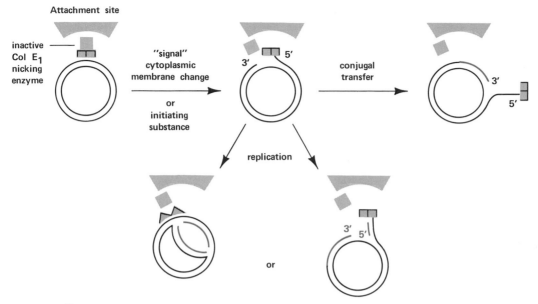

Attachment site

inactive Col E₁ nicking enzyme

"signal" cytoplasmic membrane change

or

initiating substance

conjugal transfer

replication

or

FIGURE 8-35
A scheme to illustrate participation of the "relaxation" protein in *Col* E₁ replication. Nicking action of the protein (nuclease) results in its covalent linkage to the 5' end of a distinctive strand of the DNA utilized in replication or conjugal transfer. (Courtesy of Professor D. R. Helinski.)

(iii) Or, additional ways proposed for the pilot protein—DNA complex in virus particles (Section 2 of this chapter).

Replication of the *col* E₁ plasmid does not take place in the *pol* A mutant[129], indicating that DNA polymerase I is required and that the level in mutant cells is inadequate. Although the other plasmids are maintained in *pol* A cells, their replication should not be assumed to be independent of DNA polymerase I inasmuch as the low level of enzyme known to be present in the mutant may be sufficient for multiplication of one plasmid but not another.

Initiation of replication of the *col* E₁ plasmid is a striking example of RNA priming[130]. Plasmids that accumulate in the presence of chloramphenicol, to a level of thousands per cell, were found to contain RNA in only one strand of the twisted circle but without distinction between the "heavy" and "light" strands of the duplex. Treatment with alkali or ribonuclease converted the plasmids to nicked, circular form. Synthesis of the RNA-containing plasmids was inhibited by rifampicin[130, 131]. All these findings suggest that RNA serves as a primer in the initiation of *col* E₁ DNA synthesis. The normal process of removal of the RNA, thought to involve DNA

129. Kingsbury, D. T. and Helinski, D. R. (1970) *BBRC* **41**, 1538; (1973) *J. Bact.* **114**, 1116.
130. Blair, D. G., Sherratt, D. J., Clewell, D. B. and Helinski, D. R. (1972) *PNAS* **69**, 2518; Williams, P. H., Boyer, H. W. and Helinski, D. R. (1973) *PNAS* **70**, 3744.
131. Kline, B. C. (1972) *BBRC* **46**, 2019; Sakakibara, Y. and Tomizawa, J. (1974) *PNAS* **71**, 802.

polymerase I[132] (Chapter 4, Section 8), is somehow inhibited in cells poisoned by chloramphenicol. The composition of the isolated RNA does not yet indicate the location of the primer, nor whether replication is uni- or bidirectional.

Whereas rifampicin-sensitive RNA synthesis has been implicated in replication of the F factor[133] and *E. coli* 15 minicircles[134] as well as the *col* E$_1$ plasmid, the λ dv dimer is multiplied in the presence of rifampicin[135]. Presumably an initiation mechanism like that for φX174 and ongoing *E. coli* DNA replication is used and does not require the rifampicin-sensitive event observed in the origin of the *E. coli* and λ chromosome replications (Chapter 7, Section 5, and Section 7 of this chapter).

9. Small Animal Viruses[136] SV40[137], and Polyoma[138]

Simian virus 40 (SV40) and polyoma can produce cancer in animals and are much alike in structure and biologic behavior. They are polyhedral particles containing a closed, twisted, circular, duplex DNA molecule (Form I) of about 3×10^6 daltons. An avalanche of current interest in these viruses is explained by the relative simplicity of their chromosomes and the promise for understanding malignant transformations as well as normal mechanisms of DNA replication in animal cells.

Detailed physical maps of the viral DNA[139] define regions of cleavage by specific nucleases, or recombination with the larger adenoviruses, and origins of replication and transcription (Fig. 8-36). However, mapping of the five or so genes of the virus and definition of their products has not gone beyond a recognition of regions responsible for early and late stages of virus development.

Infection takes one of two principal forms, depending on the type of cell infected: (i) viral replication in permissive cells leading to cellular death and lysis, and (ii) abortive infection in nonpermissive cells, resembling the lysogenic (latent) response (see Section 1 of this chapter). Integration of one or several copies of the viral genome into the host cell chromosome in the abortive infection results in a transformed cell showing abnormally proliferative growth.

Attempts to follow the fate of infecting particles by use of isotopes or chemistry are complicated by the need for about 100 particles, on

132. Goebel, W. and Schrempf, H. (1973) *NNB* **245**, 39.
133. Bazzicalupo, P. and Tocchini-Valentini, G. P. (1972) *PNAS* **69**, 298; Kline, B. C. (1974) *B.* **13**, 139.
134. Messing, J., Staudenbauer, W. L. and Hofschneider, P. H. (1972) *NNB* **238**, 202.
135. Hobom, B. and Hobom, G. (1973) *NNB* **244**, 265.
136. Tooze, J. (ed.) (1973) *The Molecular Biology of Tumor Viruses*, CSHL. Chapter 5; Sambrook, J. (1972) *Adv. Canc. Res.* **16**, 141; Eckhart, W. (1972) *ARB* **41**, 503.
137. Fareed, G. C., McKerlie, M. L., Salzman, N. P. (1973) *JMB* **74**, 95.
138. Winnacker, E-L., Magnusson, G. and Reichard, P. (1972) *JMB* **72**, 523; also succeeding paper, p. 539.
139. Morrow, J. F., Berg, P., Kelly, T. J., Jr., and Lewis, A. M., Jr. (1973) *J. Virol.* **12**, 653.

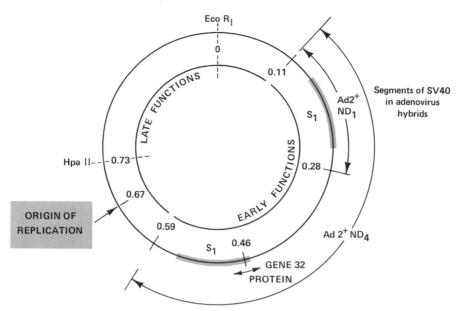

FIGURE 8-36
Genetic map of SV40. The map is oriented by the site of the *Eco* RI restriction
endonuclease cleavage. Indicated are: sites of cleavage by *Hpa* II and the single-
strand-specific S_1 nuclease which acts only on the superhelical RF, the region
of the latter bound by phage T4 gene 32 protein, the origin of replication,
regions integrated into nondefective adenovirus DNAs, and regions responsible for
early and late functions expressed in permissive cells.

the average, to produce an infected cell (plaque-forming unit). As a
result, less is known about the biochemistry of SV40 and polyoma
DNA replication than of M13 and ϕX174. However, certain important
features have become clear from recent experiments.

Replication cycle in permissive cells.   Monkey cells are permissive
for SV40, mouse cells for polyoma. Patterns of SV40 and polyoma
DNA replication have been determined by observing cultures infected for 24 to 48 hours. Replicating DNA was isolated and examined
after a several-minute pulse of [³H]-thymidine to a cell culture.
Alternatively, replicating molecules have been analyzed after incubation in vitro. In these studies, washed nuclei from infected cells
were exposed to cycles of [³H]-thymidine and α-(³²P)-triphosphates,
to which nuclei are permeable.

The sequence of events indicated by results of in vivo investigations are generally as follows: Virus particles are transported to the
nucleus where the host DNA synthesis continues, often at an accel-
erated rate. Integration of viral DNA with host DNA occurs, but its
extent and significance are uncertain. Replication starts at a single
and unique point, proceeds bidirectionally at about equal speed in
both directions, and terminates at a point on the circle opposite the
origin. Completion of progeny molecules to the final stage of a twisted

circle (Form I) takes five to ten minutes. Although this could indicate an overall rate of fork movement of about 0.5 kilobase per minute at 37°, it is likely that stages in completion of Form I rather than chain growth, limit the rate. For comparison, calculated rates of fork movements for a mammalian and an *E. coli* chromosome are 3 kilobases and 40 kilobases per minute, respectively.

The replicative intermediates (RI, Fig. 8-37) contain intact parental strands; the growing strands advance discontinuously, passing through a small-fragment stage as in bacterial systems. However the

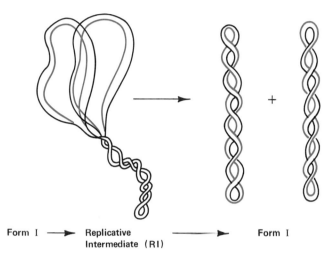

Form I ⟶ Replicative ⟶ Form I
Intermediate (RI)

FIGURE 8-37
Intermediates in the replication of supercoiled polyoma DNA (Form I). Replicative intermediates (RI) retain the unreplicated template strands in a superhelical form and the progeny strands as linear structures not attached covalently to the template strands. (Winnacker, E.-L., Magnusson, G. and Reichard, P. (1972) *JMB* **72**, 523; also succeeding paper, p. 539.) A similar pattern applies to replication of SV40 DNA. (Fareed, G. C., McKerlie, M. L. and Salzman, N. P. (1973) *JMB* **74**, 95.) (See text for further details.)

fragments are considerably smaller with a sedimentation coefficient for SV40 of only about 4S[140]. Elongation is discontinuous on both strands. The DNA chains are initiated by an RNA priming piece, as indicated by in vitro studies[141], and the RNA must subsequently be removed, because the final duplex circular product resists treatment with alkali.

Form I is isolated as a nucleoprotein; it is unlikely that replicative intermediates appear as naked DNA in the cell. Upon analysis,

140. Fareed, G. C., Khoury, G. and Salzman, N. P. (1973) *JMB* **77**, 457. Pigiet, V., Winnacker, E. L., Eliasson, R. and Reichard, P. (1973) *NNB* **245**, 203; Francke, B., and Hunter, T. (1974) *JMB* **83**, 99.
141. Magnusson, G., Pigiet, V., Winnacker, E. L., Abrams, R. and Reichard, P. (1973) *PNAS* **70**, 412; Hunter, T., and Francke, B. (1974) *JMB* **83**, 123; Pigiet, V., Eliasson, R. and Reichard, P. (1974) *JMB* **84**, 197.

replicative intermediates are observed to be from 10 to 70 percent completed and contain two loops of equal length attached to a supercoiled or twisted region of the unreplicated parental molecule (Fig. 8-37). Persistence of superhelical turns in the unreplicated portion of the circular parental duplex characterizes SV40, polyoma, and mitochondrial DNA replication as distinct from that of phage λ. However this difference may be simply a reflection of the ten-fold larger size of λ DNA and its consequent greater susceptibility to relaxation by nicking. The mechanism for relieving the twist of the parental duplex during replication is not known; it may involve transient nicking and reunion. Newly synthesized chains of greater than unit length have <u>not</u> been found and so there is no evidence to support a rolling-circle scheme as proposed for replication of φX174 and M13 duplexes.

Cell-free, nuclear preparations have been examined for further insights into the detailed molecular mechanisms of replication. Thus far, investigators have shown the following:

(i) Replicative intermediates initiated in vivo are completed by semiconservative chain elongation.

(ii) The rate approaches that observed in vivo.

(iii) Chain elongation is discontinuous on both strands involving the production of 5S chains, each initiated by an RNA priming fragment.

(iv) Completion of replication to produce replicative, intermediate molecules is aided by cytoplasmic factors[142].

(v) Initiation of replicative intermediates has not been observed.

Although discrete replicative intermediates have been readily identified (Fig. 8-37), failure to observe their initiation indicates a more complex origin. Early integration of viral DNA into the host chromosome, and acceleration of DNA synthesis suggest that initiation and subsequent viral replicative events may take place in the chromosome or may be closely linked to it.

The earliest events in replication require continuous protein synthesis[143]. The participation of a viral component also appears necessary, as judged by temperature-sensitive mutants which are unable to initiate a new round of replication, but are able to convert replicative intermediates to mature viral DNA normally[144]. No information is yet available regarding the later stages of DNA maturation and its packaging into virus particles.

<u>Replication in nonpermissive cells.</u> In cultures of hamster or mouse cells (3T3), infection with SV40 is followed not by replication of viral DNA but rather by its integration into the cell's chromosome. Transformation of a fraction of the cells to unrestrained growth is

142. DePamphilis, M. and Berg, P., personal communication.
143. Kang, H. S., Eshbach, T. B., White, D. A. and Levine, A. J. (1971) *J. Virol.* **7**, 112.
144. Tegtmeyer, P. (1972) *J. Virol.* **10**, 591.

correlated with the expression of one or two viral genes whose functioning is also needed in replication. From what is known so far, the overall patterns of SV40 and polyoma replication are similar in many ways to those described for mitochondrial DNA and certain phages such as λ. In the case of λ, the genes for integration of the viral DNA into the host chromosome and excision from it are well defined, although the enzymes they encode have not been identified. With the animal viruses, little is known about either the genetics or the detailed biochemistry.

9

Repair, Recombination, and Restriction

1. Deoxyribonucleases[1]

Extensive use of nucleases in analysis of nucleic acids has emphasized their value as reagents but has tended to obscure their physiologic importance. Nucleases do contribute significantly to biosynthesis of nucleic acids and this has already been mentioned (Chapter 7, Section 12). Equally important is their central role in repair, recombination, and restriction of DNA.

Nucleases are conveniently divided into two classes: (i) _exonucleases,_ which require a terminus from which their action is directed, and (ii) _endonucleases,_ which do not require a terminus. A convenient experimental test for distinguishing nucleases is their action on a circular DNA, either single-stranded or duplex. A previously used criterion, no longer valid, is that exonucleases produce only mononucleotides. Although exonucleases acting from the 3' end produce mononucleotides (except for exonuclease VII, the _rec B,C_ nuclease, and those exonucleases that fail in their cleavage of the

---

1. Lehman, I. R. (1967) _ARB_ **36**, 645; (1971) in _The Enzymes_ (P. D. Boyer, Ed.) **4**, 251, Academic Press, N.Y.

remaining dinucleotide at the 5′ end), some of the exonucleases attacking from the 5′ end liberate dinucleotides or even larger fragments.

Exonucleases and endonucleases discussed elsewhere in the book have for ease of reference been enumerated and briefly described in Tables 9-1 and 9-2. Many other nucleases from various sources have been at least partially purified and characterized. In a few instances, the enzyme has been available in crystalline form and studied in great detail, notably the micrococcal endonuclease (DNase II)[2]. Analysis of its structure and function is as thorough as that of pancreatic ribonuclease.

An _exonuclease_ acts either from the 3′ end (3′ → 5′) or the 5′ end (5′ → 3′) but, except for exonuclease VII, the _B. subtilis_, rec B,C, and the _M. luteus_ ultraviolet-repair enzyme (see next section), never has the capacity to do both. The second major distinction among exonucleases is their specificity with regard to single-strand or duplex structure. In no instance has a specificity toward base sequence been demonstrated.

2. Anfinsen, C. B., Cuatrecasas, P. and Taniuchi, H. (1971) in _The Enzymes_ (P. D. Boyer, Ed.) **4**, 177, Academic Press, N.Y.

TABLE 9-1
Exonucleases: 3′ → 5′ and 5′ → 3′

3′ → 5′ Exonuclease	DNA strand	Comments (products are 5′ mono-nucleotides except where noted)
Exonuclease I[a] (E. coli; xon A, sbc B)	single	Degrades up to terminal dinucleotide; degrades glucosylated DNA
DNA polymerase I[b] (E. coli; pol A)(Exonuclease II)	single	Acts on frayed or mismatched terminus of duplex; proofreading function; formerly called exonuclease II; complete degradation
Exonuclease IV[c] (E. coli)	single	Complete degradation of oligonucleotides
Exonuclease VII[d] (E. coli)	single	Produces oligonucleotides progressively from 3′- and 5′-ends; Mg^{2+} not required
Phage T4-induced[e] exonuclease IV; dex A	single	Complete degradation of oligonucleotides
Phage T4-induced[f,g] DNA polymerase	single	Gene 43 product; resembles DNA polymerase I; degrades up to terminal dinucleotide
Snake venom[h] phosphodiesterase	single	Complete degradation; also acts on RNA and nucleoside di- and tri-phosphates
rec B,C nuclease[i] (E. coli) (Exonuclease V)	single, duplex	Dependent on ATP; acts as ATPase; no action at a nick; also a 5′ → 3′ exonuclease and an endonuclease; oligonucleotide products; functions in recombination and repair
Exonuclease III[j] (E. coli); xth A	duplex	Removes phosphate from 3′-phosphate terminus; stops when chain becomes single-stranded; acts at a nick; exo III⁻ mutants have normal phenotype
B. subtilis exonuclease[k]	duplex	Extracellular; produces 3′ nucleotides; also acts 5′ → 3′ on single strand; acts on RNA

TABLE 9-1 (*continued*)
Exonucleases: 3′ → 5′ and 5′ → 3′

299

5′ → 3′ Exonuclease	DNA strand	Comments (products are 5′-mono- and oligonucleotides except where noted)
B. subtilis exonuclease[k]	single	Extracellular; produces 3′ nucleotides also acts 3′ → 5′ on double strand; acts on RNA
Exonuclease VII[d] (E. coli)	single	Produces oligonucleotides progressively from 3′- and 5′- ends; Mg^{2+} not required
Spleen phosphodiesterase[l]	single	Produces 3′ nucleotides; requires 5′-hydroxyl terminus; also acts on RNA
Neurospora exonuclease[m] (N. crassa)	single	Also acts on RNA
Phage T5-induced exo-nuclease[n]	single, duplex	Produces mono- and oligonucleotides; mutant-infected[o] cells show arrested DNA synthesis
rec B,C nuclease[i] (E. coli) (ExonucleaseV)	single, duplex	Dependent on ATP; acts as ATPase; no action at a nick; also a 3′ → 5′ exonuclease and an endonuclease; oligonucleotide products; functions in recombination and repair
Phage λ-induced exo-nuclease[p]	single, duplex	Strong preference for 5′-phosphate; no action at a nick; preference for duplex over denatured DNA, but acts on oligonucleotides[q], functions in recombination[r]
DNA polymerase I[s] (E. coli; pol A) (Exonuclease VI)	duplex	Produces mononucleotides and oligo-nucleotides of two to ten residues; acts at 5′-hydroxyl-, mono P-, tri P; may function in excision repair
Phage T7-induced[t] exonuclease	duplex	Gene 6 product; essential for replication
Phage SP3-induced[u] exonuclease (B. subtilis)	duplex	Produces dinucleotides only

[a]Lehman, I. R. and Nussbaum, A. L. (1964) *JBC* **239**, 2628; Kushner, S. R., Nagaishi, H., Templin, A., Clark, A. J. (1971) *PNAS* **68**, 824.

[b]Chapter 4, Section 9.

[c]Jorgensen, S. E. and Koerner, J. F. (1966) *JBC* **241**, 3090.

[d]Chase, J. W. and Richardson, C. C., (1974) *JBC* **249**, 4545, 4553.

[e]Warner, H. R., Snustad, D. P., Koerner, J. F. and Childs, J. D. (1972) *J. Virol.* **9**, 399.

[f]Cozzarelli, N. R., Kelly, R. B. and Kornberg, A. (1969) *JMB* **45**, 513.

[g]Huang, W. M. and Lehman, I. R. (1972) *JBC* **247**, 3139.

[h]Khorana, H. G. (1961) **in** *The Enzymes* (P. D. Boyer, H. Lardy & K. Myrbäck, Eds) **5**, 79, Academic Press, N.Y. (Second Edition); Sulkowski, E. and Laskowski, M. Sr. (1971) *BBA* **240**, 443.

[i]Wright, M., Buttin, G. and Hurwitz, J. (1971) *JBC* **246**, 6543; Goldmark, P. J. and Linn, S. (1972) *JBC* **247**, 1849.

[j]Richardson, C. C. and Kornberg, A. (1964) *JBC* **239**, 242; Richardson, C. C., Lehman, I. R. and Kornberg, A. (1964) *JBC* **239**, 251; Milcarek, C. and Weiss, B. (1973) *J. Bact.* **113**, 1086; Masamune, Y., Fleischman, R. A. and Richardson, C. C., (1971) *JBC* **246**, 2680.

[k]Okazaki, R., Okazaki, T. and Sakabe, K. (1966) *BBRC* **22**, 611.

[l]Bernardi, G. **in** *The Enzymes* (P. D. Boyer, Ed.) **4**, 271 (1971) Academic Press, N.Y.

[m]Rabin, E. Z., Tenenhouse, H. and Fraser, M. J. (1972) *BBA* **259**, 50.

[n]Frenkel, G. D. and Richardson, C. C. (1971) *JBC* **246**, 4839.

[o]Frenkel, G. D. and Richardson, C. C. (1971) *JBC* **246**, 4848.

[p]Little, J. W., Lehman, I. R. and Kaiser, A. D. (1967) *JBC* **242**, 672; Little, J. W. (1967) *JBC* **242**, 679.

[q]Radding. C. M., personal communication.

[r]See Section 3 of this chapter.

[s]Chapter 4, Sections 12, and 13.

[t]Kerr, C. and Sadowski, P. D. (1972) *JBC* **247**, 305.

[u]Trilling, D. M. and Aposhian, H. V. (1968) *PNAS* **60**, 214.

TABLE 9-2
Endonucleases

Enzyme	DNA strand	Comments (nick contains 5'-phosphate end except where noted)
Neurospora endonuclease[a,b] (*N. crassa*)	single	Also acts on RNA
S$_1$ endonuclease[c] (*Aspergillus oryzae*)	single	Also acts on RNA
rec B,C nuclease[d] (*E. coli*)	single	Partially ATP-dependent; also an exonuclease; functions in recombination and repair
Phage T7-induced[e,f] (*E. coli*)	single	Gene 3 product; essential for replication; preference for single *vs* double strand, about 150
Phage T4-induced[g] (Endo IV) (*E. coli*)	single	*den* A gene product; splits —TpC— sequences to yield 5'-dCMP-terminated oligonucleoties; average chain length of product varies with conditions
Endonuclease I[h] (*E. coli*)	single, duplex	Periplasmic location; average chain length of product is seven; inhibited by tRNA; produces double-stranded break; produces a nick when complexed with tRNA; endo I⁻ mutants grow normally
DNase I[i] (bovine pancreas)	single, duplex	Average chain length of product is four; produces double-stranded break in presence of Mn^{2+}
DNase II (*Micrococcus pyogenes*,[j] calf thymus[k], calf spleen[l])	single, duplex	Produces 3'-P terminus
Endonuclease II[m] (*E. coli*)	duplex	Few single-strand nicks in melted poly d(A-T)
Repair enzymes[n] (*M. luteus, B. subtilis,* T4-induced endonuclease V)	duplex, with lesions	Nick at 5' end of lesion 3'-P
Restriction enzymes	Unmodified duplex	Produces double-stranded break (see Section 5 of this chapter)

[a] Rabin, E. Z., Tenenhouse, H. and Fraser, M. J. (1972) *BBA* **259**, 50.
[b] Linn, S. and Lehman, I. R. (1965) *JBC* **240**, 1287, 1294.
[c] Sutton, W. D. (1971) *BBA* **240**, 522; Beard, P., Morrow, J. F. and Berg, P. (1973) *J. Virol.* **12**, 1303.
[d] Wright, M., Buttin, G. and Hurwitz, J. (1971) *JBC* **246**, 6543; Goldmark, P. J. and Linn, S. (1972) *JBC* **247**, 1849.
[e] Center, M. S., Studier, F. W. and Richardson, C. C. (1970) *PNAS* **65**, 242.
[f] Sadowski, P. D. (1971) *JBC* **246**, 209.
[g] Sadowski, P. D. and Bakyta, I. (1972) *JBC* **247**, 405; Vetter, D. and Sadowski, P. D. (1974) *J. Virol.* **14**, 207.
[h] Lehman, I. R., Roussos, G. G. and Pratt, E. A. (1962) *JBC* **237**, 819; Dürwald, H. and Hoffman-Berling, H. (1968) *JMB* **34**, 331.
[i] Matsuda, M. and Ogoshi, H. (1966) *J. Bioch.* **59**, 230; Laskowski, M. Sr., in *The Enzymes* (P. D. Boyer, Ed.) **4**, 313 Academic Press, N.Y., 3rd Edition; Zimmerman, S. B. and Coleman, N. F. (1971) *JBC* **246**, 309.
[j] Anfinsen, C. B., Cuatrecasas, P. and Taniuchi, H. (1971) in *The Enzymes* (P. D. Boyer, Ed.) **4**, 177, Academic Press, N.Y.
[k] Bernardi, G. (1961) *BBA* **53**, 216.
[l] Bernardi, G. and Griffe, M. (1964) *B* **3**, 1419.
[m] Hadi, S.-M., Kirtikar, D. and Goldthwait, D. A. (1973) *B* **12**, 2747.
[n] See Section 2 of this chapter.

Endonucleases discriminate with respect to both secondary structure and sequence. An endonuclease may show a strong preference for single-stranded DNA or for duplex DNA. In some instances, endonucleases that function in repair of lesions seek out and identify regions of distortion in the DNA helix created by a thymine dimer or some other lack of correct base-pairing. The enzyme then introduces a single-strand scission, a nick, to one side of the lesion as a first step toward its excision by an exonuclease.

A second principal characteristic of endonucleases is recognition of base sequences. It had been assumed that this feature was found only in restriction endonucleases, which recognize a sequence of about six nucleotides of a duplex in making a double-stranded break. While recognition of base sequence is most sharply defined in a restriction enzyme or in cytosine-specific cleavage by T4-endonuclease IV (Table 9-2), it also appears to characterize all endonucleases. When pancreatic DNase I, *E. coli* endonuclease I, or spleen DNase II are freed of all contaminant activities, they yield digests with different frequencies (distributions) of oligonucleotides for each nuclease[3]. The 3′ and 5′ termini of the products form a pattern distinctive for each enzyme throughout the course of digestion. For these endonucleases the recognition is limited to sequences of three to four nucleotides. The relative susceptibility of the many available sequences is determined by kinetic parameters, as the K_m of the sequences and V_{max}, and may be altered by cations, ionic strength, or other reaction conditions.

In each of the next three sections in this chapter, on repair, recombination, and restriction, endonucleolytic scissions will be seen as initiating events. Even though exonucleolytic degradations may in some instances appear to be secondary, they are nevertheless equally important in the overall operations.

2. Repair[4]

Cellular repair of damage to a macromolecule is known only for DNA. In no other instance is the integrity of a single molecule so vital to the survival of the cell. Since a bacterial gene has a 50 percent chance of remaining unaltered even after being duplicated over 100 million times, it is not surprising to find that the cell has devices for preserving the integrity of DNA and removing lesions introduced into it. The duplex complementary structure of DNA insures that information lost from one strand can be retrieved from the other. Only complete loss of a base pair, transverse scission of the duplex, or failure to proofread a replication error (Chapter 4, Section 9) are irreversible.

3. Bernardi, G., Ehrlich, S. D. and Thiery, J. P. (1973) *NNB* **246**, 36.
4. Hanawalt, P. C. (1972) *Endeavour* **31**, 83.

Damage to DNA is in the form of strand scissions, or modification or loss of bases sustained from chemical or physical agents in the environment, such as hydrogen ions, alkylating drugs, ultraviolet light and X-rays, and also from the wear and tear of metabolic events. Spontaneous depurination, attributable to H^+, occurs at a significant rate. The chief damage to DNA by ultraviolet light, the best understood of DNA lesions, is dimerization of the pyrimidines. Adjacent thymine residues, for example, are joined in a cyclobutane ring (Fig. 4-20, Fig. 9-1) and can no longer form hydrogen bonds with adenines in the opposing strand. The helix in that region is conse-

Thymine

Thymine dimer

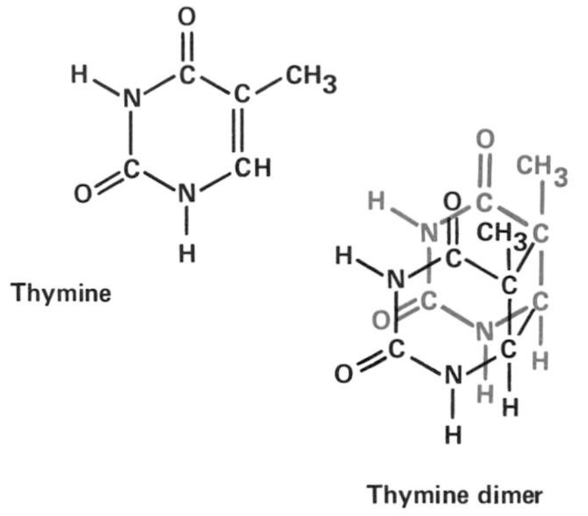

FIGURE 9-1
Structure of a thymine dimer. Adjacent thymidylate residues in a DNA chain are joined in a cyclobutane ring.

quently distorted. Unless the thymine dimer is removed, replication and transcription are prevented or disturbed.

DNA lesions are repaired by one of three kinds of mechanisms.

(i) *Photoreactivation*. The thymine or cytosine dimer is removed in situ by visible light in an enzymatically catalyzed cleavage of the cyclobutane ring.

(ii) *Excision*. The segment containing the lesion is excised in the absence of light ("dark repair") by incisions on either side of the lesion and replaced by replicative gap-filling directed by the intact complementary strand.

(iii) *Recombination*. The DNA region with the lesion is replaced, also in the absence of light, through processes that will be considered in the next section. A lesion in which the duplex strands are cross-linked (by mitomycin C, nitrogen or sulfur mustards, or x-rays) cannot be repaired by excision alone and requires a final recombinational step with homologous DNA[5].

Enzymatic photoreactivation[6], using the energy of blue light to split a pyrimidine dimer, has been observed in cells from many sources, from mycoplasma, the tiniest of bacteria, to cells from animals, such as the human leukocyte[6a]. The fact that the very simple

5. Cole, R. S. (1973) PNAS **70**, 1064.
6. Rupert, C. S. (1964) in *Photophysiology* (A. C. Giese, Ed.) Academic Press, N.Y.; **2**, 283; Cook, J. S., (1970) ibid **5**, 191.
6a. Sutherland, B. M. (1974) *Nat.* **248**, 109.

mycoplasma, with only a few hundred genes, devotes one of them to an enzyme for photoreactivation is another illustration of the importance of DNA repair.

Photoreactivation can repair any dimer that can be repaired by excision in the dark. In the first stage of a general scheme for photoreactivation (Fig. 9-2), the enzyme recognizes and binds specifically to the dimer in the dark. Upon absorption of light, energy is pro-

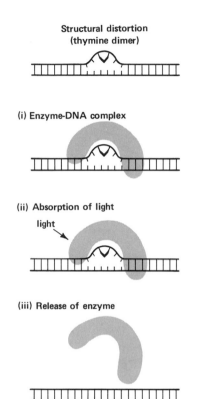

**Structural distortion
(thymine dimer)**

(i) Enzyme-DNA complex

(ii) Absorption of light

light

(iii) Release of enzyme

FIGURE 9-2
A scheme for enzymatic photoreactivation of a pyrimidine dimer (see text). (Courtesy of Professor P. C. Hanawalt.)

vided for cleavage of the dimer and the enzyme may then dissociate from the DNA. Details of enzyme structure and mechanism of action had not been studied for lack of a pure preparation. Recently, through the device of placing a gene in the chromosome of a transducing phage, the number of molecules of photoreactivating enzyme in an *E. coli* cell has been increased and the enzyme has been isolated in pure form[7].

Excision repair is the principal dark-repair mechanism for removal of the defective DNA strand and its replacement with a correctly base-paired section. This patching operation includes a replication step. At this time it is not known whether the replication step is similar to that in chromosome replication or whether it utilizes the same enzymes. There may be no fundamental or sharp distinction

7. Sutherland, B. M., Chamberlin, M. J. and Sutherland, J. C. (1973) *JBC* **248**, 4200.

among the DNA synthesis operations that fill gaps in replication, repair, or recombination.

Excision repair has been studied intensively in three experimental systems: *E. coli, M. luteus,* and T4-phage infection of *E. coli*. In none of these is there a full accounting for the physiologic events of repair by known genetic and enzymatic components. The data in each instance can be explained by a sequence of reactions generally described in the scheme shown in Fig. 9-3 with an ultraviolet lesion as an example. Four steps are indicated.

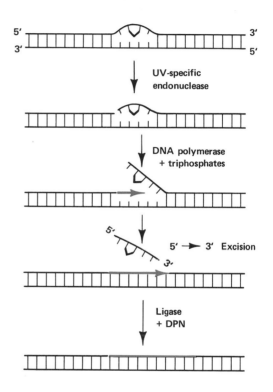

FIGURE 9-3
A possible scheme for excision repair of a pyrimidine dimer by a "cut, patch, cut and seal" operation. (Hanawalt, P. C. (1972) *Endeavour* **31**, 83.)

(i) *Incision*. A specific endonuclease that patrols the chromosome detects a distortion in the DNA helix caused by some aberration. The enzyme may be specific for a pyrimidine dimer, a gamma ray lesion[8] or may respond to the alkylation of a purine. The enzyme introduces a single-strand scission (nick) at or near the lesion, usually to the 5′ side of it.

(ii) *Repair replication*. A polymerase such as DNA polymerase I uses the intact complementary strand as template for replication. In this way information lost in the lesion is recovered from the complementary strand. The 3′ end of the nicked strand serves as primer and the 5′ end of the damaged strand is displaced.

(iii) *Excision*. The defective 5′-ended segment is removed through cleavage by an exonuclease. The latter may be part of the polymer-

8. Wilkins, R. J. (1973) *NNB* **244**, 269; Mattern, M. R., Hariharan, P. V., Dunlap, B. E. and Cerutti, P. A. (1973) *NNB* **245**, 230.

ase, as with the $5' \rightarrow 3'$ exonuclease of DNA polymerase, or may be independent of it. With a separate exonuclease, this excision step may precede repair.

(iv) *Sealing.* Ligase action joins the newly made patch to the main body of DNA, thereby restoring a normal DNA strand.

Repair in E. coli[9]. A number of mutants with increased sensitivity to ultraviolet light have been isolated. Genes whose products are required for excision repair have been mapped (Fig. 4-32) but in no instance, except for *pol* A and *lig*, have the proteins or their functions been clearly identified. The genes *uvr* A and *uvr* B appear to be responsible for endonucleolytic steps which lead to excision and *uvr* C for a step later in repair[10]. Other genes, *rec* A, *rec* B, *rec* C, and *uvr* D are involved in a repair system separate from excision repair. Neither system is capable of repairing all dimers alone. A double mutant, *uvr* A *rec* A, is exceedingly sensitive to ultraviolet light (Fig. 9-4); one dimer per genome is lethal.

---

9. Howard-Flanders, P., Boyce, R. P. and Theriot, L. (1966) *Gen.* **53**, 1119; Braun, A. and Grossman, L. (1974) *PNAS* **71**, 1838.
10. Grossman, L., personal communication.

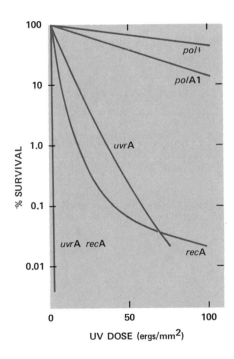

FIGURE 9-4
Comparative sensitivities of *E. coli* strains to ultraviolet. *Pol*+ sustains normal repair, whereas the *pol* A1 mutant with reduced levels of DNA polymerase I is partially deficient in excision repair. The *uvr* A strain, which appears to lack the damage-specific endonuclease, is totally deficient in this function. The *rec* A strain is completely deficient in recombinational repair; the double mutant, *uvr* A *rec* A, thus lacks both dark-repair mechanisms. (Hanawalt, P. C. (1972) *Endeavour* **31**, 83.)

An endonuclease, called endonuclease II[11], isolated for its specificity in nicking alkylated DNA, acts at a few single-stranded regions in duplex DNA that result from partial melting in A–T-rich segments. Although this enzyme is not the product of the identified uvr genes it may share in the responsibility for making incisions near lesions.

DNA polymerase I has been implicated in replicative repair and excision[12] by both in vivo[13] and in vitro[14] studies (Fig. 9-5). Yet

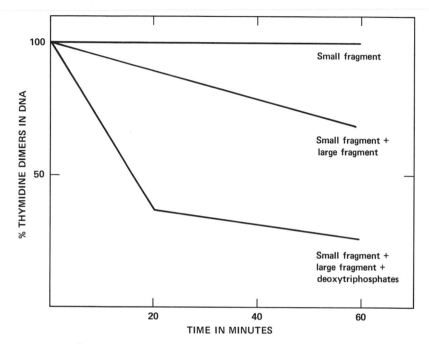

FIGURE 9-5
Thymine dimer excision by E. coli DNA polymerase.
The small fragment (5′ → 3′ exonuclease; see Chapter 4,
Sections 12, 14) is active only in the presence of the
large fragment (polymerase) and most active during
concurrent polymerization. (Source: Friedberg, E. C.
and Lehman, I. R. (1974) BBRC, 58, 132.)

pol A mutants, with lowered levels of the enzyme do manage to repair ultraviolet lesions. The size of repair patches in the pol A mutant is of the order of 1500 nucleotides, some 100 times larger than in the wild-type cell[15]. There are indications that the auxiliary replicative enzymes active in repair are DNA polymerases II and III[16]. Complementing these enzymes in excision of the fragment

11. Hadi, S.-M., Kirtikar, D. and Goldthwait, D. A. (1973) B 12, 2747.
12. Kelly, R. B., Atkinson, M. R., Huberman, J. A. and Kornberg, A. (1969) Nat. 224, 495.
13. Kato, T. and Kondo, S. (1970) J. Bact. 104, 871; Town, C. D., Smith, K. C. and Kaplan, H. S. (1971) Science 172, 851.
14. Friedberg, E. C. and Lehman, I. R., (1974) BBRC 58, 132.
15. Cooper, P. K. and Hanawalt, P. (1972) JMB 67, 1.
16. Masker, W., Hanawalt, P. C. and Shizuya, H. (1973) NNB 244, 242; Tait, R. C., Harris, A. L. and Smith, D. W. (1974) PNAS 71, 675; Youngs, D. A. and Smith, K. C. (1973) NNB 244, 240.

may be the $5' \rightarrow 3'$ exonuclease component of DNA polymerase I (Chapter 4, Section 14) or some other exonuclease.

Repair in M luteus[17]. The genetic analysis of excision repair in _M. luteus_ is not as advanced as in _E. coli,_ but similar enzymes capable of incision and excision have been implicated. Originally thought to involve an incision by a specific endonuclease to produce a 3'-phosphoryl nick, the sequence of enzymatic events now appears to be identical to that in _E. coli._ An incision creates a 3'-hydroxyl nick which is extended by the _M. luteus_ polymerase. The $5' \rightarrow 3'$ exonucleolytic component of the polymerase excises the lesion and replicative gap-filling makes sealing by ligase possible.

Repair after T4-phage infection[18]. It is significant that the selective pressures in the evolution of T4 have favored the emergence and persistence of at least two phage genes for the incision and excision steps of ultraviolet repair. The strikingly greater ultraviolet resistance of T4 as compared to T2 phage can be attributed to the ν gene, whose product is an endonuclease specific for ultraviolet-irradiated DNA, and to an additional T4-gene product, an exonuclease, with an excision function. Presumably, T4-induced DNA polymerase and ligase complete the repair and sealing.

Xeroderma pigmentosum[19]. Patients with this rare, hereditary skin disease are abnormally sensitive to sunlight and are prone to develop skin cancers. The disease has been linked to a deficiency observed in vitro. Cells cultured from the skin of afflicted patients do not carry out the excision repair which is demonstrable in fibroblasts cultured from normal skin. The experimental design from which this inference has been drawn is the same as that used in the study of repair in bacterial cells. It is based on the fact that DNA is not altered in buoyant density if the newly synthesized DNA with a density label is added in small patches. By contrast, the buoyant density is drastically increased when, in replication, the new DNA is added as long stretches.

The assay for an excision-repair defect is of interest clinically as well as scientifically. After ultraviolet irradiation of normal fibroblasts and subsequent incubation in a medium containing (^{3}H)-5-bromouracil, most of the newly synthesized DNA contains the (^{3}H) density label at an unaltered buoyant density, presumably as small patches. Incorporation that takes place in unirradiated fibroblasts is semiconservative, the label being found as long stretches at a hybrid buoyant density. Absence of excision repair

17. Kaplan, J. C., Kushner, S. R. and Grossman, L. (1971) _B_ **10**, 3315; also succeeding paper, p. 3325; Mahler, I., Kushner, S. R. and Grossman, L. (1971) _NNB_ **234**, 47; Hamilton, L., Mahler, I. and Grossman, L. (1974) _B._ **13**, 1886.
18. Sekiguchi, M., Yasuda, S., Okubo, S., Nakayama, H., Shimada, K. and Takagi, Y. (1970) _JMB_ **47**, 231; also succeeding paper, p. 243; Friedberg, E. C. and King, J. J. (1971) _J. Bact._ **106**, 500; Ohshima, S. and Sekiguchi, M. (1972) _BBRC_ **47**, 1126.
19. Cleaver, J. E. (1971) **in** _Nucleic Acid-Protein Interactions_ (Ribbons, D. W., Woessner, J. F., Schultz, J., eds.), North-Holland Publishing Co., Amsterdam p. 87.

is suspected when analysis of irradiated fibroblasts from the skin of a patient fails to show incorporation of (^{3}H)-5-bromouracil into small patches.

3. Recombination[20]

In the preceding discussion of DNA repair, emphasis was placed on the excision repair of DNA; the faulty segment is excised by incisions on both sides of the lesion and replaced by replication directed by the intact complementary strand. An important alternative is the mechanism in which a preformed patch supplied by another copy of the DNA is used to replace the defective segment[21]. An exchange of DNA in a repair process is an example of recombination, but only a minor one among the many examples in which entire genes or parts of genes are moved from one DNA molecule to another.

Recombinational assortment of genes in meiosis, and the duplication, deletion, inversion, and translocation of genes were all well known long before DNA was recognized as the genetic material. Perhaps the most dramatic demonstration of recombination is bacterial transformation, in which DNA from the external medium is assimilated into the chromosome.

Three categories of recombination are now recognized.

(i) General recombination, in which exchange between homologous segments takes place anywhere along the length of the DNA molecule.

(ii) Site-specific recombination, in which exchange occurs at a specific site, as in the integration of phage λ DNA into the E. coli chromosome, and excision of the λ DNA from it.

(iii) Nonhomologous recombination, in which foreign DNA is inserted at random, for example, phage μ DNA into E. coli DNA[22].

In view of the multiple kinds of recombination and the variety of genetic and physiologic backgrounds in which recombination takes place, we must expect considerable variation in the biochemical mechanisms employed. There may also be multiple pathways in a given cell; there are at least two in E. coli. Available evidence favors a breakage-reunion model of interacting DNA strands. Yet, among the operations of restoring intact chromosomes after breakage and reunion, there may be considerable gap-filling by replication on new templates. It is likely therefore that some of the enzymes already considered in replication and repair serve in recombination. In addition, genetic studies implicate distinctive enzymes essential to recombination but dispensable for replication; so far, all genes known to affect recombination also influence repair.

20. Clark, A. J. (1971) Ann. Rev. Microb. **25**, 437; (1973) Ann. Rev. Gen. **7**, 67; Sobel, H. M. (1972) PNAS **69**, 2483; Radding, C. M. (1973) Ann. Rev. Gen. **7**, 87.
21. Cole, R. S. (1973) PNAS **70**, 1064.
22. Daniell, E., Roberts, R. and Abelson, J. (1972) JMB **69**, 1.

Relatively little information is available about the enzymes responsible for site-specific recombination of phage λ (see Chapter 8, Section 7). Even less is known about nonhomologous recombinations. Regarding general recombination, several examples can be presented in which plausible suggestions have been made for a sequence of molecular events. In no case is a sequence reliably known nor is the identity of the enzymes proven. Events indicated as sequential may be concurrent, and operations that appear to be discrete may be nicely integrated by enzyme complexes or assemblies.

Insertion of a single strand into a gap. This nonreciprocal form of recombination occurs during assimilation of DNA after it has been taken up from the medium (transformation), or transferred from a mating cell (conjugation) (Fig. 9-6a). An attractive model for the insertion process is outlined as follows. The entering single strand anneals to a homologous strand at a gap created (Fig. 9-6b) in the recipient duplex by endonucleolytic incision followed by exonucleolytic excision. The hybrid region, which need be only a few nucleotides long, is then enlarged by displacement of the recipient strand (branch migration) in either direction (Fig. 9-6c).

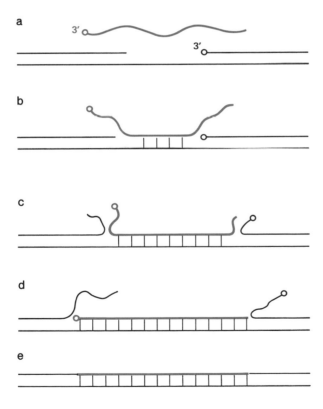

FIGURE 9-6
A recombination mechanism involving insertion of a single strand into a gap in duplex DNA (see text).

Exonucleolytic removal ($3' \rightarrow 5'$ or $5' \rightarrow 3'$) of the recipient strand may precede or follow assimilation of the donor strand (Fig. 9-6d), and may also include part of the donor strand. Properly aligned termini, prepared by replicative gap filling, if necessary, are finally joined by ligase (Fig. 9-6e).

This general scheme may also apply to recombination between the single-stranded terminus at the end of a donor duplex and an internal gap of a recipient one. A single strand may anneal at a transiently melted region of an intact recipient duplex and successfully replace a stretch of the recipient strand, the latter being excised by repair endonucleases. Melting of a recipient duplex can be facilitated by an unwinding protein (Chapter 7, Section 7) such as the gene 32 protein of phage T4, known to be essential for both replication and recombination.

Recombination between duplexes by lap-joint intermediates.
Union of two DNA molecules with overlapping, single-stranded termini is accomplished basically by the same process used in joining the cohesive ends of a linear λ phage molecule into a circle (Chapter 8, Section 7), or in forming elongated concatemers from unit-size molecules in phage T7 replication (Chapter 8, Section 5). Lap-joint intermediates have been observed in phage T4 recombination, the best understood example of general recombination, and are also postulated for phage λ.

The key enzyme in phage λ recombination is the 5'-exonuclease, product of the *exo* gene[23]. It is produced early in infection and has been shown to be essential for a major portion of recombination but dispensable for replication. It is a $5' \rightarrow 3'$ exonuclease with a preference for a 5'-phosphoryl terminus and for duplex over denatured DNA; however, oligonucleotides are also digested. The enzyme binds but does not act at nicks. A related β protein, determined by an adjacent gene, forms a complex with the exonuclease but has no demonstrable influence on its activity in vitro. Both *exo* and β are necessary for recombination in vivo. In vitro studies with the enzyme support the lap-joint scheme shown in Fig. 9-7. Redundant single-stranded branches, if present, can be removed by a cellular $3' \rightarrow 5'$ exonuclease such as exonuclease I, or by $5' \rightarrow 3'$ action of λ exonuclease on a duplex region. Alternatively branch migration may expose a redundant 5'-terminated single-stranded branch, which is susceptible to the oligonucleotidase action of λ exonuclease. Covalent union of the recombinant molecules is carried out by ligase.

Insights into phage T4 recombination have come from studies[24] of cells infected with two kinds of phage particles: heavy-atom

23. Radding, C. M. (1973) *Ann. Rev. Gen.* **7**, 87; Carter, D. M. and Radding, C. M. (1971) *JBC* **246**, 2502; also succeeding paper, p. 2513; Cassuto, E. and Radding, C. M. (1971) *NNB* **229**, 13.
24. Anraku, N. and Lehman, I. R. (1969) *JMB* **46**, 467; also succeeding paper, p. 481; Broker, T. R. and Lehman, I. R. (1971) *JMB* **60**, 131.

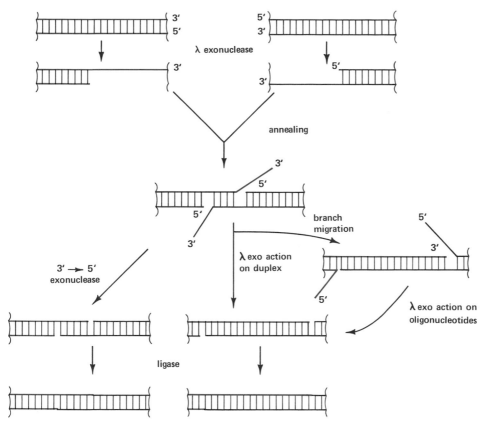

FIGURE 9-7
A recombination mechanism involving annealing of lap-joints. Subsequent
exonuclease action removes redundant single-stranded branches and prepares
the recombinant molecule for ligase action (see text).

labeled, and radioactively labeled (Fig. 9-8). DNA replication was
prevented by mutation in gene 43 (*pol*⁻) or by drugs. After 30 minutes of infection parental T4 DNA was extracted and banded in
density gradients. Some of the DNA occupied a position of hybrid
density between the heavy and light bands and represented the intermediates and final products of recombination (Fig. 9-8). The intermediates could be separated into their heavy and light components
when banded in an _alkaline_ density gradient, or after thermal
denaturation in neutral gradients. The hybrids, called "joint molecules", represent hydrogen-bonded parental units lacking a covalent
connection. When such isolated "joint molecules" were treated
with DNA polymerase I and the four deoxynucleoside triphosphates,
gaps of several hundred nucleotides were filled, enabling ligase to
make a covalent connection (Fig. 9-9).

Electron microscope examination of "joint molecules" has
indicated a variety of structures, with prominent single- and double-

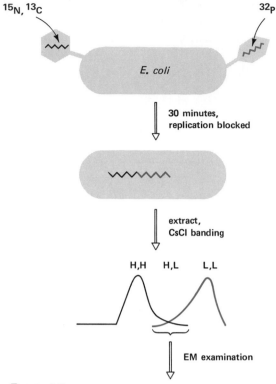

FIGURE 9-8
Procedure for studies of T4 recombination. Co-infection
with ^{32}P-labeled phage and heavy-density labeled
phage generates recombinants which can be observed
in a ^{32}P-labeled band of intermediate buoyant density.

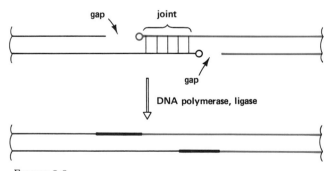

FIGURE 9-9
An isolated recombinant (joint) molecule before and after gap
filling by DNA polymerase, and sealing by ligase.

stranded branches, single-stranded ends, and gaps[25]. A high frequency of one or another kind of these recombinant molecules accumulated when phages mutant in various replication and recombination genes were examined. Such correlations suggested in some instances possible functions for the genes and plausible schemes for T4 recombination. The schemes include lap-joint intermediates analogous to the intermediate in λ recombination, or duplexes annealed through internal single-stranded segments. The sequence of events for the latter may be as follows (Fig. 9-10).

(i) Endonucleolytic nicking of the duplexes, succeeded by exonucleolytic expansion of the nicks to gaps.

(ii) Annealing of homologous single-stranded regions to form branched structures.

25. Broker, T. R. and Lehman, I. R. (1971) *JMB* **60**, 131.

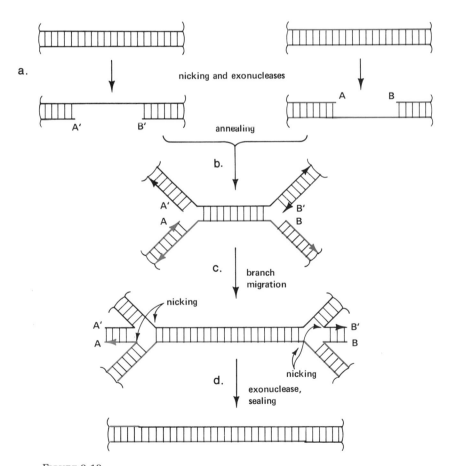

FIGURE 9-10
A recombination mechanism involving two gapped duplexes. Stages are:
(a) production of gapped duplexes, (b) annealing, (c) branch migration, and
(d) endonuclease nicking, exonucleolytic excisions, and ligase sealing.

(ii) a cell with restriction specificity, r_x, also has an m_x activity which protects its own DNA from the r_x activity.

Thus defined, three different kinds of R–M systems have been described in *E. coli*: (i) allelic host systems, distinctive for strain 15 with host specificity A[33], strain B, and strain K; (ii) two allelic phage systems, phage P1 and 15, and (iii) two drug resistance transfer factors (plasmids), RI and RII. A system in *Hemophilus influenzae* has also been characterized, but none has yet been observed in animal cells.

Modification, in all but one system, consists of methylation of the N 6-amino group of adenine residues in a specific DNA sequence; in the *Eco* RII system methylation of the 5-carbon of cytosine residues takes place. Restriction depends on an endonuclease that recognizes the same specific DNA sequence, *provided it and the opposite strand are unmodified*. The enzyme introduces successive cleavages in each strand, resulting in a double-stranded break at that sequence or, in some cases, elsewhere in the DNA[34]. A double-stranded break may mark the death of a DNA molecule as an integral unit, when repair of the break cannot take place. Extensive and nonspecific breakdown of the restricted DNA by other nucleases then follows[35].

R–M systems were first clearly recognized and defined in studies of phage infections of different *E. coli* strains[36]. The efficiency of phage infectivity, as judged by the number of plaques produced on a bacterial lawn, provided a measure of the susceptibility of the invading phage DNA to the DNA-immunity system of the host strain. As shown in Table 9-3, phage λ grown on *E. coli* C, and therefore lacking any known modification pattern, fails to plate on all strains except C and a mutant of K that lacks the restriction system. However phage λ, grown on strain K and carrying that modification pattern, does plate efficiently on strain K as well as on C. The genes for an R–M system in *E. coli* can be provided by a lysogenic phage or by plasmids and, when present, operate independently of the host system. As seen for strain C lysogenized with phage P1 (Table 9-3), all entering DNA is restricted, except that which develops in such a lysogenized cell. When two R–M systems are present in the same cell, restriction of foreign DNA is more efficient and surviving phage DNA acquires both modification patterns.

Maintenance of a characteristic modification pattern of DNA thus requires: (i) persistence of the specific DNA sequence recognized for methylation and (ii) replication in a host that has the methylation system and no restriction system that recognizes the DNA as foreign. Thus, modified DNA is an unstable genetic characteristic because it is lost within two generations when replicated in a host that lacks the proper system.

33. Arber, W. and Wauters-Willems, D. (1970) *Mol. Gen. Genet.* **108**, 203; Arber et al., (1972) ibid., **115**, 195.
34. Horiuchi, K. and Zinder, N. D. (1972) *PNAS* **69**, 3220; Horiuchi, K., Vovis, G. F. and Zinder, N. D. (1974) *JBC* **249**, 543.
35. Simmon, V. F. and Lederberg, S. (1972) *J. Bact.* **112**, 161.
36. Arber, W. and Linn, S. (1969) *ARB* **38**, 467.

TABLE 9-3
Host-specificity (hs, restriction-modification) systems in *E. coli*[a] analyzed
by efficiency of plating.

E. coli strain on which λ phage had been grown	*E. coli* strain for plating λ phage					
	C	K	B	C(P1)	K(P1)	Kr⁻
C	1	<.001	<.001	<.0001	<.0001	1
K	1	1	<.001	<.0001	<.0001	1
B	1	<.001	1	<.0001	<.0001	1
C(P1)	1	<.001	<.001	1	<.001	1
K(P1)	1	1	<.001	1	1	1

[a]*E. coli* C has no known R–M system; a lysogenized strain, K(P1) (i.e. strain K carrying phage P1) displays the R–M system of the phage as well as its own; Kr⁻ is a mutant strain lacking restriction capacity.

[b]Source: Arber, W. and Linn, S. (1969) *ARB* **38**, 467.

Genetic information for the <u>hs system</u>, as the R–M system of *E. coli* is designated, is contained in three adjacent genes, *hsr, hsm,* and *hss.* The *hsr* gene specifies the endonuclease function, the *hsm* gene a polypeptide required for both endonuclease and methylase functions, and the *hss* gene the sequence site recognition that determines the specificity of the entire system. *hsr⁻* mutants fail to restrict but do modify, whereas *hsm⁻* and *hss⁻* neither restrict nor modify.

Genetic complementation studies suggest that in the *E. coli* B system the three closely linked genes code for three diffusable products: polypeptides for restriction, modification, and specificity of site recognition. Support for this model has come from the isolation of *E. coli* B and K restriction enzymes, each of which is in fact composed of three nonidentical subunits, which may prove to be the products of the r, m, and s genes; the B modification enzyme has been shown to have two of these three subunits[37].

Only double-stranded DNA is sensitive to modification and restriction. Single-stranded phage DNAs of M13 and ϕX174 are modified or restricted only as duplex replicative forms. Modification of one strand is sufficient to confer immunity to restriction; a heteroduplex prepared from one modified and one unmodified strand is not restricted. For this reason DNA, as it is produced during semiconservative replication, is protected by the modification on the template strand.

Modification must be nearly error-free if the cell is to survive. Restriction may have this accuracy in distinguishing the correct site sequence but may miss some sequences; unmodified phages are known to escape restriction more frequently than can be explained by mutations. Such tolerance for foreign DNA may be of some evolutionary value. In the competition between modification

---

37. Lautenberger, J. A. and Linn, S. (1972) *JBC* **247**, 6176.

and restriction of DNA entering the cell, additional features of the host and the DNA may influence the outcome. For example, enzymatic activities may be affected by the physiologic as well as the genetic state of the host; and very short chromosomes, such as M13, with at most a few cleavage sites for a particular restriction enzyme may enjoy a special survival advantage.

5. Restriction Endonucleases and Modification Methylases[38]

The proteins responsible for endonucleolytic restriction and methylation modification have unusual and fascinating features that are only beginning to be understood. They are important because they operate the DNA immunity mechanisms of the cell and, by their recognition of specific DNA sequences, have become crucial reagents for the mapping and sequence analysis of DNA. Some may also prove to have a role in recombination. It is difficult to justify calling any of these proteins an enzyme, inasmuch as none in their operations as nuclease or methylase has yet been shown to carry out more than a single scission or methylation. However, the designations endonuclease and methylase are already established in common usage.

Restriction endonucleases. Because restriction activity is easier to measure, the endonucleases have been investigated more thoroughly. Still, there are many uncertainties about their purity and properties. Restriction endonucleases must display three features: (i) recognize a specific oligonucleotide sequence in duplex DNA, (ii) cleave within this sequence in some cases or at some distance away in others, and (iii) fail to cleave DNA if the recognition site is absent or modified by methylation on one or both strands.

It might be wiser at this time simply to enumerate than to try to classify these enzymes. But it is tempting to gamble that some early judgments might actually clarify a rather complex subject. Hazarding this, the restriction endonucleases are presented here as distinct from the modification methylases and divisible into two groups (Table 9-4).

The _type I enzyme_ is complex. It depends on S-adenosylmethionine and ATP, as well as Mg^{2+}. It is a potent, DNA-dependent ATPase and fails to produce distinctive cleavage termini among the products of its restriction activity. The enzyme recognizes a distinctive, but still undefined, binding site on duplex DNA. Having bound the DNA at a recognition site, the enzyme wanders far and wide on the DNA to make its cleavages. It is supposed that cleavages in the duplex strands would occur directly opposite one another rather than in a staggered manner. The fragments of DNA created

38. Meselson, M., Yuan, R. and Heywood, J. (1972) _ARB_ **41**, 447.

Restriction endonucleases from various sources[a]

Endonuclease	Source	Endonuclease	Source
Complex Type I		**Simple Type II**	
Eco K	E. coli K[c]	Eco RI	E. coli bearing RTF I[h, i]
Eco B	E. coli B[d, e]	Eco RII	E. coli bearing RTF II[j, k]
Eco P1	E. coli K12 lysogenic for P1[c, f]	Hin d II[b]	H. influenzae strain d[l, m, n]
Hin d I	H. influenzae strain d[g]	Hin d III	H. influenzae strain d[o]
		Hpa I, II	H. parainfluenzae[p, q]
		Hae[b]	H. aegyptius[o]

[a]The nomenclatural system abbreviates the organism (e.g. *Hin* d), identifies multiple restriction enzymes by Roman numerals (e.g., *Hin* d II, *Hin* d III), and indicates an enzyme specified by a virus or plasmid with a suffix (e.g., *Eco* P1, *Eco* RI). (Smith, H. O. and Nathans, D. (1973) *JMB* **81**, 419.)
[b]Cleavages are directly opposite rather than staggered.
[c]Haberman, A., Heywood, J. and Meselson, M. (1972) *PNAS* **69**, 3138.
[d]Eskin, B. and Linn, S. (1972) *JBC* **247**, 6183.
[e]Smith, J., Arber, W. and Kuehnlein, U. (1972) *JMB* **63**, 1; see also succeeding paper, p. 9.
[f]Brockes, J. P., Brown, P. R. and Murray, K. (1972) *BJ* **127**, 1; Brockes, J. P. (1973) *BJ* **133**, 629.
[g]Gromkova, R., Bendler, J. and Goodgal, S. H. (1973) *J. Bact.* **114**, 1151.
[h]Hedgpeth, J., Goodman, H. M. and Boyer, H. W. (1972) *PNAS* **69**, 3448.
[i]Mertz, J. E. and Davis, R. W. (1972) *PNAS* **69**, 3370. RTF is resistance transfer factor, a plasmid.
[j]Bigger, C. H., Murray, K. and Murray, N. E. (1973) *NNB* **244**, 7.
[k]Boyer, H. W., Chow, L. T., Dugaiczyk, A., Hedgpeth, J. and Goodman, H. M. (1973) *NNB* **244**, 40.
[l]Kelly, T. J. and Smith, H. O. (1970) *JMB* **51**, 393.
[m]Smith, H. O. and Wilcox, K. W. (1970) *JMB* **51**, 379.
[n]Roy, P. H. and Smith, H. O. (1973) *JMB* **81**, 427; 445.
[o]Middleton, J. H., Edgell, M. H. and Hutchison, C. A., III, (1972) *J. Virol.* **10**, 42.
[p]Gromkova, R. and Goodgal, S. H. (1972) *J. Bact.* **109**, 987.
[q]Sharp, P. A., Sugden, B. and Sambrook, J. (1973) *B*, **12**, 3055.

by such cleavages would be exposed to degradation, whereas recombination would be favored by overlapping ends created by some type II enzymes.

The _type II enzyme_ is simpler. It shows no requirement for S-adenosylmethionine or ATP and requires only Mg^{2+}. It makes a well-defined, double-stranded cleavage at a defined site also recognized by a corresponding modification methylase. In some instances the cleavage is staggered, making it possible for the overlapping ends to be rejoined, or recombined, with other ends created by cleavage of any DNA molecule by the same enzyme.

Eco K endonuclease[39]. This protein has been purified to homogeneity. The molecule of 420,000 daltons contains three nonidentical subunits in the stoichiometry: one γ of 55,000, two α of 125,000, and two β of 60,000 daltons. Based on genetic evidence and an in vitro interchange of subunits with those of the _Eco_ B restriction-modification proteins, it has been suggested that one type of subunit (γ) confers specificity (_hss_), the second (α) endonuclease (_hsr_), and the third (β) methylase (_hsm_); the latter function is absent from the _Eco_ B endonuclease.

---

39. Haberman, A., Heywood, J. and Meselson, M. (1972) *PNAS* **69**, 3138.

Organization of the protein into two sets of identical subunits is in keeping with the symmetry of the DNA in the restriction-modification operations required. Kinetic studies suggest a two-step mechanism, the protein dissociating between the first and second cleavages. Attempts to determine the identity of the 5′ terminus, by labeling it with γ-(^{32}P)-ATP by action of polynucleotide kinase, have not succeeded with either this endonuclease[40] or with the Eco B enzyme[41].

The function of ATP for Eco K and Eco B endonuclease activity is not understood. Most phenomenal is the turnover of ATP. Over 10,000 molecules of ATP are hydrolyzed to ADP and inorganic phosphate for every phosphodiester bond broken. The endonuclease is thus a potent DNA-dependent ATPase with a turnover number of over 1000 per minute, even though as a nuclease it may not turn over at all.

The ATPase behavior is intriguing[41,42]. It is dependent on Mg^{2+} and S-adenosylmethionine and also requires DNA with correct and unmodified sequence. After the cleavage, the enzyme forms a complex with S-adenosylmethionine and the restricted DNA, and ATPase action ensues. It continues for hours, long after the restrictive cleavages have been made in the DNA. Unlike the nuclease action which is exceedingly slow, requiring hours, and showing no turnover, the enzyme as an ATPase hydrolyzes up to 10^4 molecules per minute. The enzyme must remain tightly bound to unmodified DNA because in its absence it is known to be rapidly inactivated by ATP and Mg^{2+}. Pancreatic DNase degradation of the DNA aborts the ATPase action, presumably by dissociating the enzyme from the DNA. In vivo, the function might be performed by other DNases, such as the rec B,C nuclease. ATP in the cell may serve to maintain the oligomeric endonuclease in an active form. In vitro, in its isolated and damaged state, the protein may dissipate the ATP needlessly.

The role of S-adenosylmethionine in endonuclease action is as uncertain as that of ATP. According to one view[43] it provides an allosteric function in binding the protein to DNA. Another suggestion[44] is that it serves directly in restrictive cleavage through methylation reactions. However, attempts to identify the methylated bases or nucleotides at cleavage sites have been unsuccessful, a result perhaps of their sparsity and heterogeneity. Although Eco K endonuclease preparations are known to methylate adenine residues in unmodified DNA, it is still uncertain whether this activity represents the true modification methylase of E. coli K.

Eco B endonuclease[45]. The pure protein is essentially similar to the Eco K molecule. It has three nonidentical subunits ($\gamma = 55,000$,

40. Murray, N. E., Batten, P. L. and Murray, K. (1973) JMB **81**, 395.
41. Eskin, B. and Linn, S. (1972) JBC **247**, 6183.
42. Eskin, B. and Linn, S. (1972) JBC **247**, 6192; Linn, S., Lautenberger, J. A. Eskin, B. and Lacky, D. (1974) Fed. Proc. **33**, 1128.
43. Meselson, M., Yuan, R. and Heywood, J. (1972) ARB **41**, 447.
44. Murray, K., personal communication.
45. Eskin, B. and Linn, S. (1972) JBC **247**, 6183; Smith, J., Arber, W. and Kuehnlein, U. (1972) JMB **63**, 1; also succeeding paper, p. 9.

321

SECTION 5:
Restriction
Endonucleases
and Modification
Methylases

$\alpha = 135,000$, and $\beta = 60,000$ daltons), but only one of each, adding up to a mass of 250,000 daltons. The controlling genes are *hss* for γ, *hsr* for α, and *hsm* for β. In vivo, the *hsr* and *hsm* gene products can be interchanged in strains B and K. The absence of methylase activity in *Eco* B implies some mechanism for suppressing it not manifested by *Eco* K.

The functions of ATP and S-adenosylmethionine, essential to the endonuclease action, are as mysterious as for the *Eco* K enzyme. The dependence of the ATPase action on the topological characteristics of the DNA suggests[46] that the enzyme wanders from a specific site of recognition to multiple, nonspecific sites where cleavages are made.

Eco P1 endonuclease[47]. A partially purified preparation of this enzyme (molecular weight 200,000) also has the capacity to methylate, but this may be due to contamination with a known methylase (see below). Unlike its role with *Eco* K and *Eco* B endonucleases, S-adenosylmethionine stimulates but is not absolutely required for restriction; ATP, while required, is not hydrolyzed to a measurable extent. More sensitive assays that would detect ATP breakdown at a level equivalent to endonuclease cleavages would be needed to prove the absence of any ATP involvement at all.

Eco RI and Eco RII endonucleases[48,49,50]. These proteins are coded by the resistance transfer factors (plasmids). *Eco* RI is about 58,000 daltons, a dimer of two 29,000 dalton subunits[48,49], and appears to be the simplest and best understood of the *E. coli* restriction systems. Only Mg^{2+} is required. The specific sequence cleaved and modified has been defined (Table 9-5). The staggered cleavage produces overlapping (cohesive) 3'-hydroxyl and 5'-phosphoryl ends. The sites of cleavage are symmetrical. They exhibit 180° rotational (two fold) symmetry about an axis perpendicular to the helical axis of DNA. Cleavage of the symmetrical substrate may depend on symmetrical orientation of the two subunits of the enzyme. Cleavage of SV40 DNA is at the unique site of occurrence of the hexanucleotide sequence, and furnishes one of the prime reference points for mapping this important chromosome[51].

Eco RII has been partially purified; its estimated mass is 100,000 daltons. The sites of cleavage (Table 9-5), as with *Eco* RI, are rotationally symmetrical and staggered, although the axis of symmetry is through a base pair rather than between base pairs.

46. Horiuchi, K. and Zinder, N. D. (1972) *PNAS* **69**, 3220; Horiuchi, K., Vovis, G. F. and Zinder, N. D. (1974) *JBC* **249**, 543.
47. Haberman, A., Heywood, J. and Meselson, M. (1972) *PNAS* **69**, 3138; Brockes, J. P., Brown, P. R. and Murray, K. (1972) *BJ* **127**, 1; Brockes, J. P. (1973) *BJ* **133**, 629.
48. Hedgpeth, J., Goodman, H. M. and Boyer, H. W. (1972) *PNAS* **69**, 3448.
49. Mertz, J. E. and Davis, R. W. (1972) *PNAS* **69**, 3370.
50. Bigger, C. H., Murray, K. and Murray, N. E. (1973) *NNB* **244**, 7; Boyer, H. W., Chow, L. T., Dugaiczyk, A., Hedgpeth, J. and Goodman, H. M. (1973) *NNB* **244**, 40.
51. Morrow, J. F. and Berg, P. (1972) *PNAS* **69**, 3365; Dugaiczyk, A., Hedgpeth, J., Boyer, H. W. and Goodman, H. M. (1974) *B.* **13**, 503.

TABLE 9-5
Base sequences at sites of restriction and modification[a]

Restriction and modification enzyme[b]	Sequence at sites of restriction and modification[c]
Eco RI[d]	— C —T —T —│— A*—A—G —— 5′ ↓ 5′ —— G —A—A*—│— T — T —C —— ↑
Eco RII[e,f]	— C — G—G —A— C*—C— G —— 5′ ↓ 5′ —— G — C— C*—│— T— G — G — C —— ↑ Axis of symmetry
Hin d II[g,h,i]	— C —A*— Pu—│— Py —T —G —— 5′ ↓ 5′ —— G — T— Py—│— Pu — A*— C —— ↑
Hin d III[i,j]	— T —T— C —│— G —A— A*—— 5′ ↓ 5′ —— A*—A— G —│— C — T — T —— ↑
Hae (B. subtilis X5)[a]	— C — C —│— G — G —— 5′ ↓ 5′ —— G — G —│— C — C —— ↑
Eco K, Eco B, Eco Pl, Hin d I[a]	Indefinite

[a] Source: Murray, K. and Old, R. W. (1974) Prog. N.A. Res. **14**, 117.
[b] See Table 9-4 for sources of enzymes.
[c] A* is 6-methylaminopurine, C* is 5-methylcytosine, Pu is A or G, Py is T or C. Arrows show points of cleavage.
[d] Hedgpeth, J., Goodman, H. M. and Boyer, H. W. (1972) PNAS **69**, 3448.
[e] Bigger, C. H., Murray, K. and Murray, N. E. (1973) NNB **244**, 7.
[f] Boyer, H. W., Chow, L. T., Dugaiczyk, A., Hedgpeth, J. and Goodman, H. M. (1973) NNB **244**, 40.
[g] Kelley, T. J. and Smith, H. O. (1970) JMB **51**, 393.
[h] Smith, H. O. and Wilcox, K. W. (1970) JMB **51**, 379.
[i] Roy, P. H. and Smith, H. O. (1973) JMB **81**, 427; 445.
[j] Old, R. W., Murray, K. and Roizes, G., personal communication.

The staggered cleavages by the *Eco* RI, and RII and *Hin* d III endonucleases (Table 9-5) provide overlapping ends that facilitate the
joining of any two DNA duplexes generated by one of these enzymes.
This type of endonucleolytic cleavage may be seen as an invitation
to recombination of the cleaved DNA, as well as to degradation, and
could conceivably serve both functions in vivo (Chapter 11, Section 5).

H. influenzae and other endonucleases[52].   Restriction enzymes have
been isolated from many Hemophilus species and strains[53]. The

52. Kelly, T. J. and Smith, H. O. (1970) JMB **51**, 393; Smith, H. O. and Wilcox, K. W. (1970)
    JMB **51**, 379; Roy, P. H. and Smith, H. O. (1973) JMB **81**, 427, 445.
53. Smith, H. O. and Nathans, D. (1973) JMB **81**, 419.

323

SECTION 5:
Restriction
Endonucleases
and Modification
Methylases

first characterization of a restriction enzyme and of a specific, symmetrical, cleavage sequence came from studies of *Hin* d II. Thought originally to be the only restriction endonuclease in this organism, it was later recognized that at least two additional restriction enzymes (*Hin* d I and *Hin* d III) exist in this strain. The existence of the other two endonucleases was obscured by their failure to restrict phage T7 DNA, the substrate employed for *Hin* d II.

Whereas *Hin* d III cleaves the double strand to produce cohesive ends, *Hin* d II does not; in both cases, as with the *Eco* R I and R II cleavage sites, there is a twofold axis of symmetry.

It seems strange that the *Hin* d III cleavage site, which should appear, on a random basis, about 40 times in a DNA molecule the size of T7, should be completely absent. It remains to be explained how T7 phage grown on an *E. coli* strain without a modification system can plate with full efficiency on *E. coli* strains B and K. When examined in vitro, the T7 DNA has five sites susceptible both to cleavage by *Eco* B endonuclease and to modification by *Eco* B methylase[54]. Evidently there are important features in DNA sequence and structure recognized by restriction-modification systems that have not yet been characterized.

As the result of a need for highly sequence-specific endonucleases to serve as reagents for DNA mapping, and with improvement in assay procedures to detect their action, many restriction enzymes have been found and characterized (Table 9-4). An enzyme preparation from *H. parainfluenzae* (*Hpa* I, *Hpa* II) for example, has proved to be important in the analysis of SV40[55] and M13[56] DNAs. Limited cleavage of small genomes such as these may also be used to recognize mutants that have either lost a restriction site or acquired it.

Modification methylases. These enzymes methylate DNA at a specific site; as a result the DNA is protected against cleavage by a corresponding restriction endonuclease which recognizes that site. Methylation utilizes S-adenosylmethionine as methyl donor and duplex DNA as acceptor, producing a pair of 6-methylaminopurine residues (except in the one instance in which the product is 5-methylcytosine; see Table 9-5). The modified residues are symmetrically oriented in the specific restriction-modification sequence.

The clearest insight into methylation modification came with the isolation of *Eco* B methylase and studies of its action[57]. The protein consists of two subunits: $\beta=20,000$ and $\gamma=55,000$ daltons; they are controlled by *hsm* and *hss* genes, respectively. It was purified as an activity that enabled the unmodified, duplex form of fd DNA to infect spheroplasts of *E. coli* B. This action was shown to involve

54. Eskin, B., Lautenberger, J. A. and Linn, S. J. (1973) *Virol.* **11**, 1020; Murray, N. E., Batten, P. L. and Murray, K. (1973) *JMB* **81**, 395.
55. Morrow, J. F. and Berg, P. (1972) *PNAS* **69**, 3365; Danna, K. J., Sack, G. H., Jr. and Nathans, D. (1973) *JMB* **78**, 363.
56. Tabak, H., Griffith, J., Geider, K., Schaller, H. and Kornberg, A. (1974) *JBC* **249**, 3049.
57. Eskin, B. and Linn, S. (1972) *JBC* **247**, 6183; Smith, J., Arber, W. and Kuehnlein, U. (1972) *JMB* **63**, 1; also succeeding paper, p. 9.

the transfer of four methyl groups to fd DNA. Two restriction-modification sites in the fd genome were inferred from the isolation of a mutant that had lost one site and accepted only two methyl groups in vitro, and another mutant that had lost both sites and accepted no methyl groups.

Eco P1 methylase[58] also has been characterized as a dimer. Its two unlike subunits are 70,000 and 45,000 daltons, respectively. The enzyme by methylating unmodified phage DNA protects it from attack by the *Eco* P1 restriction endonuclease. It is difficult to decide whether a similar action carried out by the *Eco P1 restriction endonuclease* is an inherent property or results from contamination with this methylase. A similar uncertainty concerns the modification methylase activity of the *Eco K restriction endonuclease*, a separate *Eco* K methylase of smaller size, without endonuclease activity, having already been identified[59]. If *Eco* P1 and *Eco* K endonucleases can indeed express both restriction and modification activities in vivo, there must be a mechanism that determines whether their binding to an unmodified recognition sequence will be followed by cleavage or by protective methylation.

Other methylases. Methylase activities have been identified in restriction-modification systems containing the endonucleases *Eco* RI, *Eco* RII, *Hin* d II, and *Hin* d III. The sequences at sites of modification have been determined (Table 9-5) but little is known as yet about the isolated methylases in this rapidly developing field.

OTHER MODIFICATIONS

The description thus far of the *hs* system in vivo and in vitro has failed to mention three other significant kinds of modification. One is the glucosylation of T-even phage DNA known only in these phages (see Section 6 of this chapter). The other two are widespread systems for methylation by S-adenosylmethionine of adenine and cytosine residues in DNA, forming 6-methylaminopurine and 5-methylcytosine to an extent many times greater than that of the *hs* system.

Despite extensive studies there is no evidence that the 6-methylaminopurine and 5-methylcytosine modifications by potent and specific methylases are related to a restriction system, or to any other known functions. Experiments[60] in which a phage-induced enzyme destroyed the cell's S-adenosylmethionine supply, and with it the S-adenosylmethionine-mediated methylations in the cell, virtually eliminated 6-methylaminopurine and 5-methylcytosine from the DNA, but did not show up any defects or alterations in function. It is difficult to believe that methylations of adenine and cytosine, so widespread in plant and animal species, are maintained without performing some important functions.

58. Brockes, J. P., Brown, P. R. and Murray, K. (1972) *BJ* **127**, 1; Brockes, J. P. (1973) *BJ* **133**, 629.
59. Murray, K., personal communication.
60. Gefter, M., Hausmann, R., Gold, M. and Hurwitz, J. (1966) *JBC* **241**, 1995.

6. Restriction and Glucosylation Modification[61]

The first host-controlled modification of DNA was recognized in T-even phages, which were observed to adapt to better growth on a particular *E. coli* strain only after growing in that strain. The phenomenon came to be understood as resulting from glucosylation of the phage DNA, which protected it from degradation by specific nucleases in *E. coli*. These restriction enzymes recognize unglucosylated hydroxymethylcytosine (HMC) residues (Fig. 2-13) in certain sequences and degrade the DNA.

Understanding of the restriction-glucosylation modification developed from the following discoveries:

(i) T2, T4, and T6 phages possess HMC, in place of cytosine, in their DNA[62].

(ii) The HMC residues are partially or totally glucosylated in a specific pattern[63] (Table 9-6).

TABLE 9-6
Glucosylation patterns of HMC residues of T-even phage DNA.

	T2	T4	T6
	Percent of total HMC residues		
Unglucosylated	25	0	25
α-Glucosyl	70	70	3
β-Glucosyl	0	30	0
β-glucosyl-α-glucosyl (α-gentiobiosyl)	5	0	72

(iii) The extent and type of glucosylation is determined by the particular phage[64].

(iv) Glucosylation occurs by transfer of glucose from uridine diphosphate glucose (UDPG), furnished by the host cell, to newly assembled phage DNA, catalyzed by specific α- and β-glucosyl transferases[64]. The enzymes are coded by phage *gt* genes (Table 9-7), T4 α *gt* and β *gt*, located at 330° and 298°, respectively, on the circular map[65] (Fig. 8-27).

(v) Restriction of unglucosylated HMC-DNA in *E. coli* is due to its degradation by nucleases coded by *E. coli* genes *r2,4* and *r6*. Restriction affects only the entering DNA, not intracellularly synthesized DNA[66].

(vi) *E. coli* mutants defective in UDPG pyrophosphorylase fail to

61. Revel, H. R. and Luria, S. E. (1970) *Ann. Rev. Gen.* **4**, 177.
62. Cohen, S. S. (1968) *Virus-Induced Enzymes*, Columbia University Press, N.Y.
63. Lehman, I. R. and Pratt, E. A. (1960) *JBC* **235**, 3254.
64. Kornberg, S. R., Zimmerman, S. B. and Kornberg, A. (1961) *JBC* **236**, 1487; Josse, J. and Kornberg, A. (1962) *JBC* **237**, 1968.
65. Op. cit. in footnote 61.
66. Revel, H. R. (1967) *Virol.* **31**, 688.

TABLE 9-7

Glucosylation by distinctive glucosyl transferases induced in cells infected by T-even phages

Phage	Glucosylation
T2, T4, T6	UDPG + HMC-DNA $\xrightarrow{\alpha\text{-gt}^a}$ G-α-HMC-DNA + UDP
T4	UDPG + HMC-DNA $\xrightarrow{\text{T4 }\beta\text{-gt}^b}$ G-β-HMC-DNA + UDP
T6	UDPG + G-α-HMC-DNA $\xrightarrow{\text{T6 }\beta\text{-gt}^b}$ G-β-G-α-HMC-DNA + UDP

$^a\alpha$-gt = α glucosyl transferase
$^b\beta$-gt = β glucosyl transferase

make UDPG and as a result produce phages with unglucosylated HMC-DNA. These phages, called T*, cannot infect these mutants or any other cells except those defective in restriction (e.g. *E. coli* r2,4⁻,r6⁻ and *Shigella dysenteriae*)[67].

The restriction-modification system based on glucosylation has a clear resemblance to the *hs* system based on methylation; several important differences form a convenient basis for discussion.

Methylation to form 6-methylaminopurine involves only a small fraction of the adenine residues, in the neighborhood of 0.1 percent. A host-specificity methylase responsible for modification has been found widely distributed among bacteria and suggests an immunity system to protect cells of a given strain from foreign DNA. An intimate functional and structural relationship is found in the *hs* system between the modification and restriction components of the operation. Glucosylation, by contrast, modifies all or nearly all the HMC residues of the DNA. The system has been observed only in the closely related T2, T4, and T6 phages. A sharp distinction exists between the phage-induced glucosylation enzymes and the host system which restricts unglucosylated DNA. What selective advantage glucosylation offers is not clear, since growth of nonglucosylated phages (T*) in nonrestrictive host cells seems relatively unimpaired.

The *hs* restriction endonucleases that recognize and cleave sites lacking 6-methylaminopurine have, in several instances, been purified to homogeneity and the nucleotide sequences of some cleavage sites have been determined. This has not yet been achieved for the proteins that restrict T* DNA[68]. These endonucleases may be associated with the membrane, since they degrade the DNA introduced by infecting T* phage but not the DNA synthesized for developing T* phage[69].

A demonstration of T* restriction has been obtained with toluene-treated, uninfected *E. coli*. Such a preparation supported semi-

67. Revel, H. R. and Luria, S. E. (1970) *Ann. Rev. Gen.* **4**, 177.
68. Op. cit. in footnote 67.
69. Fukasawa, T. and Saito, S. (1964) *JMB* **8**, 175; Erikson, R. L. and Szybalski, W. (1964) *Virol.* **22**, 111.

conservative DNA replication when dCTP was one of the four deoxynucleoside triphosphates but not when dHMCTP was used in its place[70]. However, with mutants that are not restrictive (r2,4⁻, r6⁻) either the C or HMC deoxynucleoside triphosphate was able to support DNA synthesis. This result shows the presence of the restriction system in cells before infection and its capacity to attack even the HMC-DNA synthesized upon an *E. coli* DNA template. Presumably the genes *r2,4* and *r6* each specify a membrane-associated nuclease, one with a preference for HMC sequences in T2 and T4, and the other in T6.

Is there any relationship between the *hs* and the *r2,4* and *r6* restriction systems? Although none is apparent, the observations suggest that HMC-glucosylation protects T-even phages against restriction. These phages, unlike λ and fd, are completely insensitive to the *hs* system; the absence of HMC eliminates all but the T-even phages as targets for the *r2,4*, and *r6* system. Functions of the latter do not appear to depend on methylation. In cells deprived of their methylating capacity through unavailability of S-adenosylmethionine[71], the *hs* system is depressed, whereas the *r2,4*, and *r6* are not.

70. Fleischman, R. A. and Richardson, C. C. (1971) *PNAS* **68**, 2527.
71. Gefter, M., Hausmann, R., Gold, M. and Hurwitz, J. (1966) *JBC* **241**, 1995.

10

Transcription — RNA Polymerases

1. Comparison of RNA and DNA Polymerases

RNA synthesis directed by a DNA template is too intimately related to DNA synthesis to be omitted entirely, even from a book that narrows its focus to replication and excludes transcription. Aside from the formal similarities of the two processes, there are several persuasive reasons for paying close attention to RNA synthesis when one studies DNA synthesis.

(i) RNA polymerase is a protein that carries out a coordinated, DNA-directed synthesis of a chain from start to finish, and a great deal is known about the process.

(ii) RNA transcriptions serve as specific priming devices for initiating DNA synthesis.

(iii) RNA transcription and DNA replication are interdependent at several levels of metabolic regulation.

(iv) RNA itself may have a structural role in the organization of the chromosome.

Before considering RNA polymerases in detail it would be helpful to cite the similarities and distinctions between the RNA and DNA polymerases of *E. coli* (Table 10-1). The differences are sharpened by the comparison with DNA polymerase I in the tabulation. Until the discovery of DNA polymerases II and III, and the complex form of DNA polymerase III, the contrast between the multisubunit structure of RNA polymerase and the single polypeptide of DNA polymerase I was striking. Now, the similarities are becoming more evident (Chapter 5, Section 3).

It is a biochemical accident that the isolation of a nearly intact transcriptional complex has been relatively easy, but isolation of a physiological replicase assembly has not. As a result of this circumstance we know more about the physiology of RNA polymerase

TABLE 10-1
Comparison of *E. coli* RNA polymerase and DNA polymerase I

	RNA Polymerase	DNA Polymerase I
Similarities		
DNA template direction	yes	yes
Requirement for 4 triphosphates	yes	yes
Base-pairing analog substitution	yes	yes
Divalent cation requirement	yes	yes
Synthesis: $5' \rightarrow 3'$, antiparallel	yes	yes
Anomalous reiterative and de novo synthesis	yes	yes
Anomalous use of DNA units: triphosphate, template, primer	yes	—
Anomalous use of RNA units: triphosphate, template, primer	—	yes
Distinctions		
Function	transcriptase	replicase
Size, in daltons	480,000	109,000
Subunits	5	1
Starts chains	yes	no
Triphosphate binding sites	2	1
K_m for triphosphates:		
Initiation	high (>500 μM)	low (5 μM)
Elongation	low (5–20 μM)	low (5 μM)
Terminates chains	yes	no
Recognition of sequences and strands	yes	no
Intact duplex is template	yes	no
Product	short, free	long, duplex
Ultra-high-fidelity copying	uncertain	yes
Proofreading	no	yes
Repair function	no	yes
Linked to translation	yes	no
Turnover number, nucleotides/sec	~50	~10
Comparison with in vivo rate	same	~1%
Inhibited by rifampicin, actinomycin	yes	no

than we do about DNA polymerase. However, when attempts are made to assess the in vivo functions of the numerous factors less tightly held by RNA polymerase, the answers are as elusive as the missing parts of the replicase puzzle.

RNA and DNA polymerases have basically identical catalytic properties in the growth of polynucleotide chains. RNA polymerases require all four ribonucleoside triphosphates for complementary base pairing with a DNA template, just as DNA polymerases require all four deoxyribonucleoside triphosphates. Nucleophilic attack of the 3'-hydroxyl terminus of the growing chain on the α-phosphate of the monomeric 5'-triphosphate propagates the chain in a $5' \rightarrow 3'$ direction. Both polymerases can show the anomalous lack of template direction and dependence described for DNA polymerase in the reiterative and de novo synthesis of homopolymers and copolymers (Chapter 4, Section 15).

RNA polymerase is sharply distinguished from DNA polymerases by its capacity, with a duplex template, to start a chain and to terminate it. This property is of course needed to select for transcription certain genes among the many available on the chromosome. It is accomplished by recognition of the nucleotide sequences that identify the strand and precisely where on the strand the start is to be made. These sequences are called the promoter region for a given gene or group of genes. DNA polymerases by contrast, cannot start chains but rely on RNA polymerases to do it for them. RNA polymerases terminate chains at the end of a defined genetic sequence. Once in motion, DNA polymerases copy a template until it runs out.

The true template for RNA polymerase is a duplex that undergoes localized and transient melting during transcription. By contrast DNA polymerases are inert on such intact duplexes and require frayed regions or auxiliary proteins to unwind the helical duplex and expose relatively long single-stranded segments for replication.

DNA polymerases are exquisitely designed for ultra-high fidelity in copying any template, but show no discrimination among sequences being copied. In the interest of error-free copying, certain DNA polymerases have proofreading capacities built into their operation; DNA polymerase I can even excise lesions from the 5' end of one strand of a template duplex it is replicating (see Chapter 4, Sections 12 and 13). The proofreading function, a $3' \rightarrow 5'$ exonuclease that recognizes and excises a mismatched primer terminus, represents an enzyme conformation different from the one that polymerizes. This dual role of DNA polymerase bears some resemblance to the distinctive states of selection-initiation and of transcription which will be described for RNA polymerase.

As more is learned about the location of the polymerases in the cell and how they are regulated further distinctions will become evident. For example, RNA polymerase action must be linked closely to translation and express the need for synthesis of certain RNA and protein molecules at a particular time and place. DNA polymerase synthesis, progressing in a discontinuous and pulsatile fashion, depends on signals still unknown. There must also be factors to gear replication to membrane growth and to other cellular operations associated with cell division.

2. Structure of *E. coli* RNA Polymerase[1]

RNA polymerase of *E. coli*, the most extensively studied of bacterial RNA polymerases, is representative of the enzyme isolated from a number of bacterial genera, including *Azotobacter*, *Pseudomonas*, *Bacillus*, *Caulobacter*, and *Micrococcus*. All these RNA polymerases are remarkably similar in organization and properties. On the other hand, the RNA polymerase induced by T7 phage infection is distinctive, and will be discussed separately in Section 4 of this chapter.

In considering the structure of an RNA polymerase, the method of purifying the enzyme is especially important. Problems with persistent contaminants, lost factors, proteolytic alterations, denaturations, aggregation, and dissociation, which are difficulties plaguing studies of all complex enzymes, have been especially troublesome in studies of RNA polymerase. There have been inconsistencies in results obtained by different laboratories, or even with successive preparations of the enzyme in the same laboratory.

RNA polymerase is composed of α, β, β', and σ subunits[2] in the ratio: 2:1:1:1 (Table 10-2). When the fragments obtained from tryptic

TABLE 10-2
Subunits of RNA polymerase

Subunit	E. coli	B. subtilis
	Molecular weight $\times 10^{-3}$	
α	40	43
β	150	150
β'	160	150
σ	90	55
ω	10	

digests of the α, β, and β' subunits were chromatographed, the patterns showed little overlap, indicating their unique polypeptide compositions. The composite molecular weight is 480,000 daltons with an S value of 15. At ionic strengths below 0.1, the enzyme forms a dimer with an S value of 23.

Two ω subunits, present in some *E. coli* preparations but absent

1. Chamberlin, M. in *The Enzymes* (P. D. Boyer, Ed.) Academic Press, N.Y. (1974) *ARB* **43**, 721; Burgess, R. (1971) *ARB* **40**, 711; Von Hippel, P. H. and McGhee, J. D. (1972) *ARB* **41**, 231.
2. Burgess, R. R., Travers, A. A., Dunn, J. J. and Bautz, E. K. F. (1969) *Nat.* **221**, 43; Travers, A. A. and Burgess, R. R. (1969) *Nat.* **222**, 537.

from others, have not been found in the other bacterial polymerases[3]. The ω subunit was dissociated without loss of the activity of the intact enzyme (holoenzyme) and was not required in the reconstitution of an active holoenzyme after disaggregation of native enzyme by urea treatment. The role of ω in RNA polymerase is therefore uncertain; it might be disclosed if activity were measured under more exacting and physiological conditions.

Chromatography of RNA polymerase on phosphocellulose separates the σ subunit, which fails to bind to the adsorbent, from $\alpha_2\beta\beta'$, called the core polymerase, which does bind. The latter is distinguished from the holoenzyme in being unable to utilize an intact duplex template such as T2 DNA while being equally active on nicked or singled-stranded DNA. Poly d(A–T), frayed enough to be susceptible to the single-strand-specific exonuclease I, serves as a template for the core polymerase (Table 10-3).

TABLE 10-3
Relative activities of holoenzyme
and core polymerase

Polymerase	Template	
	poly d(A–T)	T2 DNA
	units/mg	
Holoenzyme	20,000	8000
Core	16,000	<300

The σ subunit is inert by itself. Although σ is essential for strand- and sequence-specific transcription of a duplex, its function in the holoenzyme may well be different from the adaptor role assigned to it originally (see Section 3 of this chapter). Other proteins, called rho (ρ) and kappa (κ) may influence termination of RNA polymerase action but cannot as yet be considered part of the enzyme.

Antibiotics that bind RNA polymerase and inhibit its action have been useful for gaining important insights into its structure and actions. Rifampicin, for example, which binds firmly to the β subunit, completely blocks initiation of RNA synthesis by the enzyme in vitro and in vivo[4]. Mutants resistant to rifampicin inhibition have a resistant RNA polymerase due to an altered β subunit which fails to bind the drug. Used this way, rifampicin has shown RNA polymerase to be the enzyme responsible for transcription in the cell, and provides a means of exploring the mechanism of the enzyme's action. Streptolydigin also binds the β subunit; it blocks chain growth rather than initiation.

The holoenzyme has been dissociated into its component subunits, which have been resolved and then successfully reconstituted in

3. Berg, D., Barrett, K. and Chamberlin, M. (1971) Meth. Enzym. 21, 506.
4. Wehrli, W. and Staehelin, M. (1971) Bact. Rev. 35, 290.

50–60 percent yield[5,6]. The separated polypeptides are individually inert; they become active when reassembled in this order[5]:

$$2\alpha + \beta \rightarrow \alpha_2\beta$$

$$\alpha_2\beta + \beta' \xrightarrow{\sigma} \alpha_2\beta\beta'$$

$$\alpha_2\beta\beta' + \sigma \rightarrow \alpha_2\beta\beta'\sigma$$

The presence of σ appears to stimulate assembly at an intermediate step before being itself incorporated into the holoenzyme. While mutants in the α, β', and σ subunits are not known, the requirement for the subunits in reconstituting enzymatic activity is strong evidence that they are essential.

3. Function of *E. coli* RNA Polymerase[7-10]

The equation for RNA polymerase action can be written simply:

$$(ATP)_n + (GTP)_n + (UTP)_n + (CTP)_n \xrightarrow{\text{DNA}} (AMP\text{–}GMP\text{–}UMP\text{–}CMP)_n + 4\,nPP_i$$

This equation resembles that for DNA polymerase, except that the DNA template serves a catalytic role and is neither extended as a primer nor incorporated into a stable duplex with the product. RNA polymerase initiates a chain and terminates it. Formation of the phosphodiester bonds for chain growth is thus only one stage in a remarkably complicated pathway of reactions. They include, in order: (i) *binding to template*, (ii) *initiation of chain*, (iii) *elongation of chain*, and (iv) *termination of chain and enzyme release*.

I BINDING TO TEMPLATE

The binding and initiation steps are difficult to analyze separately. Binding may be defined by an association of the enzyme with DNA tight enough for cosedimentation in velocity gradients or retention on nitrocellulose filters; proteins ordinarily adhere to these filters, but duplex DNA does not. Initiation is described by the pairing of either ATP or GTP with the first nucleotide in the template sequence to be transcribed. A valid determination of initiation requires covalent attachment of the ATP or GTP to the second ribonucleotide. Measurement of initiation is thus inseparable from the beginning of chain growth.

To separate the binding and initiation steps, use has been made of

5. Heil, A. and Zillig, W. (1970) *FEBS Lett.* **11**, 165.
6. Ishihama, A. and Ito, K. (1972) *JMB* **72**, 111; Ishihama, A. (1972) *B.* **11**, 1250.
7. Chamberlin, M. in *The Enzymes* (P. D. Boyer, Ed.) Academic Press, N.Y. (1974) *ARB* **43**, 721.
8. Losick, R. (1972) *ARB* **41**, 409.
9. Bautz, E. K. F. (1972) *Prog. N. A. Res. and Mol. Biol.* **12**, 129.
10. Schäfer, R., Zillig, W. and Zechel, K. (1972) *Polymerization in Biological Systems*, Ciba Foundation Symposium, Elsevier, N.Y., p. 41; (1973) *EJB* **33**, 207.

drugs, low temperatures, or high ionic strengths, coupled to early or delayed addition of the triphosphates. Both rifampicin and low temperatures (<20°) permit binding but block initiation. Heparin, a competitive inhibitor of DNA, and high salt concentrations (>0.1 M NaCl) block binding but permit initiation and chain elongation (Table 10-4). The binding event as defined here includes several

TABLE 10-4
Effects of agents on RNA polymerase

	Binding to DNA	Initiation of synthesis
Rifampicin	Yes	No
Low temperatures	Yes	No
Heparin	No	Yes
High salt concentrations	No	Yes

steps, sometimes called preinitiation steps, that together permit rapid initiation when the triphosphates are added.

With a T7 DNA molecule as template, two kinds of enzyme complex have been observed: a tight one with a half-life of many hours that involves up to eight enzyme molecules, and a weak one with a half-life of less than a second that may involve as many as 1500 enzyme molecules. Based on the number of promoter sites per genome, estimated independently, and the number and variety of polymerase complexes observed, the following model for template binding may be suggested.

(i) Primary binding is manifested randomly along the length of the DNA in very weak and transient complexes; in the unlikely event of enzyme excess, these complexes begin to saturate the DNA.

(ii) Tight but labile complexes then form at or near the promoter site in a region that may be considered a preinitiator zone, in effect a staging area for initiation[11]. Studies with phage λ had suggested the site of promoter mutations are separated from the starting point of RNA transcription (Fig. 10-1)[12]. The size of a preinitiator zone in phage λ was judged to be large enough for several RNA polymerase molecules to line up. More recent biochemical studies[12a] have defined the location and sequence of the operator region in DNA which is bound by the λ repressor protein; it proves to be the same site bound and competed for by RNA polymerase. The site is 33 base pairs long and immediately precedes the start of the N gene (Fig. 10-1). It is not large enough to accommodate more than one poly-

11. Schäfer, R., Zillig, W. and Zechel, K. (1972) *Polymerization in Biological Systems*, Ciba Foundation Symposium, Elsevier, N.Y. p. 41; (1973) *EJB* **33**, 207.
12. Blattner, F. R., Dahlberg, J. E., Boettiger, J. K., Fiandt, M. and Szybalski, W. (1972) *NNB* **237**, 232.
12a. Maniatis, T., Ptashne, M., Barrell, B. G. and Donelson, J. (1974) *Nat.* **250**, 394.

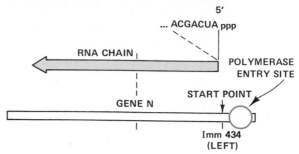

FIGURE 10-1
A map of the leftward promoter region of phage λ. This
diagram indicates the area for entry of RNA polymerase
molecules to start transcription at the initiation site with
the sequence indicated by the transcript (see text).

merase molecule. A most remarkable feature of the sequence is the
presence of three different types of 180°-rotational symmetry; these
may be devices designed to attract regulatory proteins, in this
instance repressor and polymerase, and a *Hin* d restriction endonuclease (Chapter 9, Section 5) as well.

(iii) Conformational changes in the enzyme convert a tight but
labile association with the template into a more stable one and poise
it for rapid initiation when it reaches the starting point.

(iv) Several points for initiation may be identified in a promoter
region; in T7, there appear to be two starts with ATP and one with
GTP[13].

Multiple starts in the promoter region may make for more efficient
use of the template. The region transcribed in advance of the true
initiator point would, of course, be superfluous. Such extensions
at the 5' end can be removed by cleavage or ignored in ribosomal
translation.

The promoter region refers loosely to the section of DNA that
controls initiation of RNA synthesis. This region needs to be studied
more precisely with reference to the binding sites and the sequence
for starting transcription. One approach to determining the actual
sequence of nucleotides in promoter regions is to isolate, for analysis,
the DNA covered with RNA polymerase molecules, taking advantage
of the protection against DNase degradation afforded by the enzyme
cover. This technique complements others which are described in
Chapter 11, Section 1.

Two features seem important in considering the structure of the
promoter region. One is a specific sequence recognized by RNA
polymerase; another is a secondary structure that will facilitate
opening of the duplex to expose the template strand for transcription. Particularly pertinent is evidence that the superhelical circular
form has greater template efficiency than the nicked, relaxed form,

13. Chamberlin, M. in *The Enzymes* (P. D. Boyer, Ed.) Academic Press, N.Y. (1974) *ARB* **43**, 721;
Dunn, J. J. and Studier, F. W. (1973) *PNAS* **70**, 1559; Minkley, E. G. and Pribnow, D. (1973)
JMB **77**, 255.

both in vitro[14] and in vivo[15]. The negatively twisted DNA with its deficiency of base pairing per unit length is, according to one view[16], energetically favorable to denaturation. The susceptibility of a DNA region to become unwound may be a crucial factor influencing the rate of a transcriptional start.

How does the binding of core polymerase differ from that of the holoenzyme? What is the function of σ? It had been considered when σ was discovered that there was a family of units serving as adaptors in the recognition of promoter sequences. A particular σ might seek out a specific sequence and pry that duplex region apart so that the core polymerase could initiate transcription. However, σ lacks any demonstrable affinity for DNA or DNA-like adsorbents, such as phosphocellulose. Furthermore, binding studies of the core polymerase with a T7 DNA molecule have shown that about 1500 enzyme molecules are involved in complexes with a half-life as long as 20 minutes. It would seem then that the function of σ is to prevent the core polymerase from being dissipated in tight complexes at non-specific sites all over the chromosome. It reduces this nonspecific binding a thousandfold. Instead of recognizing DNA, σ binds the core polymerase and may serve as an allosteric effector in the holo-enzyme[17]. It may provide the holoenzyme with a structure favoring formation of a tight complex specifically in the promoter staging area for initiation. Soon after chain elongation is underway, σ dissociates from the complex and may then combine with a free core enzyme to form a new holoenzyme[17].

The β' subunit is thought to function in the primary binding step. The β subunit serves in initiation and elongation and, as judged from reconstitution studies, contains sites for specific association with the α subunit. The action of the α subunits is as yet undetermined.

II, III INITIATION AND ELONGATION OF CHAIN

Experimental studies of the initiating event have depended on monitoring chain elongation. The first diester bond formed can be observed through exchange of ^{32}P-labeled PP_i into the β, γ positions of a triphosphate, for example UTP:

$$(DNA) \cdot ppp \, \frac{A}{G} + pppU \rightarrow (DNA) \cdot ppp \, \frac{A}{G} \, pU + PP_i$$

$$(DNA) \cdot ppp \, \frac{A}{G} \, pU + {}^{32}P^{32}P_i \rightarrow (DNA) \cdot ppp \, \frac{A}{G} + {}^{32}p^{32}ppU$$

Thereafter, isolation of the nascent RNA chains relies on identifying a 5' terminus containing the original triphosphate. Through alkaline

---

14. Hayashi, Y. and Hayashi, M. (1971) B. **10**, 4212.
15. Puga, A. and Tessman, I. (1973) JMB **75**, 83.
16. Bauer, W. and Vinograd, J. (1970) JMB **47**, 419; (1970) **54**, 281; Davidson, N. (1972) JMB **66**, 307.
17. Burgess, R. R., Travers, A. A., Dunn, J. J. and Bautz, E. K. F. (1969) Nat. **221**, 43; Travers, A. A. and Burgess, R. R. (1969) Nat. **222**, 537.

hydrolysis adenosine or guanosine tetraphosphate is generated from the terminus:

$$ppp \, {A \atop G} \, pXpYpZ \xrightarrow{\text{OH}^-} ppp \, {A \atop G} \, p + Xp + Yp + Z.$$

Reinitiations are prevented by one of several agents, such as high salt concentrations, low temperatures, rifampicin, or heparin; these agents do not interfere with chain elongation.

RNA polymerase has a specific binding site for the initiating triphosphate and another for those added during elongation (Table 10-5). The initiating site is specific for ATP and GTP; the only sig-

TABLE 10-5
Triphosphate binding sites of RNA polymerase

	Initiation	Elongation
Specificity	ATP, GTP	All four
Analogs	Only ITP	All that base-pair
Mono- and diphosphates	AMP, ADP	No
Dinucleotides	Yes	No
Triphosphate concentration	High; $> 500 \ \mu$M	Low; 5-20 μM
Mg^{2+} required	No	Yes
Half-life	0.2 second	0.02 second

nificant exceptions are AMP, ADP, and dinucleotides[18]. Whereas affinities of AMP and ADP are greatly reduced and these nucleotides are not extended in chain growth, dinucleotides, particularly those with A or G at the 3′ end, have a relatively strong affinity and are extended. Mg^{2+} is not required. By contrast, the elongation site depends on Mg^{2+}, absolutely requires a ribonucleoside triphosphate, and binds any one of the four. Also a variety of analogs that base-pair with the template are bound in this site, including the triphosphates of the potent antibiotics formycin and tubericidin, which compete with ATP (Chapter 7, Section 14).

The initiation site requires extraordinarily high concentrations of ATP and GTP, over 0.5 mM. For the elongation site, the apparent K_m values for all four ribonucleoside triphosphates are two orders of magnitude lower. Is there any rationalization for this remarkable distinction in affinity for the two sites? The high levels of triphosphate required for initiation may reflect the absence of a still unrecognized initiation factor, or it may be a regulatory device. Since physiological concentrations of ATP and GTP are in the range of 1 to 3 mM[19], significantly lower values should discourage RNA synthesis, and indirectly, DNA synthesis too. However, once the commitment to RNA synthesis is made with insertion of the initi-

18. Minkley, E. G. and Pribnow, D. (1973) *JMB* **77**, 255.
19. Mathews, C. K. (1972) *JBC* **247**, 7430.

ating triphosphate, then the chain is completed even at low triphosphate concentrations. This unique triphosphate initiation site is a feature that most clearly distinguishes RNA polymerase from the DNA polymerases.

It is difficult to assess the fidelity of chain initiations, but correctness of base-pairing in chain elongation can be gauged from the transcription of defined homopolymer and copolymer templates. No errors have been observed, even though the experiments would detect them in a frequency of one per several thousand. Base-pair selection by the enzyme depends on the proper size and shape of a Watson-Crick base pair. Remarkably, the base analog 7-deazanebularin[20] (Fig. 10-2), as a triphosphate, is accepted by RNA polymerase

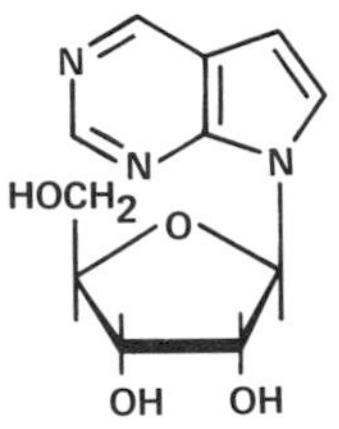

7 - Deazanebularin

FIGURE 10-2
Deazanebularin, an analog of both ATP and GTP for RNA polymerase despite the lack of amino and keto groups for hydrogen-bonded base pairing.

as substitute for both GTP and ATP, despite the failure to form hydrogen bonds with cytosine and the lack of the 6-amino group for forming one of the two hydrogen bonds for the A–U base pair. Incorporation of the analog must be ascribed to its correct size for fitting into a base-pair niche on the enzyme.

Two notable exceptions to proper template direction and template dependence are reactions reminiscent of DNA polymerase I and described as reiterative and de novo syntheses. In reiterative transcription[21], a single residue in the template is repeatedly copied to produce a homopolymer. A sequence of thymines in DNA will generate poly A if the nucleoside triphosphate needed to complement the base following thymine is not supplied in the medium. For example, TTTTTTC will generate a long chain of poly A from ATP provided GTP is excluded. Although a tract of residues in the template favors reiteration, even a single residue will suffice. The alternating copolymer d(T–C) directs the synthesis of poly G when ATP is omitted.

In de novo synthesis, RNA polymers are formed by RNA polymerase in the absence of added template. After a significant lag period and in the presence of Mn^{2+}, the homopolymer pair poly A· poly U is formed when ATP and UTP are substrates[22], and the alternating copolymer poly (I-C) when ITP and CTP[23] are substrates.

20. Ward, D. C. and Reich, E. (1972) *JBC* **247**, 705.
21. Chamberlin, M. and Berg, P. (1964) *JMB* **8**, 708; Adman, R. and Grossman, L. (1967) *JMB* **23**, 417.
22. Smith, D. A., Ratliff, R. L., Williams, D. L. and Martinez, A. M. (1967) *JBC* **242**, 590.
23. Krakow, J. S. and Karstadt, M. (1967) *PNAS* **58**, 2094.

Although the enzyme employed is pure, judged by the best available standards, rather large amounts of enzyme are required. In view of the known capacity of the enzyme to carry out reiterative synthesis, it is difficult to rule out the possibility that it is a contaminating fragment of RNA on the enzyme that serves as a "nucleating" template.

After polymerization of about ten nucleotide residues, σ is released. This event represents a drastic change in state of the enzyme. As a holoenzyme it has been geared for *site selection and initiation*. Once chain growth is underway, loss of σ and appearance of a nascent RNA chain begin a new role for the enzyme. It enters a *transcription state* with replicative functions closely resembling those of the DNA polymerases. What directs the transition between these states is not known. Competition between the nascent RNA chain and σ for a site in the active template-polymerase complex may account for the change.

The fundamental reason that transcription, unlike replication of a DNA duplex, is a conservative process may lie in the organization of the enzymes themselves or, in part at least, in the relative stability of secondary structures of polynucleotides (Fig. 10-3). The synthe-

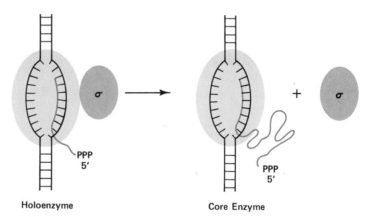

Holoenzyme   Core Enzyme

FIGURE 10-3
Diagram of a transcription bubble at an early stage in a DNA-RNA hybrid and at a later stage in which the σ subunit has dissociated from the core enzyme.

sized RNA chain may peel away from the DNA because, in folding upon itself, a more stable structure than the RNA-DNA hybrid is produced. As judged by melting properties of model polymers, an RNA duplex is more stable than a DNA duplex, and an RNA-DNA hybrid is least stable.

Streptolydigin binds the β subunit and blocks elongation[24]. From this it may be inferred that β functions both in growth and initiation of chains. The role of the α subunit which binds β is not clear. Acti-

24. Riva, S. and Silvestri, L. G. (1972) *Ann. Rev. Microb.* **26**, 199.

nomycin D[25], an important and profound inhibitor of virtually all RNA polymerases (bacterial and animal), does not bind the enzyme. It interferes with the template by intercalating into a duplex at a G–C base pair and binding to the deoxyguanosine residue. The lesser effectiveness of actinomycin D as an inhibitor of DNA polymerase may be due to the greater dependence of RNA polymerase on a duplex state of the template.

IV TERMINATION OF CHAIN AND ENZYME RELEASE

Termination of transcription requires that the elongation process stop at some point on the DNA template and that the RNA product and enzyme be released. Exactly how this is accomplished is not well understood. Termination can be brought about in vitro at a specific site by a DNA sequence, by protein factors, or by high salt concentrations.

Two small in vitro transcripts of λ DNA similar to in vivo ones have been isolated, each with the sequence pUpUpUpUpUpUpA at the 3′ terminus of the chain[26]. Secondary structure of an unusual sequence in the DNA template or the RNA product may disrupt the active ternary complex, terminating synthesis and discharging the enzyme.

Termination has also been ascribed to specific proteins, the most prominently considered being rho (ρ)[27]. This protein decreases synthesis on phage DNA templates. In its presence, the transcription of λ and T7 DNA generated RNA products that seemed to correspond to early mRNA transcripts found in vivo. However, the influence of ρ on site-specific termination of T7 DNA transcription is still uncertain. One reason is that the RNA products do not correspond to the five discrete species of early mRNA found in vivo, representing transcription of the left 20 percent of the T7 chromosome. Another reason is that in vitro transcription in the absence of ρ yields an RNA molecule of 2.5×10^6 daltons that does correspond to the complete 20 percent of the "r" strand of T7 DNA, starting at the left end of the chromosome[28]. Furthermore this large transcript, subsequent to transcription, is cleaved into five mRNA molecules. The agent responsible for this posttranscriptional processing of mRNA, sometimes known as a "sizing factor", has been identified as RNase III[28], an enzyme known to degrade duplex RNA[29].

RNase III⁻ mutants of *E. coli* accumulate the large RNA which, upon isolation and treatment with purified RNase III, is converted to the proper mRNA fragments[30]. Presumably, regions of secondary

25. Sobell, H. M., Jain, S. C., Sakore, T. D. and Nordman, C. E. (1971) *NNB* **231**, 200.
26. Dahlberg, J. E. and Blattner, F. R. (1973) *Fed. Proc.* **32**, 664 abs.
27. Roberts, J. W. (1969) *Nat.* **224**, 1168; Goldberg, A. R. and Hurwitz, J. (1972) *JBC* **247**, 5637; Schäfer, R. and Zillig, W. (1973) *EJB* **33**, 215; Minkley, E. G., Jr. (1973) *JMB* **78**, 577.
28. Dunn, J. J. and Studier, F. W. (1973) *PNAS* **70**, 1559, 3296; Hercules, K., Schweiger, M. and Sauerbier, W. (1974) *PNAS* **71**, 840.
29. Robertson, H. D., Webster, R. E. and Zinder, N. D. (1968) *JBC* **243**, 82; Robertson, H. D. (1971) *NNB* **229**, 169.
30. Dunn, J. J. and Studier, F. W. (1973) *PNAS* **70**, 1559; Nikolaev, N., Silengo, L. and Schlessinger, D. (1973) *JBC* **248**, 7967.

structure in the mRNA provide signals for the specific cleavages. This kind of processing resembles that already known to form tRNA (Chapter 11, Section 4) and is also analogous to posttranslational division of cistron-length polypeptides into component proteins: the polio virus capsid proteins, the T4 head proteins, and the familiar zymogen conversions.

A factor with ρ-like properties, recently isolated from *E. coli*, is a protein called kappa (κ)[31]. It is a dimer of identical, 17,000 dalton polypeptides, which binds DNA at many sites, but appears to terminate transcription specifically at a few. As with ρ, neither the RNA product nor the enzyme are discharged at termination. Therefore, both ρ and κ act as blocking rather than termination factors. To account for product and enzyme release in vivo, these or additional factors need properties not observed thus far in vitro.

Elevated mono- and divalent cation concentrations promote terminations that frequently generate discrete transcripts of phage DNAs. Salt may act by reducing affinity of the enzyme for the template and thus enhance the influence of DNA sequences that induce termination. Salt also makes polymerase released from synthesis available for reinitiation by reducing the capacity of the RNA product to bind and sequester the polymerase.

Single-stranded templates. Single-stranded DNA and RNA are bound by the enzyme and transcribed by it. Even though it is clear that duplex DNA is the physiological template for the enzyme, the behavior of single-stranded polynucleotides and damaged DNA is very important to the interpretation of in vitro studies. In the interesting instance of the M13 virus, replication of single-stranded viral DNA is initiated by RNA polymerase (Chapter 8, Section 3).

Compared to duplex DNA, many more starts are made on single-stranded DNA and the chains produced are shorter. Masking by a DNA-binding protein, or by a viral capsid protein, may prevent initiation of RNA chains from occurring at biologically incorrect sites. In the transcription of single-stranded polynucleotides such as poly dT and poly rU, matching errors can be detected and are in fact rather frequent. Incorporation of GMP in place of AMP may be as high as one in thirty[32].

Since single-stranded polynucleotides displace σ from the holoenzyme, it is never clear in the interpretation of an experiment which form of the enzyme is involved and whether the polynucleotide occupies the template site, product site, or both. The outcome may also be uncertain in a competition between polymerases for binding nicked and terminal regions of duplex DNA. RNA polymerase binding is very strong and has been found to prevail over that of DNA polymerase I[33].

31. Schäfer, R. and Zillig, W. (1973) *EJB* **33**, 201.
32. Strniste, G. F., Smith, D. A. and Hayes, F. N., (1973) *B.* **12**, 603.
33. Berg, P., Kornberg, R. D., Fancher, H. and Dieckmann, M. (1965) *BBRC* **18**, 932.

Factors that regulate and stimulate transcription.   In its functioning to transcribe a specific region of the chromosome, RNA polymerase must be guided and regulated by auxiliary factors; several of
them have already been recognized at this early stage of exploration
and will be mentioned here.

(i) An additional protein may function as a subunit by binding to
and directing the enzyme. This is suggested by events in phage T4
and λ infections[34] (see Section 4 of this chapter) and possibly in
sporulation[35].

(ii) Modifications of the polypeptide chains of the subunits
observed after T4 infection[36] may alter transcriptional specificity.

(iii) A positive regulatory protein promotes transcription of a
particular DNA region, as in the case of the cyclic AMP receptor
protein for the *lac* operon[37].

(iv) A negative regulatory protein blocks transcription by binding
a particular DNA region. Repressors of the *lac* operon and λ prophage
are examples.

(v) Several factors, probably proteins, increase the amount and
rate of RNA synthesis in vitro[38]. Their significance has not been
evaluated as yet.

4. RNA Polymerase in Phage Infections

Introduction of a new genome upon phage infection often requires
major adaptations to assure the specific, rapid, and temporally
correct transcription needed for production of the phages. T7, T4
and λ infections introduce novel transcriptional devices and controls. As the studies of phage DNA replication emphasize (Chapter
8), the phage genomes in their simplicity and variety can become a
showcase for host cell activities; in this instance the transcriptional
operations, endogenous and phage-specific, can be examined.

T7 infection[39].   Initially the host RNA polymerase is used to transcribe the "left" 20 percent of the T7 chromosome which includes
gene 1, the determinant of T7 RNA polymerase. Translation of this
early message produces T7-specific RNA polymerase which then
transcribes the rest of the T7 chromosome.

The most remarkable and distinctive feature about the induced
enzyme, compared to the host polymerase, is its simplicity (Table

34. Stevens, A. (1972) *PNAS* **69**, 603; (1974) *B.* **13**, 493; Snyder, L. (1973) *NNB* **243**, 131; Geor-
    gopoulos, C. P. (1971) in *The Bacteriophage Lambda* (A. D. Hershey, Ed.) CSHL, p. 639.
35. Linn, T. G., Greenleaf, A. L., Shorenstein, R. G. and Losick, R. (1973) *PNAS* **70**, 1865.
36. Goff, C. G. and Weber, K. (1970) *CSHS* **35**, 101; Schachner, M., Seifert, W. and Zillig, W.
    (1971) *EJB* **22**, 520; also preceding paper; Goff, C. G. (1974) *JBC* **249**, 6181.
37. Emmer, M., deCrombrugghe, B., Pastan, I. and Perlman, R. (1970) *PNAS* **66**, 480.
38. Ghosh, S. and Echols, H. (1972) *PNAS* **69**, 3660; Ramakrishnan, T. and Echols, H. (1973)
    JMB **78**, 675; Cukier-Kahn, R., Jacquet, M. and Gros, F. (1972) *PNAS* **69**, 3643.
39. Chamberlin, M. and Ring, J. (1973) *JBC* **248**, 2235.

TABLE 10-6
Comparison of Host and T7-induced RNA polymerases

Distinctive properties	Host	T7-Induced
Molecular weight	480,000	107,000
Subunits	5	1
Initiating triphosphate	ATP, GTP	GTP
Templates		
all bacterial DNAs	yes	no
all phage DNAs	yes	no
T7 DNA	yes	yes
region of T7 DNA	"left 20 percent"	"right 80 percent"
poly (dG), poly (dA), poly (dT)	yes	no
poly (dC)	yes	yes
Anomalous de novo synthesis	yes	no
Mn^{2+} substitutes for Mg^{2+}	yes	no
Inhibition by rifampicin, streptolydigin, streptovaricin	yes	no
Inhibition of chain growth by heparin, high salt, poly (rU)	no	yes
Efficient termination	no	yes
Turnover number, nucleotides/sec	~50	~250

10-6). Within a single polypeptide it retains the capacity to start and terminate chains, as well as to transcribe faithfully a DNA template. By exceedingly rapid chain growth and efficient termination it may account, in part at least, for the short phage growth cycle. Rifampicin insensitivity of the T7 polymerase distinguishes it from the host enzyme[40]. However, more hydrophobic rifampicin derivatives, such as AF/ABDP, containing a benzyl substituent, bind at high concentrations to the T7 enzyme and inhibit its initiation actions.

How, despite its severe streamlining, does the T7 enzyme manage the job as well as does the very complex host enzyme? The answer may be in the relatively simpler demands for intiation and termination. Studies of template preference have indicated that starts may be limited to a specific promoter with a cytosine residue in a pyrimidine-rich sequence, and the termination of transcripts seem limited to the extreme right end of the chromosome. The enzyme seems to be specially and narrowly designed for transcribing the T7 template. An analogous template specificity has been observed in the behavior of the phage T3-induced RNA polymerase for T3 DNA[41].

T4 infection. Unlike T7, T4 phage does not induce a novel RNA polymerase. Instead it exploits the host enzyme by altering it[42], as

40. Chamberlin, M. and Ring, J. (1973) *JBC* **248**, 2245.
41. Chakraborty, P. R., Sarkar, P., Huang, H. H. and Maitra, U. (1973) *JBC* **248**, 6637.
42. Goff, C. G. and Weber, K. (1970) *CSHS* **35**, 101; Schachner, M., Seifert, W. and Zillig, W. (1971) *EJB* **22**, 520; also preceding paper; Goff, C. G. (1974) *JBC* **249**, 6181.

well as through interactions with additional new protein subunits dictated by the phage[43]. Four such proteins, ranging in molecular weight from 10,000 to 22,000, were found to be associated with the isolated RNA polymerase midway in the infection. Two of the proteins have been identified as products of genes 33 and 55 which are known to be essential for late mRNA synthesis and phage maturation. It is not yet understood what properties these two proteins impart to the host enzyme that give it the specificity for transcription of late T4 messages, nor is it known what the genetic origins and enzymatic properties of the other two proteins are.

The alterations observed in the host enzyme following T4 infection include ADP-ribosylation of the α subunits[44] and changes in the peptides of a tryptic digest of all four subunits[45]. But the significance of these changes is still unknown.

λ infection.   As with T4 infection, proper transcription of λ DNA is carried out by the host polymerase directed by the phage gene N[46]. Without the gene N protein, transcription is initiated but then stops prematurely after about 1000 residues, near the site where σ was observed to act. Although the gene N protein has not been isolated, its existence and binding to the β subunit of polymerase has been inferred from the bacterial gro N mutation (see Chapter 8, Section 7). The effect of a mutation in gene N can be reversed by a mutation in the bacterium, called gro N, localized to the β subunit of RNA polymerase. Presumably an altered β subunit, as judged by rifampicin resistance, can bind an altered gene N product and carry out λ transcription.

5. Eukaryotic RNA polymerases[47]

The first instance of DNA-directed RNA synthesis from ribonucleoside triphosphates was observed with a preparation of rat liver nuclei[48]. Most of what we know about the regulation and operation of transcription is based on the isolation and characterization of RNA polymerases from bacterial systems. Are the patterns of enzyme organization and function in prokaryotes reflected in eukaryotic species?

Since a method for obtaining a useful quantity of pure eukaryotic enzymes has been available for only a short time, less is known about these. The enzymes have been purified from calf thymus[49], rat liver[50],

43. Stevens, A. (1972) PNAS **69**, 603; (1974) B. **13**, 493; Snyder, L. (1973) NNB **243**, 131; Horvitz, H. R. (1973) NNB **244**, 137.
44. Goff, C. G. (1974) JBC **249**, 6181.
45. Schachner, M. and Zillig, W. (1971) EJB **22**, 513.
46. Georgopoulos, C. P. (1971) in The Bacteriophage Lambda (A. D. Hershey, Ed.) CSHL, p. 639.
47. Kedinger, C., Gissinger, F., Gniazdowski, M., Mandel, J.-L. and Chambon, P. (1972) EJB **28**, 269; also succeeding papers; Weaver, R. F., Blatti, S. P. and Rutter, W. J. (1971) PNAS **68**, 2994; Transcription of Genetic Material (1970) CSHS **35**.
48. Weiss, S. and Gladstone, L. (1959) JACS **81**, 4118.
49. Kedinger, C., Gissinger, F., Gniazdowski, M., Mandel, J-L. and Chambon, P. (1972) EJB **28**, 269; also succeeding papers.
50. Seifart, K. H. (1970) CSHS **35**, 719.

TABLE 10-7

Comparison of calf thymus RNA polymerases

	A	B	C
Location	Nucleolus	Nucleoplasm (Euchromatin)	
Relationship to RNA polymerases I and II of rat liver nuclei	I	II	III
α-Amanitin inhibition	No	Yes	
Subunit structure:	200,000	214,000	
	126,000	140,000	
	51,000		
	44,000	34,000	
	25,000	25,000	
	16,500	16,500	
Proportions	1:1:1:1:2:2		
Molecular weight	500,000		
Elution from DEAE cellulose by ammonium sulfate	0.10–0.12 M	0.20–0.24 M	
Heat lability: 45° for 2 min percent initial activity	10	90	
Preferred DNA template	Duplex, homologous	Denatured, heterologous	
Optimal assay condition	Mg^{2+}	Mn^{2+}, 0.4 M KCl	
pH optimum	8.5	7–7.5	
Products	rRNA	mRNA	4S and 5S RNA

and from human tissue culture cell lines[51]. Calf thymus RNA polymerases have a complex set of subunits. Two types, A and B, have been distinguished (Table 10-7), both restricted to the nucleus. Based on their distribution in fractionation, type A is localized in the nucleolus and is not inhibited by α-amanitin, a drug that blocks mRNA synthesis in the cell. Type B is found in the nucleoplasm, associated with the diffuse, active chromatin; it is inhibited by α-amanitin.

The physiologic importance of the B type may be inferred from the observation that cells which gain resistance to α-amanitin have also developed a type B polymerase resistant to the drug[52]. Which of the subunits of B is the target of α-amanitin inhibition and how A and B may be structurally and functionally related are not known.

The animal RNA polymerases from all sources show the same basic requirements for a DNA template and the four complementary ribonucleoside triphosphates. Protein stimulatory factors[53] have been isolated from calf thymus, rat liver, and ascites tumor cells, and human cells in culture. These protein factors enhance transcriptional rates of only the type B enzyme acting on duplex DNA. Whether these factors will prove to be analogous to σ and other proteins that influence bacterial polymerases is not yet known. Despite

51. Sugden, B. and Keller, W. (1973) JBC 248, 3777.
52. Chan, V. L., Whitmore, G. F. and Siminovitch, L. (1972) PNAS 69, 3119.
53. Sugden, B. and Keller, W. (1973) JBC 248, 3777; DiMauro, E., Hollenberg, C. P. and Hall, B. D. (1972) PNAS 69, 2818; Lee, S-C. and Dahmus, M. E. (1973) PNAS 70, 1383.

numerous attempts to implicate histones and other agents as regulatory factors, nothing definitive can yet be said about any of the stages of the initial transcription process. What is already quite clear is that the final messenger RNA in mammalian cells, unlike that in bacteria, is pared down from a much larger transcript which is modified by addition of a poly A tail[54].

Among the known DNA-dependent RNA polymerases, the enzyme isolated from mitochondria of *Neurospora crassa*[55] is of special interest because of its size. It is a single polypeptide chain of only 64,000 daltons and the smallest RNA-polymerase known. It is sensitive to rifampicin, resistant to α-amanitin, and shows a template preference for native or denatured mitochondrial DNA over calf thymus DNA. In its simplicity, this polymerase resembles the phage T7 enzyme and may reflect the fact that both transcribe a relatively small genome of limited complexity. There is also evidence that in vivo transcription of mitochondrial DNA is not asymmetric; thus the selection of a specific strand and sequence is avoided. Instead, later processing steps are depended upon to choose those regions of the mitochondrial transcript to be retained for translation.

Once the polymerases of a cell are characterized, it becomes important to determine how their levels are regulated. A large increase in mitochondrial DNA polymerase occurs in response to treatment with ethidium bromide (see Chapter 6, Section 4). Presumably, injury to _mitochondrial_ DNA induces a response from _nuclear_ genes for increased production of mitochondrial DNA polymerase. It is important to know how the RNA polymerases are regulated and their response to external agents, such as ethidium bromide.

54. Darnell, J. E., Jelinek, W. R. and Molloy, G. R. (1973) *Science* **181**, 1215.
55. Küntzel, H. and Schäfer, K. P. (1971) *NNB* **231**, 265.

11

Synthesis of Genes

1. Determination of Nucleotide Sequence[1]

The sequence of nucleotides in a cell's DNA (i) spells out the primary sequence of all of that cell's proteins, (ii) contains the information that regulates the synthesis of proteins, and (iii) determines the survival of the DNA against the immunity defenses of the cell. Thus, the determination of DNA sequence is of major importance for understanding biology.

Progress in DNA sequencing had lagged behind that for protein, and RNA, for several reasons.

(i) Lack of small DNA molecules (the size of proteins and transfer RNA).

(ii) Lack of nucleases to produce specific cleavages and overlapping fragments.

---

1. Murray, K. and Old, R. W. (1974) *Prog. N.A. Res.* **14**, 117; Wu, R., Donelson, J., Padmanabhan, R. and Hamilton, R. (1972) *Bulletin de l'Institut Pasteur* **70**, 203; Barrell, B. G. (1971) **in** *Procedures in Nucleic Acids* (Cantoni, G. L. and Davies, D. R., Eds.) Harper and Row, New York **2**, 751.

(iii) Lack of partially digested sequences (such as RNA yields, due to secondary structures).

(iv) Limited variety of nucleotide constituents as compared to proteins.

(v) Lack of reliable chemical degradations.

(vi) Relatively low specific radioactivity attainable in DNA as compared to RNA.

Fortunately, recent discoveries of specific endonucleases with a capacity to produce defined chains of manageable size, and further development of methods for fractionation and analysis of oligonucleotides have brightened prospects for DNA sequence determination.

A. Production of Defined Chains by Specific Endonuclease Cleavage

Even the smallest DNAs, ϕX174 and SV40, are about 5500 nucleotides or base pairs long, nearly 100 times longer than the longest RNA sequence yet determined. The DNAs need to be reduced to defined fragments of about 50 to 100 nucleotides long, of "tRNA size", for convenient sequence analysis. By using endonucleases that cleave at specific sequences, large DNA molecules can be broken down first to "ϕX size" and then to still smaller fragments. The following are examples of endonucleases that have been used:

Size in   "λ"   "ϕX"   "tRNA"
base pairs: 50,000 ⟶ 5000 ⟶ 1000 ⟶ 100
                          ↑   ↑   ↑
Endonuclease: *Eco* RI   *Hin* d   T4 Endo IV

The RI restriction enzyme of *E. coli* (Chapter 9, Section 5) is a specific endonuclease that cleaves a DNA duplex to "ϕX size" by introducing two single-strand, staggered breaks in a specific hexanucleotide sequence (Table 9-5). Based on a random distribution of a specific hexanucleotide, the RI sequence should occur about once per 4000 base pairs, and this frequency of cleavage has been found in some animal and bacterial DNAs[2]. A single break is produced in SV40 DNA[3] with 5000 base pairs. However, only 5 breaks are introduced into λ DNA and none in T7[4]; presumably some chromosomes have evolved by selection to avoid cleavage by the RI enzyme.

Restriction endonucleases from *H. influenzae* and *H. parainfluenzae* (Chapter 9, Section 5) degrade ϕX size DNA to fragments of intermediate length. T7 DNA with about 40,000 base pairs can be

2. Mertz, J. E. and Davis, R. W. (1972) *PNAS* **69**, 3370.
3. Morrow, J. F. and Berg, P. (1972) *PNAS* **69**, 3365; Mulder, C. and Delius, H. (1972) *PNAS* **69**, 3215.
4. Allet, B., Jeppesen, P. G. N., Katagiri, K. J. and Delius, H. (1973) *Nat.* **241**, 120; Davis, R. W., personal communication.

split into 40 duplex fragments with an average length of 1000 base pairs. SV40 is cleaved by an *H. influenzae* preparation (containing Hin *d* II and III enzymes) at 11 sites[5] and ϕX174 at 13 or more[6], whereas the M13 circular duplex is split only once[7]. On the other hand, an *H. parainfluenzae* preparation (containing *Hpa* I and II) splits SV40 at four sites[5] but M13 at nine places[7]. In each instance the distinctive cleavage fragments are readily separable by gel electrophoresis.

For producing fragments in the tRNA size range so that total sequence may be analyzed, T4 endonuclease IV (Table 9-2) has been useful[8]. The enzyme prefers single-stranded DNA and generates pC-ended chains of varying length. To maximize the specificity and molar yields of fragments, digestion of ϕX174 DNA was carried out at low temperature and high ionic strength. Hairpin regions in the DNA under these conditions limit the number of cleavage sites. Fractionation of such digests in urea-polyacrylamide gel electrophoresis showed that over 90 percent of the DNA was in fragments larger than 600 residues, but discrete fragments from 15–150 residues long were also liberated and were well resolved on the gel. The complete sequence of one of these ϕX174 fragments, 48 residues long, was determined (see below).

The nucleases just discussed are only three examples among many possibilities for preparing small, defined DNA chains by specific endonucleolytic cleavage. Other specific endonucleases are available, and by varying incubation conditions and the sequence of successive applications of different nucleases their usefulness can be multiplied. Still other means can be used for defining specific DNA regions, and these will be cited.

B. Production of defined chains by other methods

In addition to the use of specific endonucleases, there are other approaches to obtaining defined DNA fragments.

(i) *Use of defined oligonucleotides as priming fragments for replication*[9]. Oligonucleotides of known sequence, prepared by chemical synthesis, by polymerase transcription from a DNA or RNA template, or by specific endonuclease cleavage, can be annealed to a DNA chain and can then serve as primers for extension by DNA polymerase. By labeling different deoxynucleoside triphosphate substrates the sequence copied by polymerase can be determined directly. Graded extents of synthesis can be achieved by synchro-

5. Danna, K. J., Sack, G. H., Jr. and Nathans, D. (1973) *JMB* **78**, 363; Sambrook, P. A., Sugden, B. and Sambrook, J. (1973) *B*. **12**, 3055.
6. Edgell, M. H., Hutchison, C. A. III and Sclair, M. (1972) *J. Virol.* **9**, 574.
7. Tabak, H. F., Griffith, J., Geider, K., Schaller, H. and Kornberg, A. (1974) *JBC* **249**, 3049.
8. Ziff, E. B., Sedat, J. W. and Galibert, F. (1973) *NNB* **241**, 34.
9. Sanger, F., Donelson, J. E., Coulson, A. R., Kössel, H. and Fischer, D. (1973) *PNAS* **70**, 1209.

nizing the start of replication, and regulating the time and temperature of incubation. By inserting ribonucleotides[10], or other analogs, in place of the standard deoxynucleotides, specific points are provided for subsequent cleavage of the synthetic chain by alkali or specific nucleases.

(ii) *Use of a protein to mask specific regions of DNA for protection against degradation by nonspecific endonucleases.* This technique has been used to determine the sequences of: (*a*) the *lac* operator[11] of E. coli (a specific region of DNA involved in regulation of gene expression) and phage λ immunity regions[12], which are bound and masked by their respective, highly specific repressor proteins, (*b*) the duplex DNA promoter regions of phages λ[13], fd[14], and φX174[15], bound and masked by RNA polymerase, and (*c*) a ribosome-binding region on φX174 DNA, corresponding to a coat-protein initiation site[16].

(iii) *Use of the endonucleases acting only on single strands at denatured regions in a duplex.* Specific cleavages have been introduced by S_1 nuclease (see Table 9-2) in two small regions of SV40 DNA which are susceptible in the superhelical but not in relaxed form[17]. This type of "denaturation cleavage" may prove to be specific enough for preparing defined duplex fragments. Exonuclease I digestion (Table 9-2) of single-stranded φX174 DNA, after an initial nicking of the circle by endonuclease, removes 99 percent of the DNA, leaving only one or two duplex hairpins of about 20 base pairs for examination[18].

C. Total Sequence Determination[19]

Analysis of certain φX174 oligonucleotides obtained in high yield and purity from initial digestion of ^{32}P-DNA by T4-endonuclease IV (Table 9-2) illustrates some of the methods of sequence determination[20]. Characterization of a 48-residue oligonucleotide isolated by gel electrophoresis (Fig. 11-1) depended on depurination and on analysis of the products of partial digestion with exonucleases. Oligonucleotides, 5 to 30 residues long, are displayed as a finger-

10. Jackson, J. F., Kornberg, R. D., Berg, P., RajBhandary, U. L., Stuart, A., Khorana, H. G. and Kornberg, A. (1965) *BBA* **108**, 243; Salser, W., Fry, K., Brunk, C. and Poon, R. (1972) *PNAS* **69**, 238.
11. Gilbert, W., and Maxam, A. (1973) *PNAS* **70**, 3581.
12. Maniatis, T. and Patshne, M. (1973) *PNAS* **70**, 1531; Pirrotta, V. (1973) *NNB* **244**, 13; Maniatis, T., Patshne, M., Barrell, B. G. and Donelson, J. (1974) *Nat.* **250**, 394.
13. Le Talaer, J.-Y. and Jeanteur, Ph. (1971) *PNAS* **68**, 3211.
14. Okamoto, T., Sugiura, M. and Takanami, M. (1972) *NNB* **237**, 108; Heyden, B., Nüsslein, C. and Schaller, H. (1972) *NNB* **240**, 9.
15. Chen, C.-Y., Hutchison, C. A. III, and Edgell, M. H. (1973) *NNB* **243**, 233.
16. Robertson, H. D., Barrell, B. G., Weith, H. L. and Donelson, J. E. (1973) *NNB* **241**, 38.
17. Beard, P., Morrow, J. F. and Berg, P. (1973) *J. Virol.* **12**, 1303.
18. Schaller, H., Voss, H. and Gucker, S. (1969) *JMB* **44**, 445.
19. Ziff, E. B., Sedat, J. W. and Galibert, F. (1973) *NNB* **241**, 34; Sanger, F., Donelson, J. E., Coulson, A. R., Kössel, H. and Fischer, D. (1973) *PNAS* **70**, 1209.
20. Ziff, E. B., Sedat, J. W. and Galibert, F. (1973) *NNB* **241**, 34.

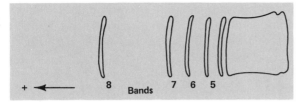

Polyacrylamide Gel Electrophoresis

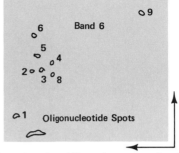

Ionophoresis and Homochromatography of Band 6

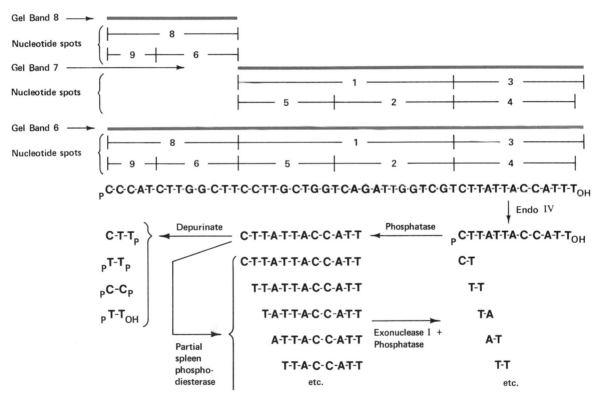

FIGURE 11-1

A scheme for total sequence determination of a 48-residue fragment of φX174 DNA. Fragments are first generated by endonuclease IV digestion, and then are fractionated into bands by polyacrylamide gel electrophoresis. Bands of oligonucleotides are eluted and digested more vigorously with endonuclease IV, and then are fractionated into oligonucleotide spots by ionophoresis (for base composition) and homochromatography (for size). Oligonucleotide spots are sequenced by the indicated procedures and others mentioned in the text. (Courtesy of Professor K. Murray.) (Ziff, E. B., Sedat, J. W. and Galibert, F. (1973) NNB **241**, 34.)

print (Fig. 11-1) using a two-dimensional, thin-layer system which had already been successfully applied to analysis of RNA sequences: first ionophoresis separated the oligonucleotides by their charge; next a chromatographic step separated them by their size.

DNA, depurinated by acid treatment, is readily hydrolyzed at the phosphodiester bonds between the apurinic sites and pyrimidine nucleotides, leaving clusters of pyrimidine-containing oligonucleotides. Analysis of these pyrimidine tracts was one of the first and remains one of the best analytic methods for DNA sequencing[21]. This procedure, applied to band 6 of Fig. 11-1, furnished essential information about overlaps of the component oligonucleotides as well as their pyrimidine composition.

Oligonucleotides recovered from the two-dimensional separation were analyzed by graded digestions with exonucleases acting in the $5' \rightarrow 3'$ direction (spleen phosphodiesterase) and the $3' \rightarrow 5'$ direction (exonuclease I, or snake venom phosphodiesterase).

The sequence determination of the ϕX174 fragment[22] just described was by a direct analysis of a ^{32}P-labeled DNA. Analyses of even greater sensitivity can be performed on *unlabeled* DNA in two different ways: (i) use of the DNA as a template, with DNA polymerase and labeled deoxynucleoside triphosphates, for replication, and a synthetic oligonucleotide as primer[23], and (ii) enzymatic ^{32}P-labeling of the termini, followed by graded exonuclease digestions (described later in this section).

D. Terminal Sequences of Defined Chains

The 3'- and 5'-terminal residues, and stretches of successive subterminal residues, have been analyzed using enzymatic polymerization[24] and phosphorylation[25]. *The 3'-terminal region can be analyzed by three methods.*

ANALYSIS OF 3'-TERMINAL REGION

(i) *Turnover catalyzed by T4 DNA polymerase*[24]. Nucleotides at the 3' terminus of a duplex are removed by the $3' \rightarrow 5'$ exonuclease action of the enzyme and immediately replaced with a labeled nucleotide by polymerase action if the suitable deoxynucleoside triphosphate is provided. This "turnover technique" has been successful in determining the terminal sequences at the cleavage site of an *H. influenzae* restriction endonuclease[24] and the terminal sequences of T7[26] and λ[27] DNA. Such labeled 3' termini can be used

21. Burton, K., Lunt, M. R., Petersen, G. B. and Siebke, J. C. (1963) *CSHS* **28**, 27; Mushynski, W. E. and Spencer, J. H. (1970) *JMB* **52**, 91, 107.
22. Ziff, E. B., Sedat, J. W. and Galibert, F. (1973) *NNB* **241**, 34.
23. Sanger, F., Donelson, J. E., Coulson, A. R., Kössel, H. and Fischer, D. (1973) *PNAS* **70**, 1209.
24. Englund, P. T. (1971) *JBC* **246**, 3269.
25. Weiss, B. and Richardson, C. C. (1967) *JMB* **23**, 405.
26. Price, S. S., Schwing, J. M. and Englund, P. T. (1973) *JBC* **248**, 7001.
27. Brezinski, D. and Wang, J. C. (1973) *BBRC* **50**, 398; Weigel, P. H., Englund, P. T., Murray, K. and Old, R. W. (1973) *PNAS* **70**, 1151.

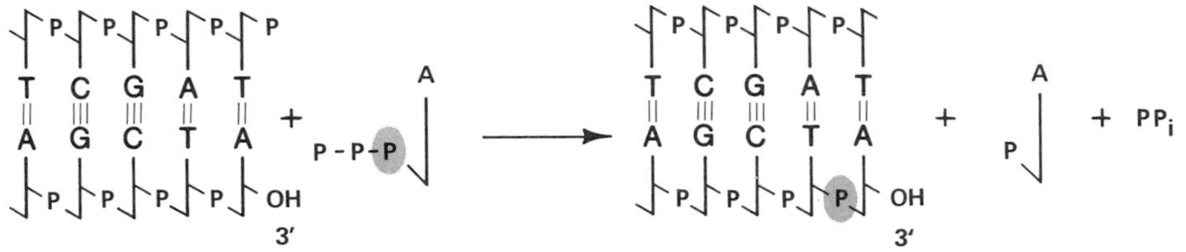

FIGURE 11-2
Turnover of the 3′ terminal sequence by phage T4 DNA polymerase. This result is the consequence of a 3′ → 5′ exonucleolytic removal of the terminus, followed by its replacement with an α-³²P-labeled dATP by polymerase action.

to determine longer, subterminal sequences by digestion with an endonuclease (e.g. pancreatic DNase) that produces a series of oligonucleotides, which overlap in size.

As illustrated in Fig. 11-2, a terminal deoxyadenylate of a sequence is released in the presence of $\alpha(^{32}P)$-dATP in the reaction mixture, and is replaced by (^{32}P) pA. If $\alpha(^{32}P)$-dTTP were present instead, pT would be incorporated in the subterminal position. By nearest-neighbor analyses (Chapter 4, Section 17), pA would be found linked to T, and pT to C. As expected from the sequence, the incorporation of pT would be inhibited by dATP and not by dCTP or dGTP (Table 11-1). From these facts, the 3′-terminal sequence pCpTpA in the DNA could be deduced. To distinguish between the two 3′ termini of a duplex, it may be necessary to separate the two strands for analysis of their complementary 5′ ends (see below).

TABLE 11-1
Determination of 3′-terminal sequences by enzymatic turnover, using controlled triphosphate additions[a]

Labeled triphosphate	Unlabeled triphosphate	Effect of unlabeled triphosphate	Sequence determined
A	none		TpA
A	T, G, or C	no effect	
T	none		CpT
T	A	inhibits	
T	G or C	no effect	
C	none		GpC
C	A or T	inhibits	
C	G	no effect	
G	none		ApG
G	A, T, or C	inhibits	

[a] Source: Englund, P. T. (1971) JBC 246, 3269.

(ii) *DNA polymerase extension after exonuclease digestion*. A string of nucleotides removed from the 3′ end of a *duplex* by exonuclease III exposes a template for replication by polymerase. The potential of this method, however, for sequence determination has not been realized because of the difficulty, compared to the turnover method, in obtaining graded and limited exonuclease digestion.

(iii) *Addition of ribonucleotides to 3′ ends of single strands*[28]. Deoxynucleotidyl transferase (Chapter 6, Section 8) can catalyze the addition of only one or two riboadenylate residues to the end of an oligodeoxynucleotide. With α-(^{32}P)-ATP, the sequence analysis of a synthetic octadeoxynucleotide proceeded in this way.

(a) The octanucleotide was partially degraded from the 3′ end with snake venom phosphodiesterase to give a spectrum of smaller-sized fragments.
(b) ^{32}P-riboadenylate was attached to each fragment.
(c) Fragments were separated on basis of size.
(d) Digestion with spleen phosphodiesterase released the 3′ terminus of each fragment as a 3′-deoxynucleotide.

Labeling a terminus also provides the opportunity for determining subterminal sequences, as mentioned for the turnover method, and 5′-terminal labeling (described next).

The structure of a *5′-terminal region*, if not inferred from a complementary 3′ end, can be determined in several ways.

ANALYSIS OF 5′-TERMINAL REGION

(i) *Labeling of the terminal nucleotide with ^{32}P*. This method, involving transfer from γ-(^{32}P)-ATP by polynucleotide kinase[29], can be helpful in all cases and is essential in some. The labeling identifies the 5′-terminal residue; subterminal sequences can be determined, as described for the 3′-labeled termini, by endonuclease cleavages that produce a series of overlapping oligonucleotides.

The successes with this method warrant attention to its special advantages. By using intensely labeled γ-(^{32}P)-ATP at a specific radioactivity in excess of 10^{11} counts per minute per micromole, the need for in vivo labeling can be avoided; or, double labeling with ^{33}P, and ^{3}H or ^{14}C may be used. Furthermore, labeling by this method can be delayed until the end of a series of degradative reactions.

This approach requires the 5′-hydroxyl group to be available and a preliminary phosphatase treatment may be needed. The known values for specific radioactivity of the ATP, and the mass of DNA phosphorylated provide an immediate determination of chain length; complete nuclease digestion of the DNA permits direct identification of the labeled terminal nucleotide. Polynucleotide kinase

28. Kössel, H. and Roychoudhury, R. (1971) *EJB* **22**, 271; see also (1971) *EJB* **22**, 310; (1971) *BBRC* **45**, 430.
29. Weiss, B. and Richardson, C. C. (1967) *JMB* **23**, 405; Murray, K. (1973) *BJ* **131**, 569; Bernardi, G. (1974) *Anal. Biocn.* **59**, 501.

acts on ribo- and deoxyribonucleotide chains of all sizes, down to and including a 3'-mononucleotide.

(ii) *Terminal dinucleotide by enzymatic digestion.* The 5' terminus labeled with ^{32}P as was just described can also be determined as the only dinucleotide remaining after total $3' \rightarrow 5'$ degradation with exonuclease I or T4 DNA polymerase (Table 9-1). It is also possible to utilize the $5' \rightarrow 3'$ exonuclease induced by *B. subtilis* phage SP3, which liberates dinucleotides successively from chains as short as tetranucleotides (Table 9-1).

(iii) *Replication of protruding 5' ends of a duplex.* The cohesive ends of phage DNAs were determined by stepwise extension of the 3' end, using DNA polymerase I and labeled deoxynucleoside triphosphates. A sequence of 12 nucleotides was identified in λ and 19 in phage 186 DNA[30] (Fig. 8-31); one 5' cohesive end may be inferred from the sequence of the other by complementarity. Other examples in which sequence determination by replication has been useful are the restriction endonuclease site[31] of *Eco* RI and the regulatory regions neighboring a tRNA gene (see Section 4 of this chapter).

E. Comments and Perspectives

Progress in DNA sequencing is quickening. Specific endonuclease cleavages and the use of proteins to bind and protect regions of DNA generate well-defined fragments. Electrophoretic methods are effective for separation and two-dimensional ionophoretic and chromatographic methods can be used to analyze them. The relatively low specific radioactivity of in vivo-labeled DNA can now be circumvented by intensive labeling of the 3' and 5' termini, and this makes it possible to deduce extended subterminal sequences. Iodination with ^{125}I shows promise for fingerprinting studies and possibly for sequence determination[32]. Polymerase replication and transcription of defined regions provide independent approaches, as do others mentioned. The prospects are now bright for stating the nucleotide sequences of chromosome maps outlined initially by heteroduplex electron microscopy, denaturation analyses, and genetic techniques.

In anticipation that commercial sources of enzymes and radioactive nucleotides will hasten the widespread use of sequencing methods, some cautions in their use and interpretation should be mentioned. Enzyme reagents may be contaminated not only with other enzymes but with oligonucleotides as well. Radioactive nucleotide reagents carry impurities from their preparation and generate more on decay. Fortunately the availability of alternative methods of analysis and the inherent biologic "sense" of the DNA sequence offer major safeguards for ultimately correct solutions.

30. Murray, K. and Old, R. W. (1974) *Prog. N. A. Res. and Mol. Biol.* **14**, 117.
31. Hedgpeth, J., Goodman, H. M. and Boyer, H. W. (1972) *PNAS* **70**, 3448.
32. Robertson, H. D., Dickson, E., Model, P. and Prensky, W. (1973) *PNAS* **70**, 3260.

2. Chemical Synthesis of Repetitive DNA Duplexes[33]

The laboratory synthesis of a gene requires that its sequence be known. Paradoxically, the synthesis of defined DNA segments, recently accomplished, may itself provide a major aid to sequence analysis.

Chemical synthesis of a polydeoxyribonucleotide required the solution of two problems: the first was finding an agent for chemical activation of the phosphate group of the mononucleotide to make a phosphodiester or internucleotide bond. Successful condensing agents proved to be dicyclohexylcarbodiimide (DCC), mesitylenesulfonyl chloride and triisopropylbenzenesulfonyl chloride (Fig. 11-3). The second problem required developing suitable protecting groups for the 3′-hydroxyl, 5′-hydroxyl, the amino substituents of the purine and pyrimidine bases, and for the phosphate group itself. A protecting group must be capable of addition and removal under mild conditions that do not damage the polynucleotide. Protecting groups and the functions they serve are shown in Fig. 11-3.

33. Khorana, H. G. (1968) *Pure and Applied Chemistry* **17**, 349; (1968) *B.J.* **109**, 709; (1968) *The Harvey Lectures* **62**, 79.

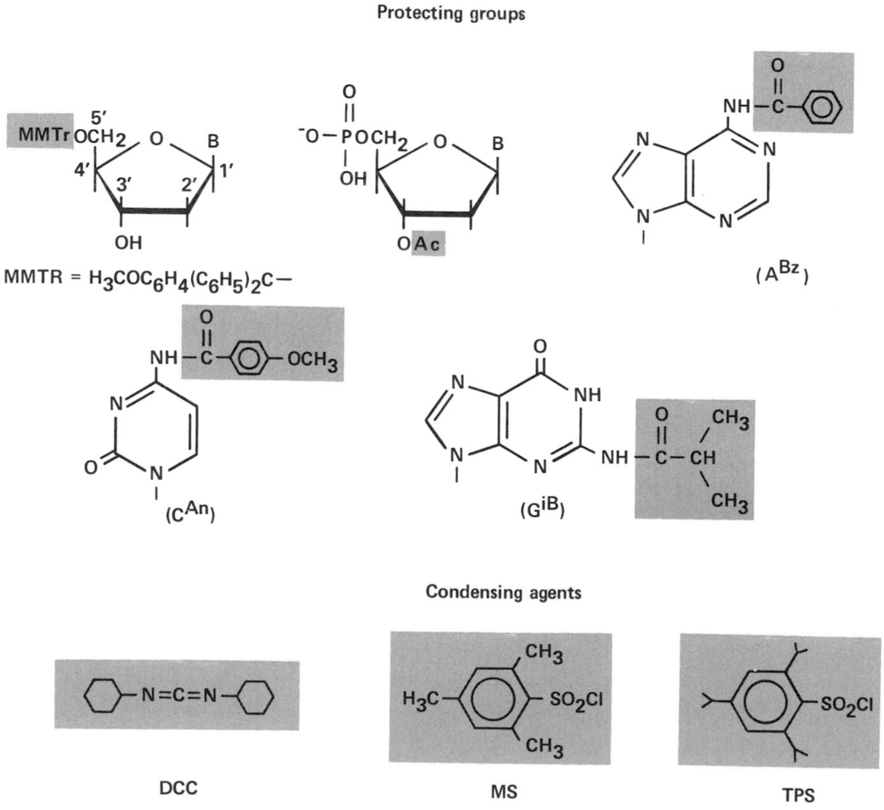

Protecting groups

MMTR = $H_3COC_6H_4(C_6H_5)_2C-$

(A^{Bz})

(C^{An})

(G^{iB})

Condensing agents

DCC

MS

TPS

In a stepwise fashion, di-, tri-, and tetranucleotides were built up (Fig. 11-4) and then condensed as oligonucleotide blocks into chains containing the di-, tri-, or tetranucleotides repeated two or three times. For each such polymer chain a polynucleotide was prepared, with a sequence that exactly matched it, by base-pairing in an anti-parallel orientation.

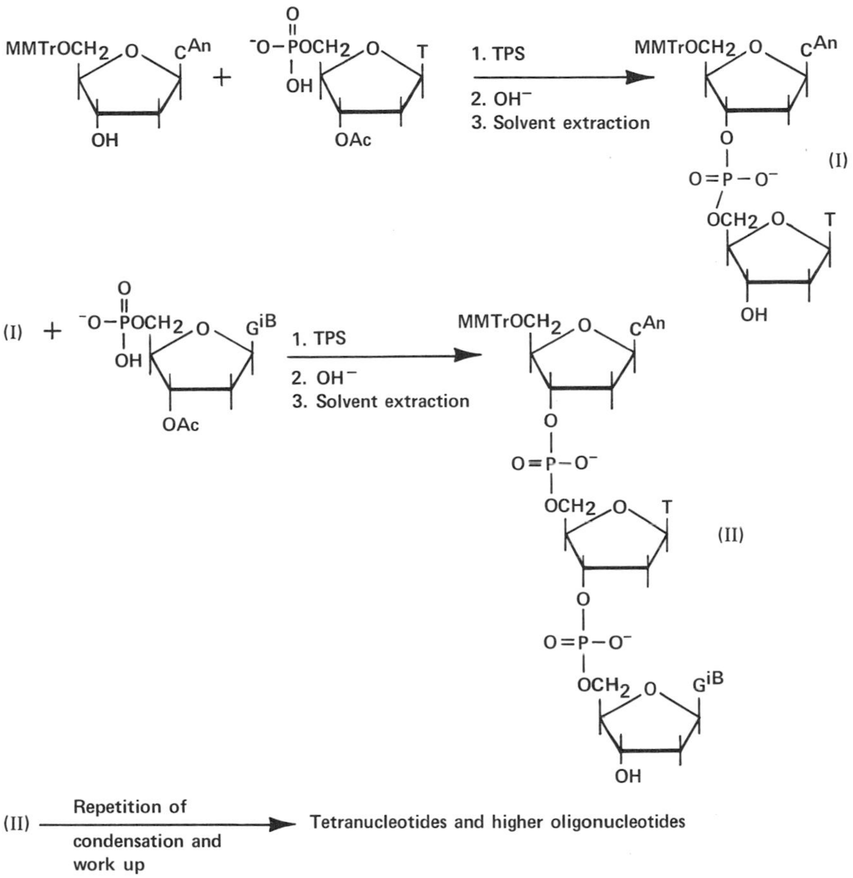

FIGURE 11-4
Condensation reactions for the synthesis of oligonucleotides. Abbreviations as in Fig. 11-3. (Courtesy of Professor H. G. Khorana.)

FIGURE 11-3 (facing page)
Protecting groups (in color blocks) for deoxynucleosides and deoxynucleotides and condensing agents (in color blocks) used in the chemical synthesis of a poly-deoxynucleotide chain. Abbreviations: MMTr, monomethoxytrityl; Ac, acetyl; Bz, benzoyl; An, anisoyl; iB, isobutyryl; DCC, dicyclohexylcarbodiimide; MS, mesitylenesulfonyl chloride; TPS, triisopropylbenzenesulfonyl chloride. (Courtesy of Professor H. G. Khorana.)

TABLE 11-2
Synthetic polydeoxynucleotides with repeating nucleotide
sequences

Repeating dinucleotide sequences	Repeating trinucleotide sequences	Repeating tetranucleotide sequences
$d(TC)_n \cdot d(GA)_n$	$d(TTC)_n \cdot d(GAA)_n$	$d(TTAC)_n \cdot d(GTAA)_n$
$d(TG)_n \cdot d(CA)_n$	$d(TTG)_n \cdot d(CAA)_n$	$d(TATC)_n \cdot d(GATA)_n$
	$d(TAC)_n \cdot d(GTA)_n$	
	$d(ATC)_n \cdot d(GAT)_n$	

These synthetic repetitive DNA duplexes served an important
function in elucidating the genetic code and other biochemical and
genetic aspects of protein synthesis[34] (Table 11-2). The short polymers, serving as template-primers, were extended by DNA polymerase with the four deoxynucleoside triphosphates to macromolecular
size. The minute quantity of polymer produced by the original
organic synthesis was thus expanded manyfold to a large polymer
which could then serve as a template and direct the synthesis by
RNA polymerase of a messenger RNA (Fig. 11-5). One or the other

34. Khorana, H. G. (1968) *B.J.* **109**, 709; (1968) *The Harvey Lectures* **62**, 79.

FIGURE 11-5
A scheme illustrating the utilization of a repetitive DNA
duplex to generate specific messengers for the synthesis of
defined polypeptides.

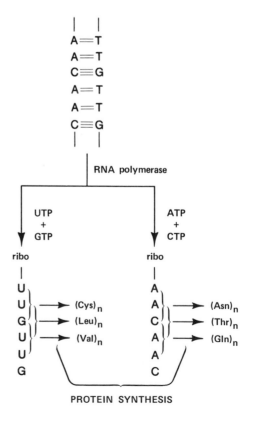

TABLE 11-3
Polypeptides synthesized in vitro
directed by messengers containing
repeating nucleotide sequences.

(System, *E. coli* B)	
Polyribo-nucleotide messenger	Polypeptides formed
REPEATING DINUCLEOTIDES	
$(UC)_n$	$(ser-leu)_n$
$(AG)_n$	$(arg-glu)_n$
$(UG)_n$	$(val-cys)_n$
$(AC)_n$	$(thr-his)_n$
REPEATING TRINUCLEOTIDES	
$(UUC)_n$	phe, ser, leu
$(AAG)_n$	lys, glu, arg
$(UUG)_n$	cys, leu, val
$(CAA)_n$	gln, thr, asn
$(GUA)_n$	val, ser, (nonsense)
$(UAC)_n$	tyr, thr, leu
$(AUC)_n$	ileu, ser, his
$(GAU)_n$	met, asp, (nonsense)
REPEATING TETRANUCLEOTIDES	
$(UAUC)_n$	$(tyr-leu-ser-ileu)_n$
$(UUAC)_n$	$(leu-leu-thr-tyr)_n$
$(GUAA)_n$	di- and tripeptides
$(AUAG)_n$	di- and tripeptides

strand of the DNA duplex could be transcribed, depending on which of the four ribonucleoside triphosphates were supplied. Thus in the presence of UTP and GTP, the mRNA produced from $d(AAC)_n \cdot d(GTT)_n$ was $r(GUU)_n$, whereas with CTP and ATP present, it was $r(AAC)_n$. Upon translation of $r(GUU)_n$, three different homo-polypeptides were obtained: polyvaline, polyleucine, and poly-cysteine; $r(AAC)_n$ led to the synthesis of polyasparagine, polythreo-nine, and polyglutamine (Table 11-3).

From the nature of the homo- and copolypeptides obtained in the translation of a variety of repeating di-, tri-, and tetranucleotide messengers, the triplet nature of the genetic code and the identity of most of the codons could be deduced or confirmed.

3. Chemical Synthesis of a Gene[35]

The successful synthesis of small DNA duplexes led to the application of these methods to the synthesis of a gene. But it was essen-

---

35. Khorana, H. G. et al. (1972) *JMB* **72**, 209; and succeeding papers.

tial to choose a gene whose sequence could be inferred directly from its product and whose size would still fall within the practical limits of organic synthesis. The gene for a tRNA with a defined sequence meets these criteria, and the total synthesis of the DNA duplex corresponding to yeast alanine tRNA has been achieved, and will be summarized here.

Since we now know that a tRNA is derived by processing a considerably larger tRNA precursor, it is evident that the true gene and the regulatory DNA segments that may surround it form a larger genetic unit; the synthesis of such a unit will be considered in the next section.

In deducing the DNA sequence corresponding to a tRNA it has been assumed that the unusual nucleotides present in tRNA have been derived by enzymatic modifications of a transcribed tRNA composed exclusively of A, U, G, and C. These modifications include methylations, deamination (adenine to hypoxanthine), and conversions of uracil to dihydro-, thio-, and 5-ribosyluracil (pseudouridine) and additions of other uncommon substituents.

The general plan of procedure relies on three crucial operations.

(i) _Chemical synthesis_ of overlapping polydeoxynucleotide segments representing the entire two strands of the DNA of chain lengths in the range of 8 to 12 nucleotides with overlapping regions of 4 to 5 nucleotides.

(ii) _Phosphorylation_ of the 5'-hydroxyl group of each segment with γ-(^{32}P)-ATP and polynucleotide kinase to enable the joining of segments to be monitored.

(iii) _Ligase_ joining of the aligned segments, monitored by esterification of the 5'-(^{32}P)-termini in internucleotide linkage. Joining renders the ^{32}P insusceptible to phosphomonoesterase and makes it a verifiable "nearest-neighbor" of the 3'-terminal nucleotide of the segment to which it was linked (see Chapter 4, Section 17).

Synthesis of the DNA duplex, corresponding to yeast tRNA (Fig. 11-6a), was accomplished through the connection by ligase of 15 segments (Fig. 11-6b). Nearest-neighbor transfer analysis verified that the head-to-tail (5'-(^{32}P)- to 3'-hydroxyl) linkage of the segments was correct. Some interesting and instructive problems were anticipated and encountered, which necessitated tactical adjustments in selecting the particular segments and the order in which they were

FIGURE 11-6 (facing page)
Sequence and structure of the major yeast alanine tRNA (top). Abbreviations: ψ, pseudouridine; T, ribothymidine; I, inosine; MeI, methylinosine; DiHU, 5, 6-dihydrouridine. A plan for the total synthesis of the DNA duplex corresponding to the yeast alanine tRNA gene (bottom). Residues 1 to 77 of the DNA, complementary to the tRNA sequence, are numbered from right to left in three lines; chemically synthesized segments are indicated by brackets. Annealing and ligase sealing of the 15 synthetic segments, as indicated, resulted in a duplex of 77 base pairs presumed to be the genetic segment which codes for this tRNA. (Courtesy of Professor H. G. Khorana.)

a.

b.

20 19 18 17 16 15 14 13 12 11 10 9 8 7 6 5 4 3 2 1

```
      ┌──────── (4) ────────┐┌──────────────── (1) ────────────────┐
     -C-T- A-A-G-G-C-C  T-G-A-G-C-A-G-G-T-G-G-T   (5') DEOXY
            | | | | | | | |  | | | | | | | | | | |
            C-C-G-G-A-C-T-C-G-T  C-C-A-C-C-A   (3') DEOXY
           └────────── (3) ──────────┘└──── (2) ────┘
```

50 49 48 47 46 45 44 43 42 41 40 39 38 37 36 35 34 33 32 31 30 29 28 27 26 25 24 23 22 21 20 19 18 17

```
      ┌──── (9) ────┐┌──────────────────── (6) ────────────────────┐
     -G-A-A-T-C  G-T-A-C-C-C -T-C-T-C-A-G-A-G-G-C-C-A-A-G-   (5') DEOXY
            | | | | | | | | | | | | | | | | | | | | | | | | | |
    -G-C-T-C-C-C-T-T-A-G-C-A-T-G-G-G  A-G-A-G-T-C-T-C-C-G-G-T-T-C-G-A-T-T   (3') DEOXY
     └──────────── (8) ────────────┘└──────── (7) ────────┘└──────── (5) ────────┘
```

77 76 75 74 73 72 71 70 69 68 67 66 65 64 63 62 61 60 59 58 57 56 55 54 53 52 51 50 49 48 47 46

```
      ┌─────────── (14) ───────────┐┌──────────── (12) ────────────┐┌──────────── (10) ────────────┐
     C-C-C-G-C-A-C-A-C-C-G-C  G-C-A-T-C-A-G-C-C-A  T-C-G-G-C-G-C-G-A-G-G-   (5') DEOXY
            | | | | | | |  | | | | | | | | | |  | | | | | | | | | |
     G-G-G-C-G-T-G  T-G-G-C-G-C-G-T-A-G  T-C-G-G-T-A-G-C-G-C-   (3') DEOXY
     └──────── (15) ────────┘└────────── (13) ──────────┘└────────── (11) ──────────┘
```

to be annealed and sealed by ligase. Two examples are cited. Because the overlap of residues 17-20 in segments 4 and 5 is very weak, involving only a tetranucleotide with three A–T base pairs, the ligase reaction had to be performed at a very low temperature, at a high Mg^{2+} concentration, and at a sacrifice in yield. Another problem may be self-complementarity. Segment 10 (Figure 11-6):

$$
\begin{array}{c}
\text{T–C–G–C–G–C–G–A–G–G–} \\
\text{| | | | | | | |} \\
\text{G–G–A–G–C–G–C–G–C–T}
\end{array}
$$

may form a tightly duplexed structure that could interfere with annealing to segments 8 and 11, and so an alternative scheme involving a segment shorter than segment 10 by two nucleotides was also tried.

The use of polynucleotide ligase in joining DNA segments has disclosed some novel properties of the T4 enzyme, such as its capacity to join the ends of DNA duplexes in the absence of any overlapping regions. In order to direct the joining of segments with greater specificity, a phosphate-blocking group has been devised[36]. It is an alkyl phosphothioate that not only prevents unwanted joinings of 5′-phosphate termini, but also facilitates the separation of these modified chains and their recovery from mixtures.

Synthesis of yeast alanine tRNA had been undertaken because it was, at the time, the only tRNA whose sequence was known. It has now become clear that the true genetic element for tRNA codes for a larger precursor molecule and that many genetic manipulations possible with E. coli are still foreclosed in yeast. To synthesize the full-length tRNA gene, and to discover its regulation, it was necessary to undertake the synthesis of the gene for a precursor tRNA of E. coli.

4. Chemical Synthesis of a Regulated Gene[37]

One of the major rewards from the chemical synthesis of the tRNA gene is the promise of a transcript to serve as substrate by which to introduce the many modifications that characterize mature tRNA. This kind of substrate would not only open a field of biochemistry of interest in itself, but also would provide a means to make novel tRNA molecules. For example, molecules with deletions, additions, or substitutions are powerful tools to study the structure-function relationships of tRNA: recognition by specific aminoacyl-tRNA synthetases, interaction with the nucleotidyl transferase that repairs the 3′-CCA end, binding to ribosomes and factors involved in protein synthesis, and pairing with codons in mRNA. The importance of determining the sequence of the adjacent promoter and terminator regions for understanding regulation of the gene needs no emphasis.

36. Harvey, C. L., Wright, R., Cook, A. F., Maichuk, D. T. and Nussbaum, A. L. (1973) B. **12**, 208.
37. Besmer, P. et al. (1972) JMB **72**, 503; Loewen, P. C., Sekiya, T. and Khorana, H. (1974) JBC **249**, 217.

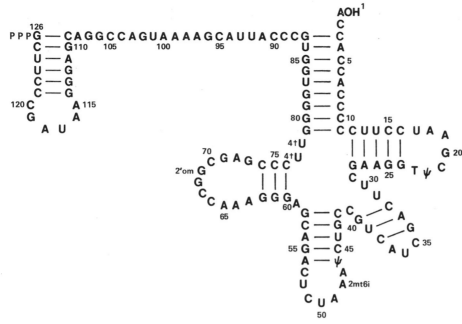

FIGURE 11-7
Structure and sequence of the *precursor* of E. coli tyrosine tRNA. Modified bases are indicated as they appear in mature tyrosine tRNA, but they are absent in the tRNA precursor. Abbreviations: 2mt6iA, N^6-isopentenyl-2-methylthioadenosine; 2′omG, 2′-methoxyguanosine; 4tU, 4-thiouridine; others as in Figure 11-6.

Tyrosine tRNA of E. coli is derived from a much longer precursor, which itself comes closer to being the initial transcription product. The isolated precursor contains 126 nucleotides and carries a 5′-triphosphate end group[38] (Fig. 11-7). There is recent evidence that this precursor is derived from a still earlier transcript containing several additional residues on the CCA 3′ end[39]. Isolation of the tyrosine tRNA precursor was made possible by working with a closely related tyrosine tRNA, the amber suppressor su$_{III}$. The gene for this tRNA has been inserted into the phage ϕ80, and therefore is more easily available for analysis.

The total synthesis of a DNA duplex corresponding to the tyrosine tRNA precursor, according to the plan in Fig. 11-8[40], has been nearly completed[41]. Already, segments 2, 4, 6, 8, and 10, representing the 51 nucleotides from the 3′ end, have been specifically annealed to the "r" strand of ϕ80 DNA containing the su$_{III}$ gene (ϕ80 psu$_{III}$). This establishes the true template strand of the gene and immediately makes it possible to determine the sequence of DNA in the 5′ → 3′ direction beyond the 3′ end of the tRNA.

38. Altman, S. and Smith, J. D. (1971) *Nat.* **233**, 35; Vickers, T. G. and Midgley, J. E. M. (1971) *Nat.* **233**, 210.
39. Schedl, P. and Primakoff, P. (1973) *PNAS* **70**, 2091.
40. Besmer, P. et al. (1972) *JMB* **72**, 503.

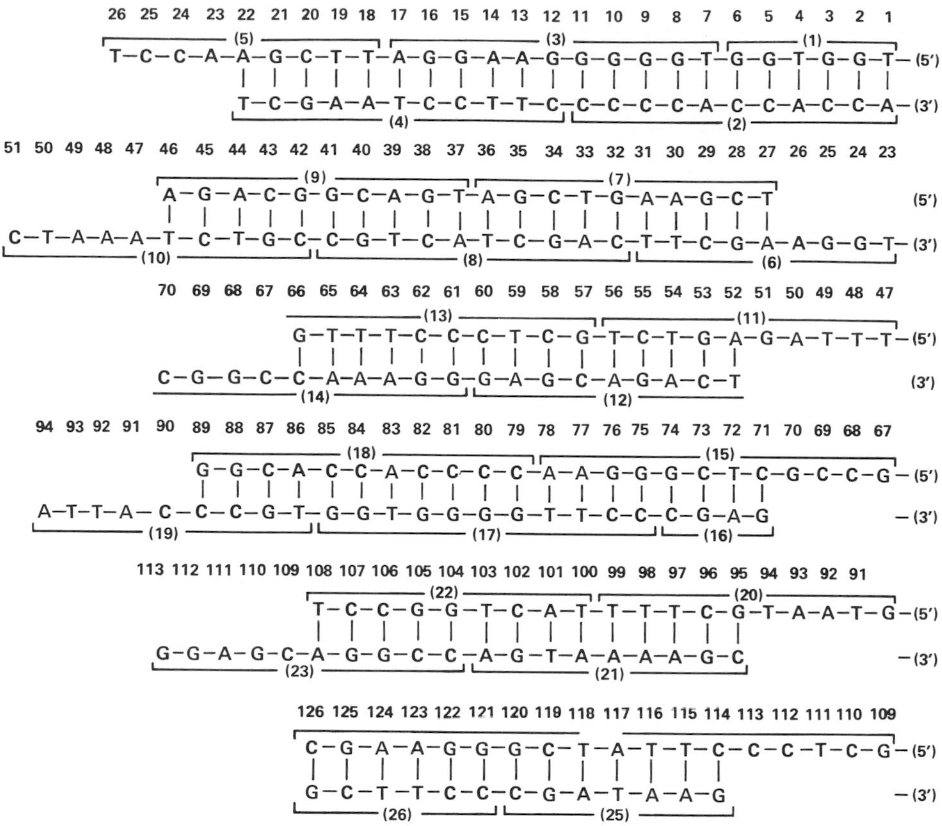

```
 26 25 24 23 22 21 20 19 18 17 16 15 14 13 12 11 10  9  8  7  6  5  4  3  2  1
      ┌──── (5) ────┐           ┌────────── (3) ──────────┐      ┌── (1) ──┐
      T—C—C—A—A—G—C—T—T—A—G—G—A—A—G—G—G—G—G—T—G—G—T—G—G—T— (5')
      | | | | | | | | | | | | | | | | | | | | | | | | | |
      T—C—G—A—A—T—C—C—T—T—C—C—C—C—C—A—C—C—A—C—C—A— (3')
      └──────── (4) ────────┘   └──────── (2) ────────┘

 51 50 49 48 47 46 45 44 43 42 41 40 39 38 37 36 35 34 33 32 31 30 29 28 27 26 25 24 23
          ┌────────── (9) ──────────┐      ┌────── (7) ──────┐
          A—G—A—C—G—G—C—A—G—T—A—G—C—T—G—A—A—A—G—C—T         (5')
          | | | | | | | | | | | | | | | | | | | | |
C—T—A—A—A—T—C—T—G—C—C—G—T—C—A—T—C—G—A—C—T—T—C—G—A—A—G—G—T— (3')
└──── (10) ────┘   └──────── (8) ────────┘   └──────── (6) ────────┘

 70 69 68 67 66 65 64 63 62 61 60 59 58 57 56 55 54 53 52 51 50 49 48 47
          ┌────────── (13) ──────────┐   ┌──────── (11) ────────┐
          G—T—T—T—C—C—C—T—C—G—T—C—T—G—A—G—A—T—T—T— (5')
          | | | | | | | | | | | | | | | | | | |
C—G—G—C—C—A—A—A—G—G—G—G—A—G—C—A—G—A—C—T (3')
└──── (14) ────┘   └──────── (12) ────────┘

 94 93 92 91 90 89 88 87 86 85 84 83 82 81 80 79 78 77 76 75 74 73 72 71 70 69 68 67
      ┌────────── (18) ──────────┐       ┌──────── (15) ────────┐
      G—G—C—A—C—C—A—C—C—C—C—A—A—G—G—G—C—T—C—G—C—C—G— (5')
      | | | | | | | | | | | | | | | | | | | | | |
A—T—T—A— C—C—C—G—T— G—G—T—G—G—G—G—T—T—C—C—C—G—A—G   (3')
└──── (19) ────────┘   └──────── (17) ────────┘   └── (16) ──┘

113 112 111 110 109 108 107 106 105 104 103 102 101 100 99 98 97 96 95 94 93 92 91
          ┌──────── (22) ────────┐       ┌──────── (20) ────────┐
          T—C—C—G—G—T—C—A—T—T—T—T—C—G—T—A—A—T—G— (5')
          | | | | | | | | | | | | | | | | | |
G—G—A—G—C—A—G—G—C—C—A—G—T—A—A—A—A—G—C   (3')
└──────── (23) ────────┘   └──────── (21) ────────┘

126 125 124 123 122 121 120 119 118 117 116 115 114 113 112 111 110 109
      C—G—A—A—G—G—G—C—T—A—T—T—C—C—C—T—C—G— (5')
      | | | | | | | | | | | | | | | |
      G—C—T—T—C—C—C—G—A—T—A—A—G   (3')
      └──────── (26) ────────┘   └──────── (25) ────────┘
```

FIGURE 11-8

A plan for the total synthesis of the DNA duplex corresponding to the gene that directs the synthesis of E. *coli* tyrosine tRNA precursor. Arrangement is indicated in Fig. 11-6.

Using the undecanucleotide segment 2 as primer, DNA polymerase to extend it, the restricted addition of deoxynucleoside triphosphates and ribonucleoside triphosphate analogs, a sequence of 23 nucleotides beyond the 3' terminus of the tRNA gene has been analyzed[41] (Fig. 11-9a). Upon examining this sequence, one fails to find the heptanucleotide U_6A that terminates certain phage λ transcripts (Chapter 10, Section 3). However, there are two noteworthy features (Fig. 11-9b). One is the 180° rotational symmetry of a pentanucleotide, which makes it possible for a distinctive secondary structure to be present in the DNA and in its RNA transcript. This structure may impart information for termination. Another feature of undetermined significance is the presence of two palindromes, sequences that read the same from either direction.

An approach similar to that used for analyzing the terminator

41. Loewen, P. C., Sekiya, T. and Khorana, H. (1974) *JBC* **249**, 217.

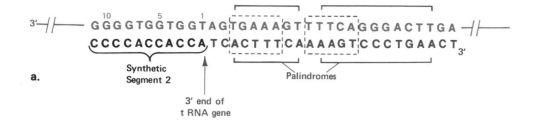

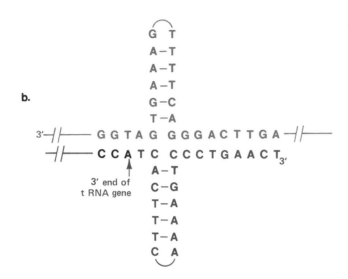

FIGURE 11-9
The sequence beyond the 3′ end of a tyrosine tRNA gene. (a) Synthetic segment
2 was annealed to the isolated r strand of phage ϕ80 psu$_{III}$ DNA and then extended
stepwise by DNA polymerase action. The absence of G simplified the determination
of the dodecanucleotide sequence to the right of the tRNA gene. Pentanucleotide
regions of 180° rotational symmetry are shown in dashed boxes and palindromes
in brackets. (b) This secondary structure in the terminator region, demonstrating
the base-pairing potentialities of its symmetry, is purely hypothetical. (Loewen, P.,
Sekiya, T. and Khorana, H. G., (1974) *JBC* **249**, 217. Courtesy of Professor
H. G. Khorana.)

region can also be applied at the other end for extension into what
should be the promoter region. The synthetic segment 24 could be
used for annealing to the "l" strand of ϕ80 psu$_{III}$ DNA in order to
serve as a primer for replication of the promoter sequence. This
sequence, analyzed for 29 nucleotides from the transcriptional
start, contains remarkable regions of 2-fold symmetry[41a].

Determination of sequences of the neighboring regulatory regions
about a gene by using them as templates in the directed elongation
of synthetic primer segments can be pursued systematically and
extensively (Fig. 11-10). Enzymatic and chemical syntheses can
follow each other, step after step, to assure fidelity and complete
unambiguity; both time synchronization of primer chain extension,
as well as restricted synthesis by omission of a deoxyribonucleoside

41a. Sekiya, T. and Khorana, H. G., personal communication.

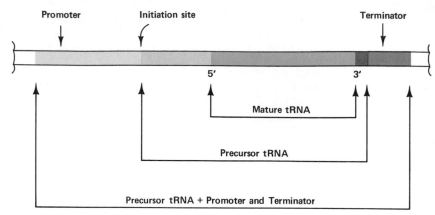

FIGURE 11-10
A diagram of the tRNA gene indicating transcriptional control elements on
either side: promoter region to the left and terminator region to the right (Courtesy
of Professor H. G. Khorana.)

triphosphate, and incorporation of ribonucleoside triphosphate analogs should be useful.

The transcription signals for operator and promoter regions and for termination of a specific gene should become clearer from studies of this kind. In other instances, where certain messenger RNAs specified by phage λ are readily isolated, the same technique should also be applicable. One especially interesting question concerns the sequences between genes that appear as tandem repetitions, i.e. a given sequence repeated consecutively. Such duplication occurs with the tyr su$_{III}$ gene· in φ80 and lends itself to analysis by the methods described.

5. Assembling Genes into a Chromosome[42]

Total synthesis of a gene that codes for a protein of average size will be far more difficult than synthesis of a tRNA gene. Assembly of operons (a cluster of genes coding for related functions) and chromosomes will be even more so. Yet the importance of the objectives argues for continued refinement of methods, and multiplication of research effort in this field. Chromosome assembly by what might be called "enzymatic tailoring," or "genetic engineering," exploits the successes of nucleic acid enzymology as well as of organic chemistry and genetics. A discussion of some current and projected efforts in chromosome assembly makes an appropriate epilogue for this book.

It has been possible to produce a covalently closed dimer circle of phage P22 DNA from the linear monomer[43]. It has been possible also to insert λ*dvgal,* an *E. coli* plasmid containing certain genes of

42. Jackson, D. A., Symons, R. H. and Berg, P. (1972) *PNAS* **69**, 2904.
43. Lobban, P. E. and Kaiser, A. D. (1973) *JMB* **78**, 453.

phage λ and the genes of the *E. coli* galactose operon, into the chromosome of a tumor virus[44]. The virus is SV40, a circular duplex (Chapter 8, Section 9) which infects and transforms animal cells in culture, and produces a variety of tumors in tissues.

The technique for assembling the chromosomes of λ*dv* and SV40 into a circular unit that may be infectious for bacterial or animal cells was based on a general plan tested first in the conversion of SV40 into double-length circles. The steps were as follows (Fig. 11-11).

(i) SV40 was converted from a circular to linear duplex by the double-stranded scission of a restriction endonuclease, *Eco* RI.

(ii) λ exonuclease degraded small portions of the 5′-ended chains in order to make the 3′-ended chains single-stranded for the next step.

---

44. Jackson, D. A., Symons, R. H. and Berg, P. (1972) *PNAS* **69**, 2904.

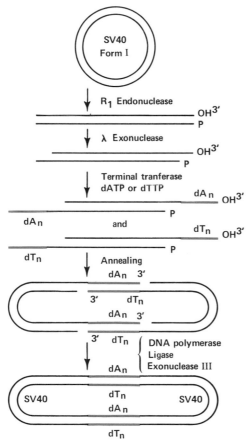

FIGURE 11-11
A scheme for producing a dimer SV40 circle from the monomer. The scheme entails conversion of the closed circle to a linear form by a restriction endonuclease, attachment of short homopolymers which form cohesive ends, and use of the latter to anneal, and then join, two such circles. The four deoxynucleoside triphosphates and DPN are included for the DNA polymerase and ligase reactions. (Courtesy of Professor P. Berg.)

(iii) Terminal nucleotidyl transferase (see Chapter 6, Section 8) extended the 3′-ended chains with relatively short lengths (50–100 nucleotides) of poly dA with one sample, and of poly dT, with another.

(iv) The two samples were annealed through the pairing of their poly dA and poly dT segments. At relatively low concentrations, formation of circular structures was favored over linear concatemers.

(v) The final gap filling, removal of loose ends, and other tailoring was carried out by DNA polymerase I in the presence of the four deoxynucleoside triphosphates; sealing of aligned ends to form a covalent duplex circle was achieved by E. coli ligase and DPN. Exonuclease III was also present to assist in the removal of any 3′-phosphoryl termini that might have been introduced by non-specific nicking and would interfere with polymerase and ligase action.

Application of this plan to some particular circular DNA requires the judicious choice of endonuclease for cleavage. The λdvgal plasmid is a supercoiled DNA molecule 6μ long; SV40 DNA is near 1.5μ. Complete digestion of λdvgal DNA with Eco RI endonuclease produces two equal-length linear molecules which, on genetic grounds, should each contain an intact galactose operon. Addition of poly dT tails to the λdvgal segments enabled them to anneal to SV40 molecules bearing poly dA tails and to form circles. When covalently closed, these hybrids were near 4.5μ long, as anticipated[44]. These molecules should be useful in determining whether bacterial genes can be introduced into a mammalian cell chromosome, and whether their information can be expressed in this novel location. Similarly, any capacity of bacterial cells to respond to SV40 DNA would yield unique information about the products of SV40 genes, and their actions in E. coli might be easier to recognize.

The technique just outlined incorporates dA · dT homopolymer duplexes as spacer regions of somewhat variable length. Such regions may interfere with some DNA functions but may enhance others, such as recombinational events. In any case, there are at least two additional approaches for connecting DNA segments that do not require this kind of linker. One relies on the overlapping ends produced as the result of staggered cleavage by a restriction endonuclease, such as Hin d III of H. influenzae, or Eco RI or Eco RII nucleases of E. coli (Table 9-6). The other exploits the capacity of T4 ligase to join the ends of any two DNA duplexes[45]. We can anticipate that refinement of these techniques and development of new ones will enlarge the scope and increase the precision of assembling genetic segments into a chromosome.

Two questions of key importance, pertaining to eukaryotic chromosomes, are their genetic content and organization on the one hand, and the control of their expression on the other. The tech-

44. Jackson, D. A., Symons, R. H. and Berg, P. (1972) PNAS 69, 2904.
45. Sgaramella, V. (1972) PNAS 69, 3389.

niques enabling the enzymatic assembly of diverse genetic segments and the highly developed knowledge of phage and plasmid structures and functions promise to be helpful in answering these questions.

Phage λ with up to 20 percent of the chromosome deleted[46] is available; packaging of a still smaller chromosome reduces its viability. Such deletions can become convenient points for incorporation of other DNA. It seems possible that DNA from a eukaryotic source such as mitochondrial DNA, inserted enzymatically, would be accepted in a viable phage package. Powerful selective techniques can be applied to identify and multiply phages which carry a deletion or insertion of a particular kind or size. Here may be a method[47] for at once obtaining transcripts of defined segments of eukaryotic DNA and determining whether factors in *E. coli* regulate their production as well as the proteins they encode. It is also immediately apparent that a phage carrying a segment of eukaryotic DNA, nuclear or mitochondrial, is the source of a large quantity of this unique DNA segment. This DNA could then be used, in one of the several hybridization methods available, to determine the frequency and specific location of this genetic segment on one or another of the animal's chromosomes.

Incorporation of eukaryotic DNA into *E. coli* and its replication and transcription have been achieved[48] recently with use of a restriction endonuclease to facilitate the joining of the DNA to an *E. coli* plasmid. *Drosophila* DNA, enriched for ribosomal RNA genes and a resistance factor plasmid[49], were each cleaved by a restriction endonuclease that produces staggered breaks with overlapping ends. The *Drosophila* genes were readily annealed to the plasmid DNA and could then be incorporated into *E. coli* where they were replicated and transcribed many times over.

6. Social Concerns about Genetic Chemistry[50]

Those interested or curious enough to read this book need not be told about the beauty or importance of DNA. Yet many among them may have a nagging concern about the use to which knowledge about DNA will be put. Isolation and synthesis of genes, assembly of genes into chromosomes, and the application of this knowledge to human problems have been suspiciously labeled "genetic engineering."

Most people do not distinguish an atom from a molecule, a virus

46. Davis, R. W. and Davidson, N. (1968) *PNAS* **60**, 243; Parkinson, J. S. and Huskey, R. J. (1971) *JMB* **56**, 369.
47. Hogness, D. S., personal communication.
48. Morrow, J. F., Boyer, H. W., Goodman, H. M., Helling, R. B., Chang, A. C. Y. and Cohen, S. N. (1974) *PNAS* **71,** 1743.
49. Cohen, S. N. and Chang, A. C. Y. (1973) *PNAS* **70**, 1293; Cohen, S. N., Chang, A. C. Y., Boyer, H. W. and Helling, R. B. (1973) *PNAS* **70**, 324.
50. Kornberg, A. (1973) *Prism* **1** (#7) 12; Hellman, A., Oxman, M. N. and Pollack, R. (Ed.) (1973) Biohazards in Biological Research, CSHL.

from a cell, or a cell from an organism; and so, through lack of under-standing may also be fearful about genetic chemistry. They might even want to enact laws which will regulate genetic research deemed to have a harmful *potential*. But how could any national group regu-late research that can be conducted all over the world?

Public fear of genetic chemistry is a threat both to the future of research, and to the application of genetic knowledge to human problems. This threat to the support of science from a public hap-hazardly informed about scientific progress is a major concern for all who regard basic science a social responsibility. It demands a clear, strong and thoughtful response from the scientific community.

All knowledge has a potential for both evil and good. The dis-coveries of fire, explosives, and even the most benign medicines have brought tragedy as well as comfort. Polio vaccines have virtu-ally eliminated one of man's most dreadful diseases. Yet these, and other life-saving vaccines, kill and maim a small fraction of those inoculated. Smallpox vaccine was used for years on a vast scale before it was recognized that _routine_ prophylactic use was more harmful than good. The first batches of polio vaccine administered contained SV40, a virus with cancer-producing potential.

Genetic research that produces exotic organisms, viruses, and bio-chemicals imposes on the scientist a special responsibility to assess their hazards in order to anticipate and avoid their accidental or purposeful misuse. Testing programs and diligent reporting are needed to protect both the laboratory people who handle such agents, and the community where these agents might accidentally spread[51].

Medicine is the application of biology and biological chemistry to problems of human disease, and could easily be called "biological engineering." Yet there is little objection to the practices of our "biological engineers," whom we call physicians and surgeons. How absurd then that treatment of disease due to defective or missing genes draws more concern about adverse effects that are remote, than do standard medical treatments whose hazards are manifest. There is an urgent and compelling need to _cure_ the genetic defect of diabetes rather than to palliate it by endless insulin treatments. The possibility that advances in genetic chemistry might conceivably enable a future tyrant to clone a race of robots should not divert attention from the difficult problems presented by the genetic diseases which plague men today.

In rebutting exaggerations of the dangers in applying chemistry to correct genetic defects, there is a risk of a comparable extravagance in evaluating the prospects of this approach. Cure of genetic disease by gene treatment is not at hand. Other approaches offer attractive alternatives and more direct application[52]. Genetic lesions may be

51. Hellman, A., Oxman, M. N. and Pollack, R. (Ed.) (1973) Biohazards in Biological Research, CSHL.
52. Lederberg, J. (1972) **in** The Challenge of Life, Biomedical Progress and Human Values (Kunz, R. M. and Fehr, H. eds.) Birkhäuser Verlag, Basel, p. 231.

avoided by counseling and by elective abortions based on accurate antenatal diagnosis; genetic defects may be relieved by appropriate transplantation treatment. In the long term, however, a basic understanding of DNA structure and function is of prime importance for all considerations of genetic disease.

Perhaps an even greater problem than human genetic disease is malnutrition. Food supply is already inadequate for a large fraction of the earth's population and threatens to become worse. There are enormous opportunities for the application of "genetic engineering" in agriculture. Development of superior seed proteins and of drought- and disease-resistant plants, perfection of methods for inserting bacterial genes into plants (e.g. nitrogen fixation genes to eliminate the need for nitrogen fertilizers), and for fine genetic control of animal breeding are all realistic projects in which a sound knowledge of DNA structure and synthesis will be of great service.

Author Index

Subject Index

(Definitions and formulas are noted in **bold** type)

Watch It Grow

Pillbugs

by Martha E. H. Rustad
Consulting editor: Gail Saunders-Smith, PhD

Consultant: Laura Jesse
Plant and Insect Diagnostic Clinic
Iowa State University, Ames, Iowa

Mankato, Minnesota

Pebble Books are published by Capstone Press,
151 Good Counsel Drive, P.O. Box 669, Mankato, Minnesota 56002.
www.capstonepress.com

1 2 3 4 5 6 14 13 12 11 10 09

Library of Congress Cataloging-in-Publication Data
Rustad, Martha E. H. (Martha Elizabeth Hillman), 1975–
    Pillbugs / by Martha E. H. Rustad.
    p. cm. — (Pebble books. Watch it grow)
    Includes bibliographical references and index.
    Summary: "Simple text and photographs present the life cycle of pillbugs" —
Provided by publisher.
        ISBN-13: 978-1-4296-2229-5 (hardcover)   ISBN-10: 1-4296-2229-6 (hardcover)
        ISBN-13: 978-1-4296-3446-5 (softcover)   ISBN-10: 1-4296-3446-4 (softcover)
        1. Wood lice (Crustaceans) — Life cycles — Juvenile literature. I. Title.
QL444.M34R87 2009
595.3′72 — dc22   2008026941

Note to Parents and Teachers

The Watch It Grow set supports national science standards related to life science. This book describes and illustrates pillbugs. The images support early readers in understanding the text. The repetition of words and phrases helps early readers learn new words. This book also introduces early readers to subject-specific vocabulary words, which are defined in the Glossary section. Early readers may need assistance to read some words and to use the Table of Contents, Glossary, Read More, Internet Sites, and Index sections of the book.

Table of Contents

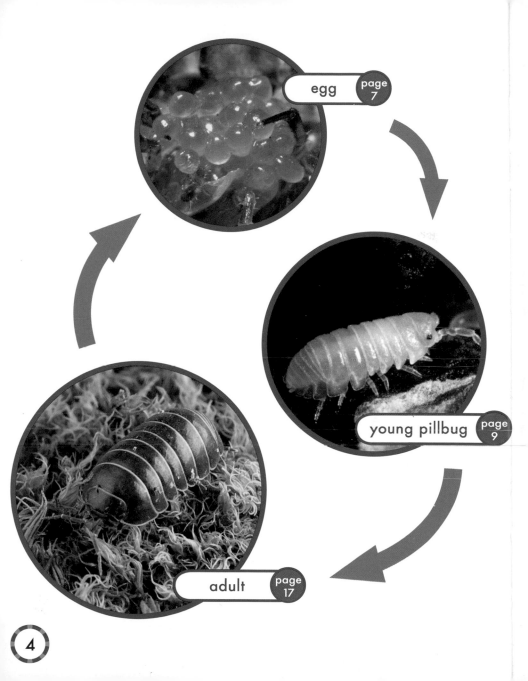

egg **page 7**

young pillbug **page 9**

adult **page 17**

4

Metamorphosis

Pillbugs are isopods.
Isopods have 14 legs
and a hard outer shell.
As they grow, pillbugs go
through metamorphosis.

eggs

From Egg to Young Pillbug

Pillbugs begin life as eggs.
Female pillbugs
lay eggs in a pouch.
The pouch rests between
the female's legs.

Young pillbugs hatch
after a few weeks.
They stay in the pouch for
another six to eight weeks.

Young pillbugs look like tiny adults. But they have 12 legs instead of 14.

As they grow, pillbugs eat dead plants and animals. Pillbugs live in damp places under wood, rocks, or leaves. They breathe through gills.

Molting

Young pillbugs molt
as they grow.
Their hard shell breaks off
in two halves.
The new shell is underneath.

Young pillbugs have 14 legs after their first molt. Young pillbugs are adults after five molts.

Adult Pillbugs

Adult pillbugs continue
to molt as they grow.
Pillbugs live up to five years.

Pillbugs can roll into a ball.
People sometimes call them
roly-poly bugs.
Their hard outer shell
keeps them safe.

Glossary

breathe — to take air in and out

damp — slightly wet

gill — a body part used for breathing

hatch — to break out of an egg

isopod — an animal that has seven pairs of legs, three main body sections, and a hard outer shell.

metamorphosis — the series of changes some animals go through as they develop from eggs to adults

molt — to shed skin or an outer shell so a new covering can be seen; when this process happens once, it is also called a molt.

pouch — a pocket between the legs of female pillbugs that holds eggs

Read More

Hughes, Monica. *Pill Bugs.* Creepy Creatures. Chicago: Raintree, 2004.

Tokuda, Yukihisa. *I'm a Pill Bug.* La Jolla, Calif.: Kane/Miller Book Publishers, 2006.

Internet Sites

FactHound offers a safe, fun way to find educator-approved Internet sites related to this book.

Here's what you do:

1. Visit *www.facthound.com*
2. Choose your grade level.
3. Begin your search.

This book's ID number is 9781429622295.

FactHound will fetch the best sites for you!

Index

Word Count: 161
Grade: 1
Early-Intervention Level: 17

Editorial Credits
Erika L. Shores, editor; Alison Thiele, designer; Marcie Spence, photo researcher

Photo Credits
Capstone Press/Karon Dubke, 4 (adult)
Dwight R. Kuhn, cover (all), 4 (egg and young pillbug), 6, 8, 10, 12, 14, 16, 20
iStockphoto/Nancy Nehring, 1
James P. Rowan, 18

Structures

ADP, etc.	5′-(pyro)-diphosphates of adenosine, etc.	P or p	phosphate (in compounds)
AMP, GMP, IMP, UMP, CMP	5′-phosphates of ribonucleosides of adenine, guanine, hypoxanthine, uracil, cytosine	^{32}P	a radioactive isotope of phosphorus with a half-life of 14.3 days
		pA, pG, pI, pU, pC	AMP, GMP, IMP, UMP, CMP
2′-AMP, 3′-AMP (5′-AMP), etc.	2′-, 3′- (and 5′-, where needed for contrast) phosphates of the nucleosides	P_i, PP_i	orthophosphate, inorganic pyrophosphate
Ap, etc.	3′-AMP, etc.	phage	bacterial virus
A-T, G-C	base pairs of adenine with thymine, and of guanine with cytosine, or their respective deoxyribonucleotides	phosphodiester	$$R-O-\overset{\overset{\displaystyle O}{\|}}{\underset{\underset{\displaystyle O^-}{\|}}{P}}-O-R$$
ATP, etc.	5′-(pyro)-trisphosphates of adenosine, etc.	poly dA, etc.	homopolymer chain of deoxy-riboadenylate of undefined length, generally over 1000 residues
^{14}C	a radioactive isotope of carbon with a half-life of 5730 years	poly dA·poly dT	homopolymer chains of poly dA and poly dT associated by base-pairing
capsid	a coat protein unit of a virus		
concatemer	a chain of an unstated number of repeated units of duplex DNA	poly d (A-T)	copolymer chain of alternating A and T residues associated with another such chain by base-pairing
core enzyme	principal part of holoenzyme; activity modified for lack of other components	poly rA·poly rU	homopolymer chains of adenylate and uridylate associated by base-pairing
d	2′-deoxyribo (nucleoside, nucleotide)	prophage provirus	provirus stage of a lysogenic phage state of a virus, integrated into the host cell chromosome; by being multiplied along with the chromosome it is transmitted from one cell generation to the next
dA_{4000}	homopolymer chain of about 4000 deoxyriboadenylate residues		
dAMP, etc.	5′-phosphates of 2′-deoxyribosyladenine, etc.		
d-A_pG	deoxyadenosine linked 3′ to 5′-phosphodeoxyguanylate (see Figure 1-2)	r	ribo (nucleoside, nucleotide)
		RF	double-stranded, circular, replicative form of DNA
DEAE-cellulose	diethylaminoethyl-cellulose (ion exchanger)	RF I	covalently closed RF
D-loop	a DNA replicative intermediate in which only one chain of a duplex region is replicated leaving the other single-stranded (see Chapter 7, Section 3)	RF II	RF with discontinuity in at least one strand
		σ	sigma subunit of RNA polymerase; also the lariat-shaped DNA replicative intermediate in a rolling circle mechanism (see Chapter 8, Section 4)
dna_{ts}	genetic locus for a replication protein in a temperature-sensitive (conditionally lethal) mutant		
		SS	single-stranded, circular DNA
dNMP (dXMP, dYMP)	5′-phosphate of a 2′-deoxy-ribonucleoside, in general	Sephadex	a porous gel of cross-linked dextran beads which separates molecules of different sizes by filtration
^{3}H	tritium, a radioactive isotope of hydrogen with a half-life of 12.26 years	θ	theta structure, a DNA replicative intermediate containing two segments (already replicated) of equal length as distinct from the third (unreplicated) (see Chapter 7, Section 2)
holoenzyme	complete enzyme including cofactors, small or large		
oligo dA	homopolymer chain of deoxy-riboadenylate of undefined length, generally under several hundred residues	vegetative phage	free state of a phage
		virion	a free virus particle

Henry Ford invented the first modern car.

To "invent" something means to be the first person to create it.

He was born in Michigan
on July 30, 1863.

His father wanted him to be
a farmer when he grew up.

Ford did not like farm work.
He was more interested in
machines (muh-SHEENS).

A lawn mower is a machine.

9

Detroit

Ford went to Detroit, Michigan, when he was sixteen. He got a job fixing and building machines there.

Ford loved engines (EN-juhns). Engines are what make many machines work.

Farm tractor

Ships, trains, and farm
tractors could not run
without their engines.

A long time ago, people
went from place to place
in a carriage (KA-rij).

A carriage was like a big
chair with wheels on it.
It was pulled by horses.

15

16

Ford wanted to build a carriage that used an engine, instead of a horse. He worked on it in a shed behind his house.

In 1896, he built his first car.
He called it a quadricycle
(KWAHD-ruh-SYE-kuhl)
because it had four bicycle wheels.

It also had a doorbell for a horn!

FIRST · CAR

20

For many years, only rich people could buy cars.

Ford did not like this. He wanted everyone to have a car.

In 1903, Ford started
his own company.

Later that year, he built another car. He called it the Model A. He made sure it did not cost a lot of money.

Then he built another car.
He called it the Model T.

He sold millions (MIL-yuhnz)
of Model Ts. People called them
"Tin Lizzies."

25

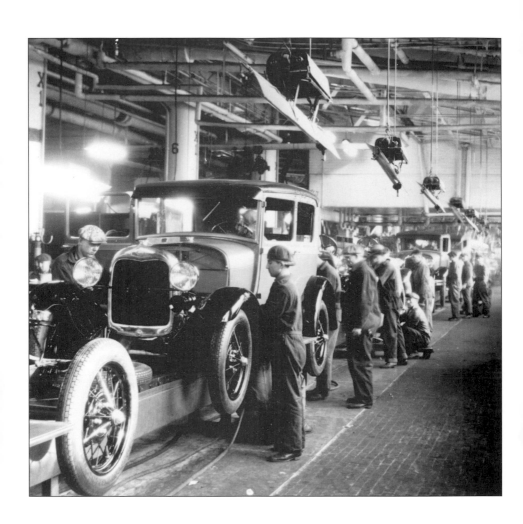

To make more Model Ts,
Ford set up assembly
(uh-SEM-blee) lines.

Workers stood in one place.
Each time a car passed, a worker
would add a part. Cars were
made faster and cheaper.

Ford died in 1947.

If it weren't for him, we might still be riding around in carriages pulled by horses!

Words You Know

assembly line

carriage

engine

Henry Ford

30

machine

Model A

quadricycle

Tin Lizzie

31

Index

About the Author

More than fifty published books bear Wil Mara's name. He has written both fiction and nonfiction, for both children and adults. He lives with his family in northern New Jersey.

Photo Credits